T0137396

Natural Computing Series

Series Editors: G. Rozenberg
Th. Bäck   A.E. Eiben   J.N. Kok   H.P. Spaink

Leiden Center for Natural Computing

More information about this series at http://www.springer.com/series/4190

Peter R. Lewis • Marco Platzner • Bernhard Rinner
Jim Tørresen • Xin Yao
Editors

Self-aware Computing Systems

An Engineering Approach

Springer

Editors
Peter R. Lewis
School of Engineering & Applied Science
Aston University
Birmingham, United Kingdom

Marco Platzner
Department of Computer Science
Paderborn University
Paderborn, North Rhine-Westphalia
Germany

Bernhard Rinner
Institute of Networked and Embedded Systems
Alpen-Adria-Universität Klagenfurt
Klagenfurt am Wörthersee
Austria

Jim Tørresen
Department of Informatics
University of Oslo
Oslo, Norway

Xin Yao
School of Computer Science
University of Birmingham
Birmingham, United Kingdom

ISSN 1619-7127
Natural Computing Series
ISBN 978-3-319-81937-2   ISBN 978-3-319-39675-0   (eBook)
DOI 10.1007/978-3-319-39675-0

Printed on acid-free paper

This Springer imprint is published by Springer Nature
The registered company is Springer International Publishing AG Switzerland

Foreword

This book considers the design of new computation systems that are in some ways more responsive to the environment and their own state than current system designs and aim to be more reliable through the creation of self-aware and self-expressive systems. One of the driving forces of this work is the realisation of the growth in system complexity and the difficulty of using current "standard" methods and designs to continue to create working systems. This is certainly relevant, as the interest in the design and understanding of complex computing systems in technical applications has been growing significantly in various research initiatives like autonomic, organic, pervasive or ubiquitous computing and in the multi-agent system community. There have been many novel applications demonstrating a wide range of self-* properties, as well as studies looking also at emerging global behaviour due to self-organised local interaction. The authors of this book present the results of a large European cooperative project focusing specifically on self-awareness, which may be seen as one of the essential backgrounds for developing and supporting the other self-* properties, which is addressed here by the term "self-expression".

Ever since researchers have realised that machines could be programmed to have increasingly adaptive behaviors, there has been much research on how to introduce adaptive behaviour and more biological like capabilities into systems – more types of reasoning, more types of awareness, and more types of intelligent processing. Particularly important in adaptation is that the system has the knowledge and the capabilities that allow it to do these adaptations in novel situations and at runtime. There are many examples of large-scale programmes to foster the understanding of the necessary attributes and architectures of systems capable of these adaptations. Hence there were programs on adapting routers and networks in real time (e.g., DARPA's Active Networks), platforms and other plug and play architectures with robust real time services (e.g., DARPA's META program; Europe's AUTOSAR (AUTomotive Open System ARchitecture), programs that worked to understand emergent behaviour and make use of it (Europe's Organic Computing), systems with computational reflection used for resource management (e.g., reflective architectures), and, of course, an enormous amount of work on multi-agent systems and autonomous computing.

In this landmark EU project, these slowly developing themes, drawn from a wide diversity of fields, have been brought together and further developed with both thoughtful discussions on foundations and new research and developments in the engineering of several application areas.

One particularly important aspect of this book is the way in which it builds up our repertoire of engineering methods for self-awareness by purposely drawing its concepts for self-awareness from a diversity of fields and its examples from a diversity of applications. Most importantly, these applications span across different levels of computational systems, from agents and applications (interactive music systems in Chapter 14) to middleware services (Chapter 11) to adaptive networks (Chapter 10) and even hardware (Chapters 8, 9 and 12.)

Starting from insights into "self-awareness" achieved by other disciplines like psychology and philosophy, the notions of "computational self-awareness" and "self-expression" are systematically developed. The majority of the book focuses on computational systems that require some form of anticipation where the new algorithms and methods are needed to provide the appropriate anticipatory behaviour. In practice, these methods can include different forms of self-awareness (such as awareness of goals, of the current state and readiness of system resources, of one's planning process and of the ordering of events), such that the system is not simply reacting to events and changes, but can anticipate them. The ideas and mechanisms outlined are applied to a number of interesting applications: Computational finance applications using heterogeneous computing clusters are investigated and include self-adaptive algorithms that are supported by hardware; low-latency adaptive network processing; run-time reconfigurable hardware acceleration; heterogeneous computing and hardware/software co-processing for algorithmic trading and reconfigurable hardware acceleration of self-optimisation of reconfigurable hardware designs. Self-awareness in distributed smart camera networks is considered for both single cameras at a node level and multiple camera systems within a network. Interesting bio-inspired methods are aimed at the network level, artificial pheromones are employed to construct a local neighbourhood graph, allowing adaptation in the network as topologies change. A hypermusic demonstrator is considered as a third application. This considers various methods and techniques to enable adaptability (self-expression) in musical output. Three methods within this application are the focus of this work, each providing different input information and overall levels of information: *SoloJam* provides rather overarching rhythmic shaping; *Funky Sole Music* provides what might be considered more specific, lower level, inputs such as walking tempo, movement types and foot activity; *PheroMusic* considers more links between musical soundscapes.

Thus, the book provides a comprehensive introduction to self-aware computation providing a broad range of new theoretical background and foundation before moving on to consider details of architectures and techniques to help design self-aware computational systems, from nodes to networks. Many of the problems that have been addressed in this book will continue to be timely for many years to come and could well provide the focus of research strands within many research fields. Particular challenges remain with respect to performance, safety and security properties of such systems. Although self-awareness is supposed to improve the performance of computational systems in complex environments, there is still a lack of formal frameworks for rigorously arguing about the behaviour of such systems.

The authors are all well known in this research area and the editors, Lewis, Platzner, Rinner, Tørresen and Yao have done an excellent job in pulling together what is an excellent book.

Los Angeles   *Kirstie Bellman*
Karlsruhe   *Hartmut Schmeck*
York   *Andy Tyrrell*

March 2016

Preface

Self-aware computing is an emerging field of research. It considers systems and applications able to proactively gather and maintain knowledge about aspects of themselves, learning and reasoning on an ongoing basis, and finally expressing themselves in dynamic ways, in order to meet their goals under changing conditions. The aspects they might be aware of include their own internal state, capabilities, goals, environment, behaviour and interactions. The presence of gathered knowledge permits advanced intelligent decision making leading to self-expression: that is, effective, autonomous and adaptive behaviour, based on self-awareness. Self-awareness and self-expression capabilities are key to designing and operating future computing systems that will inherently and autonomously deal with high levels of dynamics and uncertainty, heterogeneity, scalability, resource constraints and decentralisation. Concepts of self-awareness have been established in psychology, philosophy and cognitive science but are relatively new to computing. In computing systems, our concepts of self-awareness and self-expression integrate and enhance a number of recent approaches dealing with systems with so-called self-* properties, e.g., self-adaptation, self-organisation and self-healing.

This book is the first ever to focus on the emerging field of self-aware computing from an engineering perspective. It first comprehensively introduces fundamentals for self-awareness and self-expression in computing systems, proposing the new notion of *computational self-awareness*. It then focuses on architectures and techniques for designing self-aware computing systems at the node and network levels. Finally, the effectiveness of these techniques is demonstrated on a variety of case studies. While a number of books on related topics such as self-adaption and self-organisation, and even self-awareness concepts in computing, have already been published, this book is unique as it provides a holistic view of self-aware computing including its relationship with self-expression, and the process of engineering such systems, i.e., a thorough understanding of how to model and build self-aware computing systems based on design patterns and techniques.

This book targets graduate students and professionals in the fields of computer science, computer engineering, and electrical engineering, but also practitioners and scientists from other fields interested in engineering systems with advanced properties relying on their ability to reason about themselves in a complex environment. The authors and editors of this book are active researchers in various aspects related to self-aware computing systems. They have a strong track record in successfully collaborating on this topic, for example, through the European FET project "Engineering Proprioception in Computing Systems (EPiCS)". The extensive joint experience of the contributors makes this edited book consistent and well integrated. Therefore, we specifically recommend this book as reading material for the graduate level or for self-study on self-aware computing systems.

The book reports some of the latest results in self-aware and self-expressive computing, and we hope it serves as a launchpad for further research discussions and new ideas in the future.

Birmingham   *Peter R. Lewis*
Paderborn   *Marco Platzner*
Klagenfurt   *Bernhard Rinner*
Oslo   *Jim Tørresen*
Birmingham   *Xin Yao*

March 2016

Acknowledgements

The research leading to many results in this book was conducted during the EPiCS project (Engineering Proprioception in Computing Systems) and received funding from the European Union Seventh Framework Programme under grant agreement no. 257906.

The contributors would like to acknowledge additional support for research performed in individual chapters of this book.

- Chapters 6 and 7 were also supported by EPSRC Grants (Nos. EP/I010297/1, EP/K001523/1 and EP/J017515/1).
- Chapter 8 was also supported by the German Research Foundation (DFG) within the Collaborative Research Centre "On-The-Fly Computing" (SFB 901) and the International Graduate School on Dynamic Intelligent Systems of Paderborn University.
- Chapter 9 was also supported in part by HiPEAC NoE, by the European Union Seventh Framework Programme under grant agreement numbers 287804 and 318521, by the UK EPSRC, by the Maxeler University Programme, and by Xilinx.
- Chapter 12 was also supported in part by the China Scholarship Council, by the European Union Seventh Framework Programme under grant agreement numbers 287804 and 318521, by the UK EPSRC, by the Maxeler University Programme, and by Xilinx.
- Chapter 13 was also supported by the research initiative Mobile Vision with funding from the Austrian Institute of Technology and the Austrian Federal Ministry of Science, Research and Economy HRSMV programme BGBl. II no. 292/2012.
- Chapter 14 was also supported by the Research Council of Norway under grant agreement number 240862/F20.
- Peter Lewis would like to thank the participants of the Dagstuhl Seminar "Model-Driven Algorithms and Architectures for Self-aware Computing Systems", Seminar Number 15041, for many insightful discussions on notions of self-aware computing.

Contents

List of Contributors

Andreas Agne
Paderborn University, Germany. e-mail: agne@upb.de

Rami Bahsoon
University of Birmingham, UK. e-mail: r.bahsoon@cs.bham.ac.uk

Tobias Becker
Imperial College London, UK. e-mail: tobias.becker@imperial.ac.uk

Arjun Chandra
Studix, Norway. e-mail: arjun@studix.com

Renzhi Chen
University of Birmingham, UK. e-mail: rxc332@cs.bham.ac.uk

Tao Chen
University of Birmingham, UK. e-mail: t.chen@cs.bham.ac.uk

Stewart Denholm
Imperial College London, UK. e-mail: stewart.denholm10@imperial.ac.uk

Bernhard Dieber
Alpen-Adria-Universität Klagenfurt, Austria. e-mail: bernhard.dieber@counity.at

Lukas Esterle
Alpen-Adria-Universität Klagenfurt, Austria. e-mail: lukas.esterle@aau.at

Gustavo Fernández Domínguez
Austrian Institute of Technology, Austria. e-mail: gustavo.fernandez@ait.ac.at

Funmilade Faniyi
University of Birmingham, UK. e-mail: f.faniyi@gmail.com

Andreea-Ingrid Funie
Imperial College London, UK. e-mail: andreea.funie09@imperial.ac.uk

Kyrre Glette
University of Oslo, Norway. e-mail: kyrrehg@ifi.uio.no

Ce Guo
Imperial College London, UK. e-mail: ce.guo10@imperial.ac.uk

Markus Happe
ETH Zurich, Switzerland. e-mail: markus.happe@alumni.ethz.ch

Maciej Kurek
Imperial College London, UK. e-mail: mk306@imperial.ac.uk

Peter R. Lewis
Aston University, UK. e-mail: p.lewis@aston.ac.uk

Achim Loesch
Paderborn University, Germany, e-mail: achim.loesch@upb.de

Wayne Luk
Imperial College London, UK. e-mail: w.luk@imperial.ac.uk

Leandro L. Minku
University of Leicester, UK. e-mail: leandro.minku@leicester.ac.uk

Georg Nebehay
Austrian Institute of Technology, Austria. e-mail: gnebehay@gmail.com

Xinyu Niu
Imperial College London, UK. e-mail: niu.xinyu10@imperial.ac.uk

Kristian Nymoen
University of Oslo, Norway. e-mail: kristian.nymoen@imv.uio.no

Roman Pflugfelder
Austrian Institute of Technology, Austria. e-mail: roman.pflugfelder@ait.ac.at

Marco Platzner
Paderborn University, Germany. e-mail: platzner@upb.de

Christian Plessl
Paderborn University, Germany. e-mail: christian.plessl@uni-paderborn.de

Bernhard Rinner
Alpen-Adria-Universität Klagenfurt, Austria. e-mail: bernhard.rinner@aau.at

Mark Salmon
University of Cambridge, UK. e-mail: mhs39@cam.ac.uk

Jennifer Simonjan
Alpen-Adria-Universität Klagenfurt, Austria. e-mail: jennifer.simonjan@aau.at

Stephan C. Stilkerich
Airbus Group Innovation, Germany. e-mail: stephan.stilkerich@airbus.com

Tim Todman
Imperial College London, UK. e-mail: timothy.todman@imperial.ac.uk

Jim Tørresen
University of Oslo, Norway. e-mail: jimtoer@ifi.uio.no

Ariane Trammell-Keller
ETH Zurich, Switzerland. e-mail: ariane.trammell@alumni.ethz.ch

Shuo Wang
University of Birmingham, UK. e-mail: s.wang@cs.bham.ac.uk

Xin Yao
University of Birmingham, UK. e-mail: x.yao@cs.bham.ac.uk

Acronyms

ACO   Ant Colony Optimisation
AES   Advanced Encryption Standard
ALA   Ant Learning Algorithm
API   Application Programming Interface
BSD   Berkeley Software Distribution
CDC   Concept Drift Committee
CDT   Correct Detected Track
CMT   Consensus-Based Matching and Tracking
CPU   Central Processing Unit
CUDA   Compute Unified Device Architecture
CV   Computer Vision
DDD   Diversity for Dealing with Drifts
DDM   Drift Detection Method
DPS   Dynamic Protocol Stack
DWM   Dynamic Weight Majority
EA   Evolutionary Algorithm
EDDM   Early Drift Detection Method
EGO   Efficient Global Optimisation
FAT   False Alarm Track
FB   Functional Block
FF   Flip-Flop
FPGA   Field-Programmable Gate Array
FPS   Frames per Second
FMC   FPGA Mezzanine Card
FOV   Field of View
FSR   Force-Sensitive Resistor
GA   Genetic Algorithm
GPSP   General Purpose Sensor Platform
GP   Gaussian Process
GPU   Graphics Processing Unit
H2S   Hardware-to-Software

HLS   High Level Synthesis
HMM   Hidden Markov Models
HPC   High Performance Computing
ICAP   Internal Configuration Access Port
IDP   Information Dispatch Point
ILP   Integer Linear Programming
IP   Internet Protocol
IPC   Inter-process Communication
LUT   Look-up Table
MAC   Media Access Protocol
MLO   Machine Learning Optimiser
MOEA/D   Multi-objective Evolutionary Algorithm Based on Decomposition
MOP   Multi-objective Optimisation Problem
MPI   Message Passing Interface
MTBF   Mean Time Between Failures
NoC   Network-on-Chip
OSC   Open Sound Control
OT   Object Tracking
PE   Processing Element
RAP   Redundancy Allocation Problem
RAM   Random Access Memory
RTM   Reverse Time Migration
S2H   Software-to-Hardware
SA   Self-aware
SACS   Self-aware Computing Systems
SDRAM   Synchronous Dynamic Random Access Memory
SE   Self-expression
SIMD   Single Instruction, Multiple Data
SMT   Satisfiability Modulo Theories
SVM   Support Vector Machine
SoC   System-on-Chip
SOP   Single-Objective Optimisation Problem
SSE   Streaming SIMD Extensions
STEPD   Statistical Test of Equal Proportions
TCP   Transmission Control Protocol
TDF   Track Detection Failure
TPOT-RL   Team-Partitioned Opaque-Transition Reinforcement Learning
Todi   Two Online Classifiers for Learning and Detecting Concept Drift
TLD   Tracking-Learning-Detection
UDP   User Datagram Protocol
VHDL   Very High Speed Integrated Circuit Hardware Description Language

Glossary

This glossary lists important terms used in this book, in particular in Part I "Concepts and Fundamentals", with accompanying descriptions or definitions. The glossary is organised into four sections: concepts of self-awareness and self-expression, engineering self-aware systems, related approaches, and general terms. The terms in each of the sections are listed alphabetically.

Concepts of Self-awareness and Self-expression

self-awareness Self-awareness is a broad concept which describes the property of a system (typically a human) which has knowledge of "itself", based on its own senses (perceptual) and internal models (conceptual). This knowledge may take different forms (cf. levels of self-awareness), and be based on perceptions of both internal and external phenomena (cf. public vs. private self-awareness). It can be a property of single systems (e.g., agents) and collective systems.

collective self-awareness Collective self-awareness refers to the self-awareness property of a collective system, i.e., as opposed to a single agent. Levels of, and public/private self-awareness apply also at this abstraction. This means that a self-aware system is not required to have a central "knowledge" component (though it may have, if desired).

computational self-awareness Computational self-awareness is a notion we have developed to refer to a computational interpretation of self-awareness. Since much of the literature on self-awareness does not readily make sense to engineers or applies directly to technical systems, aspects of computational self-awareness are designed to describe self-awareness properties of computational systems, inspired by self-awareness in humans.

emergent self-awareness This is a special case of collective self-awareness, when the collective self-awareness properties are present, but it is not obvious how this comes about by simply examining the behaviour of individual nodes within a collective.

level(s) of self-awareness A very common theme in self-awareness theory is the distinction between several levels of self-awareness, to describe different aspects or capabilities which comprise a system's complex self-awareness. There are many examples of "sets of levels" to be found in the literature. In developing our notion of computational self-awareness, we have based a set of levels of computational self-awareness on the set of levels (for humans) proposed by Ulric Neisser. Note that our levels are not hierarchical, do not build on each other, nor are they in any particular order, save that the ecological self/stimulus awareness is the most basic, and the conceptual self/meta-self-awareness is typically the most complex.

meta-self-awareness Meta-self-awareness is one of the levels of computational self-awareness we propose, indeed the highest one in our framework. It refers to the capability of a system to be aware of its own self-awareness. This can be very useful, since it means a system has knowledge, obtained at run time, about its own self-awareness processes, including, for example, how effective its learning is at present, or how much resource is being spent to maintain its knowledge. Meta-self-awareness is closely related to, and permits, meta-reasoning. It is a concept inspired directly from human psychology.

private self-awareness Private self-awareness refers to a system's ability to obtain knowledge based on phenomena that are internal to itself. A system needs internal sensors to achieve this. Again, this is a notion which exists in human self-awareness theory, and also features in computational self-awareness.

public self-awareness Public self-awareness refers to a system's ability to obtain knowledge based on phenomena external to itself. Such knowledge depends on how the system itself senses/observes/measures aspects of the environment it is situated in, and includes knowledge of its situation and context, as well as (potential) impact and role within its environment. This is a notion which exists in human self-awareness theory, and also features in computational self-awareness.

scope of self-awareness The scope of self-awareness refers to the domain of phenomena able to be sensed and modelled by the self in question. For a system which is only privately self-aware, the scope may be the same as the span (i.e., it has no perception of its environment). For a system which has some private and some public self-awareness, the scope would be larger than the span, and include external social or physical aspects of the environment. The term scope can be useful to avoid having to use the word "level" to mean multiple things simultaneously in a passage of text.

self-aware system We do not formally define this in the book, however we generally consider a self-aware system to be one which (at least) obtains and maintains knowledge relating to itself (including its perspective of its environment), without external control.

self-awareness capability When a particular level of self-awareness is present in a system, we refer to this as the system having that particular self-awareness capability. For example, a node may have a time-awareness capability, indicating that it implements the time-awareness level. Levels may be realised in different ways simultaneously in the same system, meaning that, for example, a system may have several time-awareness capabilities.

self-explanation Another form of self-expression when based on self-awareness, self-explanation is the ability of a system to explain/justify its behaviour to an entity on the outside (such as a user or another system).

self-expression Self-expression is, in the general sense, behaviour based on self-awareness. It may include a wide range of different actions, enacted through a system's actuators, including self-adaptation, self-explanation, or just normal system behaviour. Self-expression can also be considered as a property of a collective, since a collective's behaviour can also be based on collective self-awareness. Examples of this might include the adaptive behaviour of a flock of birds in response to an external (to the flock) stimulus.

self-expression capability As with self-awareness capabilities, self-expression capabilities refer to the presence of an implementation of self-expression in a system. For example, a system which adapts its parameters in response to its goal-awareness, would have a self-expression capability. Again, multiple self-expression capabilities may be present simultaneously.

self-knowledge Self-knowledge is a general term for knowledge (usually held in a learnt model) concerning the system itself, which typically is produced as part of a self-awareness process. Note that this can include objective self-knowledge (i.e., about the system as an object in the world, how it interacts with others, how its internal state changes, etc.) and also subjective self-knowledge (i.e., about its experiences, sensor data, changing context, etc.).

self-optimisation Self-optimisation is a form of self-expression; self-optimisation is the ability of a system to optimise itself by improving metrics such as performance or power consumption.

span of self-awareness We use this term to refer to the domain of the subject of the self-awareness, i.e., it is the answer to the question: who is the self here? For example, if a single agent is self-aware, then the span is the agent. If we are considering the collective self-awareness of a network of smart sensors, then the span would be the network. The term span can be useful to avoid having to use the word "level" to mean multiple things simultaneously in a passage of text.

Engineering Self-aware Systems

(architectural) pattern We produced eight architectural patterns, which are derived from the reference architecture and describe how various capabilities (such as levels of self-awareness, etc.) can be included or excluded as appropriate to the application need.

methodology for engineering self-aware systems We developed a methodology for engineering self-aware systems, based on the reference architecture and the derived architectural patterns.

primitive A primitive is a particular block in the reference architecture, representing, for example, a level of self-awareness, self-expression and a sensor. They are instantiated for particular applications.

reference architecture We developed a reference architecture which captures the core aspects of computational self-awareness. The aim is to provide a common, principled basis on which researchers and practitioners can structure their work. We have argued that the psychological foundations, while not strictly necessary, can provide a means of channelling a wide range of ideas, which would perhaps otherwise not have occurred to engineers, acting to inspire the design of future computing systems. The architecture can also be used as a template for identifying common ways of implementing self-awareness capabilities. Different implementations of the same capability can thereby be compared and evaluated. Further, we have derived a set of architectural patterns from the reference architecture.

(self-aware) node We use the term self-aware node to refer to various types of system that are self-aware, e.g., an agent, a robot and a camera. Agent is an alternative term, but node can be used when not wanting to be specific about a particular system being an agent. We also claim that self-aware collectives (see next entry) can be viewed as self-aware nodes, at a higher level of abstraction. A node may or may not correspond to a physical system—this is not a requirement, but it may often make sense to make it correspond.

tactic/algorithm/technique A tactic is a particular instantiation of a primitive in the reference architecture, typically referred to as a particular algorithm, technique, etc. These are application specific. Multiple tactics may be suitable for a particular primitive, and some tactics may implement multiple primitives simultaneously.

Related Approaches

autonomic (computing) Autonomic computing is a vision originally pioneered by IBM, of engineered systems which manage themselves. This self-management is stated to include: self-configuration, self-optimisation, self-healing and self-protection. The aim is to reduce the need for human involvement in the management of complex computing systems. Some autonomic computing literature mentions the need for self-awareness as a characteristic to support self-management, though the literature on autonomic computing does not significantly expand on this. (Not to be confused with autonomous.)

autonomous (system) Autonomy is a broad notion with much disagreement surrounding it. However, in general, an autonomous system is one which acts without any external direction. Examples include robots, vehicles and software agents. In many cases, this ability to make decisions is based on a method of decision making pre-programmed into the system, in other cases it is learnt online at run time. The types of systems we are concerned with in this book are ones which would typically be considered to be autonomous to a greater or lesser extent. (Not to be confused with autonomic.)

metacognition/metareasoning Metareasoning is reasoning about reasoning, and has been the topic of a significant amount of research primarily in the US, where it has been primarily led by DARPA. Metareasoning relies on meta-self-awareness, and again the metareasoning community has discussed self-awareness as being important, but not expanded on the notion significantly.

organic computing This is a vision from a long-running (primarily) German research project to create "life-like" engineered systems, in which self-organising emergent behaviour is controlled (by an observer/controller component), to ensure desirability in the self-organisation. The Organic Computing literature also mentioned self-awareness as beneficial, but again does not expand on this significantly.

General Terms

adaptability In high level terms, this is similar to adaptivity, but describes a system's potential for adaptation, rather than actual realised adaptivity.

adaptivity In high level terms, this concerns the amount to which a system adapts, e.g., in the presence of a changing environment, or as a result of its learning.

collective We use the term collective to refer to various types of distributed systems, typically without central control. Examples include swarms, systems-of-systems, populations, multi-agent systems, interwoven systems, etc. The term can be used when there is a need to talk generally of these types of systems, without restricting the discussion to a specific one.

learnt model A learnt model is a model which has been induced through a process of (typically online) learning, based on data from sensors and other existing models. Learnt models hold the conceptual knowledge a self-aware system has concerning itself, its interactions, history, expectations, goals, etc.

model We use the term model in a very general way, to refer to a conceptual representation of some knowledge, typically obtained through sensors. A model could simply be a direct representation of some data, or could be abstractions of that data, or further data synthesised from sensory input.

online learning Online learning is the process of learning a model from data on an ongoing basis. Typically, not all data is available in advance (e.g., it arrives in a streaming fashion from sensors), and the concept being learnt may change over time (i.e., concept drift). In online learning, models are often used (e.g., through self-expression in this case) before learning "completes", if indeed it ever does. Hence most online learning algorithms also need to be anytime algorithms, implying that models are used and improved continuously as time goes by.

self-adaptive system A system which adapts (typically its behaviour) in response to external or internal changes, but without external control. We have argued that self-awareness is an enabling property for effective self-adaptation. When self-adaptation behaviour is based on self-awareness, it is a form of self-expression.

self-organising system   A system which changes its organisation (e.g., its structure, architecture, topology), without external control.

Chapter 1
Self-aware Computing: Introduction and Motivation

Peter R. Lewis, Marco Platzner, Bernhard Rinner, Jim Tørresen, and Xin Yao

1.1 Self-aware Computing: A New Paradigm

Designing and operating computing and communication systems are becoming increasingly challenging tasks, due to a multitude of reasons. First, compute nodes are evolving towards parallel and heterogeneous architectures to realise performance gains while minimising their power consumption. Progress in micro(nano)-electronics allows us to integrate more and more functionality on a single compute node, but at the same time requires us to deal with increasing numbers of faulty and unreliable components. Second, distributed systems are growing in the numbers and heterogeneity of nodes and must be able to cope with an increasing level of dynamics. The network topology and the collective resources of a distributed system can vary strongly during runtime since nodes may leave and enter the network dynamically. The position, functionality and available resources of each node may also change dynamically. Third, future challenging application domains have quite divergent requirements with respect to functionality and flexibility, performance, resource usage and costs, reliability and safety, and security. Fuelled by technological progress, applications with exciting levels of user interaction will be possible, and these dynamic socio-technical systems bring with them numerous additional runtime trade-offs to consider. Fourth, the size and complexity of decentralised com-

Peter R. Lewis
Aston University, UK, e-mail: p.lewis@aston.ac.uk

Marco Platzner
Paderborn University, Germany, e-mail: platzner@upb.de

Bernhard Rinner
Alpen-Adria-Universität Klagenfurt, Austria, e-mail: bernhard.rinner@aau.at

Jim Tørresen
University of Oslo, Norway, e-mail: jimtoer@ifi.uio.no

Xin Yao
University of Birmingham, UK, e-mail: x.yao@cs.bham.ac.uk

puting systems have grown at an increasingly fast rate, posing new challenges in terms of scalability and complexity. Based on our experience, we believe that current design and operation principles and methods will neither be able to scale with future systems, nor efficiently handle the variety and changing nature of requirements and optimisation goals. Novel design and operation principles and methods, such as those incorporating self-awareness (SA) and self-expression (SE), are needed.

Self-aware computing describes a new paradigm for systems and applications that proactively gather information; maintain knowledge about their own internal states and environments; and then use this knowledge to reason about behaviours. This paradigm is well suited for advanced intelligent decision making in dynamic and uncertain environments, which can in turn support effective and explainable autonomy and self-adaptation. Self-expression describes behaviours that are based on the knowledge acquired through system self-awareness, such as self-adaptation and self-explanation. A self-expressive system can adapt to its environment including its users, and thus limits the need for users to adapt to fixed system behaviours [390].

Self-awareness is not a new concept and has been studied for a long time in the fields of psychology and cognitive science [275]. However, a clear understanding and interpretation of self-awareness in computer science and engineering is lacking. There has been no universally agreed and accepted definition of self-aware computing in spite of frequent use of the word "self-aware" in different contexts. Although there are systems that are declared to be self-aware in one sense or another, little has been said about engineering methodologies that can help to build such systems. It has not previously been clear what properties a self-aware system could and should have, and what capabilities such a system might have.

This book attempts to bridge the gaps in the literature related to self-aware computing, in the spirit of recent work to translate concepts of self-awareness from psychology to computing [236]. It focuses on key ideas from self-awareness theory, leading to working definitions and pragmatic principles of *computational self-awareness* that can be used in engineering self-aware systems and understanding their behaviours. While it surveys a range of views concerning self-awareness, it does not attempt to engage in more philosophical debates on the ability of machines to achieve so-called "true" self-awareness, or what that might mean. As an example of the pragmatic approach taken in this book, building on notions from psychology, different levels of self-awareness as they apply to computing systems are described, examples are then used to illustrate what capabilities a system could have with which level(s) of self-awareness, how these might be implemented, and what benefits and costs are associated with such functionality. An engineering methodology is then introduced to facilitate the design of self-aware systems with different required capabilities.

This book builds on and extends earlier work in the related fields including autonomic computing [105] and organic computing [277] and architectures like MAPE-K and Kramer and Magee's three-layered architecture [225]. It takes an engineering approach to the design of self-aware computing systems, that includes considering different levels of self-awareness and the introduction of a reference architecture for designing self-aware and self-expressive computing systems. As a more concrete guide to engineering self-aware and self-expressive systems, this book gives detailed architectural patterns and primitives for systems with different levels of self-awareness, to facilitate the design of such systems.

Many techniques and algorithms are needed to support and implement self-awareness and self-expression at different levels. For example, proactively acquiring knowledge about oneself (e.g., as a compute node in a large decentralised system) and then building a model(s) of one's internal state and environment based on such knowledge requires online learning, i.e., learning while the system is running. Online learning algorithms are one of the key ingredients employed in self-aware and self-expressive systems. These algorithms try to make the most appropriate (trade-off) decision among several conflicting goals in a dynamic and uncertain environment, including what information to acquire, which level of abstraction to use to capture and model such information, and what resources to use for such information acquiring and modelling so as to maximise the expected performance gain and minimise the resource usage, etc. Such online learning must be able to deal with concept drifts because both the computing system and the environment it is operating in are changing and have uncertainty. The system needs to learn new concepts as they appear and forget old concepts as they become obsolete. In this book we will describe how some of the existing online learning algorithms could be used or adapted for self-aware and self-expressive systems. Examples will be given to illustrate how online learning algorithms could be implemented to support self-aware and self-expressive systems with desired capabilities.

Importantly, we do not propose a preferred online learning paradigm to support self-aware and self-expressive systems. There are many techniques and algorithms that can be used specifically for either self-awareness or self-expression. In fact, one of the key observations from our own studies revealed that using different learning techniques or strategies at different compute nodes in a large decentralised system can often lead to enhanced performance or efficiency gain compared to that achieved by a system where all nodes use the same learning technique/strategy. This book presents one such example to illustrate how heterogeneous learning strategies for self-expression in a smart camera network could help to achieve better system performance.

Performance has often been a primary concern of many studies. For example, fast learning to recognise a face from a video accurately is always of great interest. However, there might be some additional factors that a system should consider in the real world. For example, fast and accurate learning might imply heavy consumption of computing resources, which might not be available or necessary. It might also lead to higher energy consumption due to heavy computing, which is undesirable for any battery-driven autonomous system. As a result, fast and accurate learning might not be the most appropriate choice for a resource-constrained system/node. However, the availability of resources is not fixed in the real world. It is changing. A self-aware system needs to learn such changing conditions, learn the changing importance of different goals (performance vs. energy consumption), and learn the best trade-off among conflicting goals at run-time. This is one of several key issues that will be discussed in this book. Other issues that will be discussed and often considered in self-aware and self-expressive systems include robustness, decentralisation, and multi-objectivity of the systems.

1.2 Organisation of This Book

The book consists of four parts. Part I motivates the concepts of self-awareness and self-expression for engineering computing systems by looking into other disciplines and related concepts. It introduces a reference architecture for describing and engineering computational self-awareness and self-expression in computing systems. The architecture provides a common language which paves a way for identifying architectural patterns influencing the engineering of computational self-awareness and self-expression capabilities across a range of applications. Chapter 2 translates concepts from psychology to the domain of computing, introducing key ideas in self-aware computing. Chapter 3 relates our concepts of computational self-awareness and self-expression to other efforts in computer science and engineering under the self-awareness label. Depending on the fields, the term self-awareness may have different meanings. Chapter 4 concludes the first part by presenting our reference architecture for describing self-aware and self-expressive computing systems.

Part II outlines some common architectural primitives and guidelines for engineering self-aware systems using design patterns and different knowledge representation techniques, respectively. Chapter 5 provides design patterns and primitives on how to design self-aware and self-expressive computing systems in a principled way. It discusses how the proposed patterns and primitives can be used in real software system projects. Chapter 6 explains issues which may be present in self-aware and self-expressive systems such as adaptivity, robustness, multi-objectivity and decentralization, and discusses their implications in terms of knowledge representation and modelling choices. Finally, Chapter 7 concludes the second part by introducing common techniques that could be used in self-aware and self-expressive systems, including classical online learning, nature-inspired learning and socially-inspired learning in collective systems.

Part III presents the design of nodes and networks which provide self-aware and self-expressive capabilities. Many modern compute nodes are heterogeneous multi-cores that integrate several CPU cores with fixed function or reconfigurable hardware cores. In Chapter 8 we present a node architecture, programming model and execution environment for heterogeneous multi-cores, and show how the components of the reference architecture can be implemented on top of the operating system ReconOS. Chapter 9 describes how to build a heterogeneous cluster that can adapt to application requirements. Chapter 10 presents flexible protocol stacks as a promising alternative to today's static Internet architecture. Self-aware and self-expressive network nodes cooperate to select the protocol stacks that fulfil all communication requirements at the minimal cost at run-time. Finally, Chapter 11 compares different middleware paradigms and their suitability to support self-awareness in distributed applications and briefly describes a dedicated middleware implementation.

Part IV demonstrates how self-awareness and self-expression are useful in the three widely different application domains of hardware acceleration of financial computation, object tracking in multi-camera networks, and active music systems, respectively. Chapter 12 demonstrates how complex financial models can be speeded up using reconfigurable hardware combined with optimisation algorithms. Object tracking in multi-camera networks is the topic of Chapter 13, where autonomous monitoring of each camera in a network is combined with learning mechanisms to adapt its behaviour to changing conditions. Finally, Chapter 14 illustrates how persons without musical skills can influence music in interactive music systems using nature and socially-inspired methods.

The book is a result of extensive teamwork through the EU-funded research project "Engineering Proprioception in Computing Systems (EPiCS)". Eight research groups in five different countries collaborated to develop concepts and foundations for self-awareness and self-expression in computing systems and tested and demonstrated their usefulness in highly different domains. While the book includes some of the research results, it also, and more importantly, serves the purpose of stimulating further research into the field of self-aware and self-expressive computing systems.

Part I
Concepts and Fundamentals

Part I motivates the concepts of self-awareness and self-expression for engineering computing systems by looking into other disciplines and related concepts. It introduces a reference architecture for describing and engineering computational self-awareness and self-expression in computing systems. The architecture provides a common language which paves a way for identifying architectural patterns influencing the engineering of computational self-awareness and self-expression capabilities across a range of applications. Chapter 2 translates concepts from psychology to the domain of computing, introducing key ideas in self-aware computing. Chapter 3 relates our concepts of computational self-awareness and self-expression to other efforts in computer science and engineering under the self-awareness label. Depending on the fields, the term self-awareness may have different meanings. Chapter 4 concludes the first part by presenting our reference architecture for describing self-aware and self-expressive computing systems.

Chapter 2
Self-awareness and Self-expression: Inspiration from Psychology

Peter R. Lewis, Arjun Chandra, and Kyrre Glette

Abstract Self-awareness concepts from psychology are inspiring new approaches for engineering computing systems which operate in complex dynamic environments. There has been a broad and long-standing interest in self-awareness for computing, but only recently has a systematic understanding of self-awareness and how it can be used and evaluated been developed. In this chapter, we take inspiration from human self-awareness to develop new notions of computational self-awareness and self-expression. We translate concepts from psychology to the domain of computing, introducing key ideas in self-aware computing. In doing so, this chapter therefore paves the way for subsequent work in this book.

2.1 Introduction to Self-awareness

The *Oxford English Dictionary* defines *awareness* as "knowledge or perception of a situation or fact." Informally, we might typically consider that humans build up knowledge, or become aware of things, by perceiving the world around them. We observe interactions, listen to other people, watch television, read books, and, particularly in early life, learn through play. When considering awareness in humans, it is common to consider that all the knowledge we possess, all of our awareness, is acquired through perception. This idea was first postulated by Hume [187], who argued that all human knowledge is induced from experience. What then does it mean for a human to be *self*-aware? For Hume, the "self" is not a defined physical entity, but instead describes the bundle of experiences or perceptions unique to an

Peter Lewis
Aston University, UK, e-mail: p.lewis@aston.ac.uk

Arjun Chandra
Studix, Norway, e-mail: arjun@studix.com

Kyrre Glette
University of Oslo, Norway, e-mail: kyrrehg@ifi.uio.no

9

individual. A Humean form of self-awareness might then be considered to consist of an individual's knowledge of its experiences. Kant [210] criticised Hume's view, extending the scope of the self significantly, arguing that there is some entity which is the subject of these experiences, and is common through space and time. This Kantian self synthesises information from experiences with concepts held in the mind and with the imagination. Kant further argued that as an individual performs actions within the world, since its actions are based on its synthesised knowledge, they represent its self, giving rise to the self also as an object. This object in turn is something which can be perceived and experienced.

Though there is a long history of analysis of the nature of the self in philosophy, more recently, psychology has made a more pragmatic attempt to develop an understanding of the varieties of knowledge individuals possess concerning themselves. The notion of self-awareness first appears in the literature around the turn of the twentieth century [25, 382], perhaps most importantly with James [197] making the distinction between two forms of self based on the differences between the Humean and Kantian views described above. First, the *implicit* self, often referred to as the self-as-subject, or the "I", is the self which is the subject of experiences. These experiences are unique to the individual, and they are from the individual's own point of view, determined by factors such as their sensing apparatus, their situation within the world, and other factors associated with their own state. Second, the *explicit* self, or self-as-object, can be discerned. Here the self is an object of knowledge. It is a thing which can be recognised, modelled and reasoned about, including in relation to other objects in the world. An individual's awareness of its explicit self is often considered the more advanced form of self-awareness in this distinction, building on implicit self-awareness. Indeed, implicit self-awareness emerges much earlier in the lives of human infants than its explicit counterpart does [231].

One commonly considered form of self-awareness is that as measured by the so-called *mirror test* [140]. A subject being evaluated is presented with a mirror, to which it is then allowed to get accustomed. The subject is then distracted and, without its knowledge, a visible change is made to its appearance. This is usually done by marking its face, e.g., putting a spot on its cheek or forehead. The subject is then presented with the mirror again. Any behaviour directed towards this marker by the subject implies self-recognition, which is seen as being enabled by a mental representation of oneself (also known as a secondary representation). As Asendorpf et al. [18] put it:

> "[secondary representation] is not a perception of oneself but rather a constructed mental model of oneself that can be manipulated in fantasy. Therefore, the ability to recognise oneself in a mirror that requires linking a mirror image (a primary representation) with one's self marks the capacity for secondary representation."

Explicit self-awareness requires a subject to possess the capacity to construct such a secondary conceptual representation of itself. What then does the mirror test tell us about self-awareness? Humans, primates and some other animals have "passed" the mirror test [140, 18], however, Haikonen [155] showed that very little sophistication in computing machinery can enable a computational system with visual sensors to also pass. Haikonen therefore goes on to suggest that the ability or inability to self-recognise may not prove the presence or absence of self-awareness more generally at all. Since a computing system can specifically and easily be designed to pass the mirror test, using the test as a yardstick by which to tell whether or not the system is self-aware may thus be a misleading notion. Indeed, even if we accept Asendorpf's [18] claim that self-recognition requires some form of secondary representation (i.e., a conceptual model) of oneself, it is clear that the mirror test is concerned with a quite specific aspect of what might be considered self-awareness, based on a conceptual model of one's appearance. As this chapter will go on to discuss, we are concerned in this book with a broader treatment of self-awareness.

As should now be apparent, and is highlighted in recent work (e.g., by Morin [275], Legrain et al. [231] and Rochat [344]), there is much ongoing discussion about what might and what might not constitute self-awareness and various observed forms of it. In some cases, more "primitive" aspects of self-awareness are stated to fall instead into consciousness, upon which self-awareness builds. For example, Morin's definition considers self-awareness as distinct from, but building upon consciousness, as *"the capacity to become the object of one's own attention."* Further, he also addresses the conceptualisation of subjective experience, describing a self-aware organism as one that *"becomes aware that it is awake and actually experiencing specific mental events."*

In other cases, this perceptual (or pre-conceptual) subjective experience is itself also included in the scope, effectively presenting self-awareness and consciousness as overlapping concepts. In this chapter, our concern is not with attempting to settle these debates, but instead with understanding theories of self-awareness as presented in the literature, with the aim of developing concepts inspired by self-awareness to benefit the design of computing systems. Therefore, some of what is presented and discussed here is considered by some literature, but not all, to be forms of consciousness rather than self-awareness proper.

2.2 Key Concepts for Self-aware Systems

In drawing inspiration from psychology, in this book we utilise three key concepts which appear prominently in the self-awareness literature, and which have shown promise as useful concepts in the design of self-aware computing systems. These three key concepts are (i) public and private self-awareness, (ii) the extent of self-awareness capabilities can be characterised by various levels, and (iii) self-awareness can be a property of collective systems, where knowledge need not be present in a single central source, and may instead be distributed. Further, we argue that self-awareness properties alone are of limited value in computational systems, or indeed any system at all, unless accompanied by associated behaviour. In this book we use the term *self-expression* to refer to *behaviour based on self-awareness*. The remainder of this section introduces these key concepts.

2.2.1 Public and Private Self-awareness

As discussed above, Morin's definitional introduction to self-awareness contains two aspects. His first is centred on the idea of being the *"object of one's own attention"*, and establishes a subject-object view of self-awareness, where aspects of oneself are objects within a conceptual mental model. This *explicit* self-awareness (as presented, e.g., by Legrain et al. [231]) permits an individual to focus its attention on itself, to consider itself as an object within the world, and to observe and consider its own behaviour. However, Morin's second passage, that the individual *"becomes aware that it is awake and actually experiencing specific mental events"* reveals another facet of self-awareness, that which is *implicit*. This is concerned not with the self-as-object "me", but rather with "I", the self-as-subject of experiences. Here, the individual is aware of its experiences within the world, and that these are its own experiences, subjective and unique.

This distinction was first expounded in detail by Duval [111], who defined two classes of self-awareness: *subjective* and *objective*. Objective self-awareness is described as being *"focused exclusively upon the self and consequently the individual attends to his conscious state, his personal history, his body, or any other personal aspects of himself"* [111]. Subjective self-awareness by contrast is described as *"a state of consciousness in which attention is focused on events external to the individual's consciousness, personal history, or body"* [111].

Many authors have since developed this distinction further [131, 55, 58, 137, 147], in whose literature a slightly different distinction between *public* and *private* self-awareness develops. Private self-awareness is concerned with obtaining knowledge of internal phenomena, typically externally unobservable and accessible only to the individual. Such knowledge might include, for example, being hungry or having a headache. More complex private self-awareness might include an individual's knowledge of its values, opinions, goals or thought processes. Public self-awareness, on the other hand, is more concerned with how the individual can be (or is) perceived externally. This might include knowledge of how the individual appears to others, its social relationships or the effects of its interactions with the physical environment. Froming et al. [137] describe this as awareness of oneself as a *social object*.

In a minimal form, we may consider public self-awareness only insofar as it is present in individuals capable of implicit (or subjective) self-awareness. In this case, the individual would be capable of subjective perception of its environment. Given the unique situated nature of the individual, these experiences are themselves uniquely related to itself. However, in an explicit (or objective) form, public self-awareness may include knowledge of how the individual itself is or could be perceived by others, for example, how it looks (recalling the mirror test discussed above), whether it is a member of a particular group, or whether it has impacted on a shared environment. This requires knowledge not only of its environment, but also of itself as an object within that environment, an object of which others may be aware.

In summary, an individual's public self-awareness is concerned with knowledge obtained from experiences of the perceived environment in which the individual is situated. This knowledge may include both social aspects (e.g., other individuals, its own appearance) and purely physical aspects (e.g., the world). An individual's private self-awareness is by contrast concerned with knowledge obtained by experiences solely concerning the individual. This knowledge is obtained internally and is typically not available to others, unless the individual communicates it. An individual which is self-aware in both the public and private senses therefore has the capability to acquire and possess knowledge of its external environment and its internal state.

In taking inspiration from self-awareness for computing systems, we could choose (and some have previously chosen) to consider self-awareness only in its explicit, private sense, as the ability to build conceptual models of (part of) its own internals. Alternatively, we can broaden this further, to include the consideration that a system's sensing of its environment and itself provides a unique subjective experience, which can be modelled and reasoned about. Regardless of how we choose to draw the boundary around a definition of self-awareness, the literature does agree that self-awareness even within both its implicit and explicit forms is a multi-level phenomenon, where increasingly complex levels describe an individual's self-awareness capabilities.

In considering the self-awareness capabilities of systems in general and computing systems in particular, we are interested in the benefits (or not) of increased self-awareness, relative to current state-of-the-art systems. For example, to what extent can endowing a system with conceptual models of itself, its interactions, its goals, its past and its future enable more effective self-expression? Can this provide increased potential for adaptation? How important is a system's ability to learn such models for itself? What are the costs of maintaining and learning such models? And can a self-aware system build models of these costs too, taking account of them when deciding how to best conceptualise itself and its world? On the implicit side, what are the benefits of considering the subjective nature of a system's sensory input? To what extent does the consideration of a system's own state affect how that input is collected, stored, and reasoned about? Is there a benefit to considering differences between systems' own subjective experiences of a shared environment, or does this simply add complexity to a system's description, for little gain? These questions give a flavour of the reasons why, in this book, we take a broad view of self-awareness, including both implicit and explicit, and public and private forms. The chapters throughout the book engage with different aspects of self-awareness, from runtime learning and trade-off modelling, to exploring the heterogeneity of different systems' sensed experiences in a shared problem domain.

2.2.2 Levels of Self-awareness

As noted in much of the self-awareness literature (e.g., [231, 275]), it is widely believed that self-awareness is not a singular, *all or nothing* phenomenon, rather it can be thought of as a spectrum, where the capabilities of an individual are associated with one or more levels of self-awareness. Accordingly, several attempts have been made to define levels of self-awareness. Morin [275] provides a review of several of these classifications, highlighting similarities and differences between them. Some sets of levels, such as those by Rochat [344], focus on classifying self-awareness capabilities according to the way in which they are observed to develop in human children. Legrain et al. [231] provide a classification of three levels, considering only explicit self-awareness. One classification, due to Neisser [284], describes five levels which offer what is perhaps one of the most broad treatments of self-awareness, from the most minimal to the most advanced. Importantly from our perspective, Neisser's model also includes aspects of implicit self-awareness. In line with our approach of taking a broad view of self-awareness in our mission to translate concepts from psychology to computing, we have focussed on Neisser's levels as a concrete source of inspiration. These five levels now follow.

1. **Ecological self**
   The ecological self is the most minimal form of self-awareness. It permits sufficient knowledge only for basic stimulus-response behaviour, as the individual has a basic awareness of stimuli. The ecological self can be thought of as the minimum requirement for the individual to not be unconscious.

2. **Interpersonal self**
   The interpersonal self enables the individual to possess a simple awareness of its external interactions, permitting limited adaptation to others in the performance of tasks.

3. **Extended self**
   The extended self extends the interpersonal self to permit reflection of interactions over time. The individual is aware of the existence of past and future interactions.

4. **Private self**
   The private self allows the individual to process more advanced information concerning itself, such as thoughts, feelings and intentions.

5. **Conceptual self**
   The conceptual self, or self-concept, is the most advanced form of self-awareness, representing that the individual is capable of constructing and reasoning about an abstract representation of itself.

This final, most advanced level of self-awareness also permits what is sometimes termed **meta-self-awareness** [276, 366]. This is an awareness on the part of the individual that it is itself self-aware. Meta-self-awareness may consist of complex analysis and reasoning of both public and private self-awareness processes at any of the preceding levels. Examples of meta-self-awareness might include the individual's being aware that it is angry about something, or that it has recently learnt a lot about a particular topic. Smallwood et al. [366] discuss awareness at the meta level extensively, arguing that a lack of such a capability can lead to excessive mind-wandering in humans. They argue that the absence of meta level awareness leads to an individual being unable to direct its attention, thus exhibiting a lack of awareness more generally.

2.2.3 Self-awareness in Collective Systems

So far, we have just considered self-awareness in the context of a single individual. However, Mitchell [272] notes that self-awareness can also be observed in collective systems, where there is no central point at which such self-knowledge is located. Examples of these collective systems include those comprised of individuals that might normally be considered either organisms in their own right (e.g., ants in a colony) or constituent cells of a larger organism (e.g., neurons in the brain). In these cases, it appears from an external perspective that such biological collective systems are self-aware at the level of the collective, even though this property may not be present at the level of the individual component. This awareness, Mitchell describes as being concerned with

> "*information about the global state of the system, which feeds back to adaptively control the actions of the system's low-level components. This **information about the global state is distributed** and statistical in nature, and thus is difficult for observers to tease out. However, the system's components are able, collectively, to use this information in such a way that **the entire system appears** to have a coherent and useful sense of its own state.*" [272]

We have added the emphasis here, in order to highlight that a system which behaves as if it were self-aware is not necessarily required to possess a single "mind-like" component[1]. Indeed, in many cases, the entire system appears self-aware, despite only local knowledge being present at constituent parts of the collective. Self-awareness might be considered the product of emergence.

This is a key observation which can contribute to the design of self-aware systems: one need not require that such a system possess a global omniscient controller. Indeed, many natural systems appear to have been favoured by evolution which do not have such a central point of control, and rely upon relevant knowledge being available at appropriate locations within the system. It is highly likely that this idea can improve the robustness and adaptability of such systems; these are desirable properties for natural and artificial systems alike.

2.2.4 Self-expression

As we have seen, self-awareness is concerned with knowledge synthesised and held by an individual about itself and its experiences. This knowledge may be centrally held, or else distributed in nature. However, in studying the self-awareness properties of natural and computational systems, we have found it advantageous to explicitly and separately consider the related process of an individual determining its behaviour as a result of this knowledge. This process we call self-expression. In

---

[1] Indeed, while the brain has long been known to be a collective system composed of neurons, consciousness in the human mind is itself thought by some [98] to also be a distributed phenomenon, with nothing like what we might call global knowledge.

social psychology, self-expression has been defined as *"the assertion of one's individual traits"* [217]. Similarly, Chen et al. [66] define self-expressive individuals as ones which behave *"in line with their states and traits,"* where a trait is *"a genetically determined characteristic or condition"*,[2] which may be either physical or behavioural.

In this book, we therefore distinguish between self-awareness, which is the property concerned with an individual's knowledge, and self-expression, which is the property concerned with an individual's resulting behaviour, behaviour based on or informed by its knowledge and characteristics. This also helps to highlight that action or behaviour is not a requirement for self-awareness. It is, however, typically highly useful, especially in purposeful systems such as those which are engineered. For this reason, we typically consider self-awareness and self-expression together.

2.3 Computational Self-awareness

In this section, we propose that human self-awareness can serve as a source of inspiration for a new notion of *computational self-awareness* and associated *self-expression*. We introduce a general framework for the description of the self-awareness properties of computing systems. In later chapters of this book, this framework is developed into a reference architecture (Chapter 4) and a series of derived architectural patterns (Chapter 5). Together, these can be used to determine whether, how, and to what extent to build self-awareness capabilities into a system. This framework provides a common, principled basis on which researchers and practitioners can structure their work, and indeed is used throughout this book. The psychological foundations, while not strictly necessary, can provide a means of channelling a wide range of ideas – which would perhaps otherwise not have occurred to engineers – acting to inspire the design of future computing systems.

While the concepts of public and private self-awareness transfer in a fairly straightforward manner to computing systems, Neisser's [284] five levels of self-awareness lend themselves to being easily misinterpreted if discussed only in their psychological context. Therefore, we have found it useful to make these psychological concepts tangible from an engineering perspective. We do this by expressing these concepts in computational terms or processes, as part of our proposed notion of computational self-awareness.

2.3.1 Private and Public Computational Self-awareness

The first key idea of public and private self-awareness can be summarised as follows:

---

[2] *American Heritage Science Dictionary*, Houghton Mifflin Harcourt, 2005

- **Private self-awareness**: This is a system's ability to obtain knowledge based on phenomena that are internal to itself. A system needs internal sensors to achieve this.

- **Public self-awareness**: This is a system's ability to obtain knowledge based on phenomena external to itself. Such knowledge depends on how the system itself senses/observes/measures aspects of the environment it is situated in, and includes knowledge of its situation and context, as well as (potential) impact and role within its environment.

Some prior work in self-aware computing has considered only private self-awareness, i.e., a system's awareness of its own internal state. However, the importance of the availability of external sources of information to self-awareness processes should be emphasised: self-awareness is not only concerned with sources of information internal to the individual. We argue that a full consideration of computational self-awareness also includes public aspects, where a system's knowledge of itself in relation to its social and physical environment can be synthesised with private self-awareness, in order to produce integrated conceptual models. In this book, we hope to demonstrate that the distinction, inclusion and synthesis of both public and private self-awareness raise many important questions for engineers of self-aware computing systems.

2.3.2 Levels of Computational Self-awareness

We now describe our computational framing of the levels of self-awareness, which were first presented by Faniyi et al. [127] and elaborated upon by Lewis et al. [236]. It is possible to relate the levels of self-awareness to the concepts of private and public self-awareness, and hence the sources of the relevant knowledge (i.e., based on internal or external sensors). The relevance of each level to these concepts is also described. The relationship provides an indication of the architecture that will be required in order to realise each of the levels. In each case, Neisser's level of self-awareness is given on the left, and our corresponding level of computational self-awareness is on the right.

1. **Ecological self** ⟶ **Stimulus-aware**
   A system is stimulus-aware if it is able to obtain knowledge of stimuli acting upon it, enabling the ability to respond to events. It does not have knowledge of past/future stimuli. Since stimuli may originate both internally and externally, stimulus-awareness can be **private**, **public** or **both**.

2. **Interpersonal self** ⟶ **Interaction-aware**
   A system is interaction-aware if it can obtain knowledge that stimuli and its own actions constitute interactions with other systems and the environment. It is able to obtain knowledge via feedback loops that its actions can provoke or cause

specific reactions from the social or physical environment. Simple interaction-awareness may just enable a system to reason about individual interactions. More advanced interaction-awareness may involve the system obtaining knowledge of social structures such as communities or network topology. Interaction-awareness is typically based on external phenomena, whereby it is a form of **public** self-awareness. However, a system which learns about causality in internal interactions with itself would constitute a form of **private** self-awareness.

3. **Extended self** $\longrightarrow$ **Time-aware**
   A system is time-aware if it can obtain knowledge of historical and/or likely future phenomena. Implementing time-awareness may involve the system using explicit memory, time series modelling and/or anticipation. Since time-awareness can apply to both internal and external phenomena, it can be **private**, **public** or **both**.

4. **Private self** $\longrightarrow$ **Goal-aware**
   A system is goal-aware if it can obtain knowledge of current goals, objectives, preferences and constraints. It is important to note the difference between goals existing implicitly in the design of a system, and the system having access to its goals such that it can reason about or manipulate them. The former does not describe goal-awareness; the latter does. Example implementations of goal-awareness include state-based goals (i.e., knowing what may or may not be a goal state) and utility-based goals (i.e., ability to obtain a utility or objective function). Goal-awareness permits adaptation to changes in goals. When coupled with interaction-awareness or time-awareness, goal-awareness permits the ability to reason about goals in relation to other individuals or likely future goals, respectively. Since goals may exist privately to the system, or collectively as a shared or externally imposed goal, goal-awareness can be **private**, **public** or **both**.

5. **Conceptual self** $\longrightarrow$ **Meta-self-aware**
   A system is meta-self-aware if it is able to obtain knowledge of its own level(s) of awareness and how the level(s) are exercised. Such awareness permits meta-cognitive processes [84] to reason about the benefits and costs of maintaining a certain level of awareness. It further allows the system to adapt the way in which the levels of self-awareness are realised (e.g., by changing algorithms realising the levels), thus changing the degree of complexity of realisation of the levels). Since meta-self-awareness is concerned only with knowledge of internal processes, it is a form of **private** self-awareness.

Figure 2.1 shows an illustration of the relationship between private and public self-awareness concepts and these levels.

Although possession of (self-)knowledge is important to achieve computational self-awareness, we argue that it is the *ability to obtain this knowledge on an ongoing*

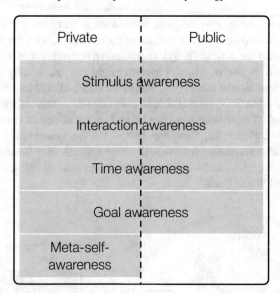

Fig. 2.1 Levels of computational self-awareness, and how they relate to private and public aspects in this framework. While the first four levels can involve perception and conceptualisation of either internal (private) or external (public) factors, meta-self-awareness is concerned with awareness of other self-aware processes, which are private to the individual.

basis throughout the system's lifetime that enables self-awareness. We do not consider a system with knowledge but no means to update or add to that knowledge during its lifetime to be computationally self-aware: it has instead been programmed by a domain expert. While the levels are considered from an architectural perspective in Chapters 4 and 5, Chapters 6 and 7 consider how online learning and knowledge representation techniques can realise these characteristics of self-aware systems on an ongoing basis.

2.3.3 Collective and Emergent Computational Self-aware Systems

Finally, self-awareness can be, and indeed is often, found in complex systems composed of many interacting subsystems. This is important for the design of self-aware computing systems. In building self-aware systems, we do not need to require a global "knowledge base" or common awareness mediator. Indeed, in many natural systems, such components are absent [272]. Whether a self-aware system is organised in a centralised or a decentralised manner is, we argue, not a conceptual difference, but an architectural one. We anticipate that, and indeed several of the examples given in this book show that, decentralised self-awareness can provide increased robustness and adaptability in the face of change.

The third key concept inspiring our framework is therefore the notion that self-awareness can be a property of a collective system, possibly even emerging from interactions within the collective. Systems within a collective that interact with each other only locally as part of a bigger system might not individually possess knowledge about the system as a whole (i.e., the global state). Global knowledge is instead distributed [272], but the system is able to collectively use this information such that it has a sense of its own state and thus be self-aware at one or more of the aforementioned levels. Figure 2.2 illustrates that, from the conceptual point of view, a collective system comprising several self-aware systems may itself be viewed as a self-aware system, and considered using the computational self-awareness framework introduced in this book.

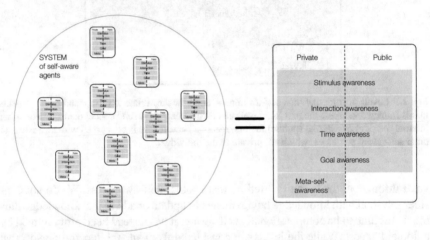

Fig. 2.2 Aspects of computational self-awareness can be considered when taking a single system view, and at subsystem level when composing a collective system. The relationship between these two may be complex, and exhibit emergent self-awareness properties.

When we talk of self-aware computing systems, we may therefore be referring to several different types of self. First, it should by now be clear that self-awareness may be a property of an autonomous agent, which is capable of obtaining and representing knowledge concerning itself and its experiences. Indeed, much of the literature on autonomous and intelligent agents is concerned with techniques for agent learning, knowledge acquisition and representation, and architectures to support these capabilities. Second, according to this third key idea of collective self-awareness, the boundary of a self-aware entity, indeed the "self", is not limited to a single agent. Chapter 4 revisits this idea, in the context of architecture, and considers both centralised and decentralised approaches.

2.4 Summary

This chapter presented a brief introduction to self-awareness concepts as they are understood in humans. Further, we have begun to consider how such concepts can be translated to the domain of computing systems. The new notion of *computational self-awareness* is proposed, based on three key concepts:

- Public and private self-awareness: A self-aware system can have the ability to obtain knowledge based on phenomena that are external and internal to itself, respectively.

- Levels of self-awareness: The self-awareness of a system can be described in terms of one or more levels of self-awareness, which characterise self-awareness capabilities. We presented a set of levels inspired by psychology, comprising stimulus-awareness, interaction-awareness, time-awareness, goal-awareness and meta-self-awareness.

- Collective self-awareness: Self-awareness may be present at the level of a single system, or at the level of a collective system, where interactions between components provide distributed learning.

Additionally, we introduced the notion of *self-expression*, behaviour based on self-awareness. Self-expression typically provides highly dynamic behaviour, able to adapt to a system's changing self-awareness as it learns. Self-expression may include behaviours considered "normal" functional system behaviour, as well as self-adaptation, self-reconfiguration, self-explanation, or any other forms of individual or collective behaviour informed by a system's self-awareness.

These concepts are used as the basis for much of the work in subsequent chapters in this book, where architectures, algorithms, platforms and case studies expand on and demonstrate their use.

Chapter 3
Relationships to Other Concepts

Kyrre Glette, Peter R. Lewis, and Arjun Chandra

Abstract This chapter will relate our concepts of computational self-awareness and self-expression to other efforts in computer science and engineering under the self-awareness label. Depending on the fields, the term self-awareness may have different meanings and may be more or less defined. Considering mainly disciplines which explicitly cover self-awareness, we present a selection of clusters of research, and their interpretation of the term. The examples range from basic, but efficient, electronic communication systems, through self-awareness in robotics and large IT systems, to more abstract and formal concepts of self-awareness emerging collectively through interaction of simple nodes. While there are many examples of work addressing self-awareness at different levels, there still seems to be a lack of general definitions and frameworks for working with self-awareness and self-expression in a computing context.

3.1 Introduction

Various research initiatives have used the term *self-awareness* explicitly to describe a property of their computing machinery within computer science and engineering. In some cases, the literature goes further and attempts to define what self-awareness might mean in the context of that research. From these initiatives, a number of clusters stand out as significant efforts to incorporate self-awareness into computing systems. However, in our view, there is a lack of a shared understanding of self-awareness concepts between these clusters. Additionally, these existing efforts have

Kyrre Glette
University of Oslo, Norway, e-mail: kyrrehg@ifi.uio.no

Peter R. Lewis
Aston University, UK, e-mail: p.lewis@aston.ac.uk

Arjun Chandra
Studix, Oslo, Norway, e-mail: arjun@studix.com

not in our view been sufficiently based on real concepts of self-awareness from psychology; rather the term self-awareness has been used as an informal description for desirable properties which might appear to fit it. In other cases, self-awareness has been identified in the literature as a likely beneficial property for computing systems to possess, though this has been elaborated on little and the term is left largely undefined. One of the potential causes for this lack of a systematic treatment of the term is likely the apparent disagreement on a universal definition for it within neuroscience and psychology literature. Computer science and engineering, where researchers tend to have a more practical approach to specifying concepts, have therefore banked on intuition for a definition, driven more by the solutions to the immediate computational problems of interest than by philosophical considerations.

Even though there is little agreement on the definition of the term both in psychology and computer science, there is a general agreement that self-awareness in computing systems, howsoever intuitively defined, can be beneficial. Some of these benefits have been summarised by Schaumeier et al. [351]. In general, the problem that self-awareness tends to address in a major portion of computing literature can primarily be characterised as the increasing difficulty we face in managing systems with physically or logically distributed autonomous components. Crucially, it is commonly believed that there is a limit to which any component may be able to dictate the behaviour of another. Decision making towards optimising system-level goals in such decentralised systems is therefore distributed across the components, each having access to partial information, which may further be unreliable in terms of the extent to which it is causally relevant to globally optimal decision making. It has been argued that components that possess self-awareness as some form of introspection capabilities [351] may aid the system towards globally efficient decisions. Intuition suggests that introspection may allow components to get fairer deals (e.g., fair allocation of tasks) for themselves out of interactions with their non-dictating and potentially unreliable neighbourhoods. Thus, if all components were to benefit from such interactions, it is likely for the system to benefit as a whole. Self-organisation of this kind may further involve components ascertaining and interacting towards preventing malicious behaviour, whilst managing their own degree of participation and role in the interactions. Additionally, systems which have to deal with humans, e.g., robotic assistants, intelligent tutors and music recommendation systems, also have to manage the degree of interaction necessary over time, so as to do "just enough" for users, thereby preventing disengagement resulting from too much or too little interaction. Such systems can benefit from having a computational notion of how they influence user behaviour.

We now introduce and discuss various clusters of research, each containing a large body of work using some notion of self-awareness. Early work surveying these efforts has been carried out by Lewis et al. [237].

3.2 Self-awareness in Artificial Intelligence

The higher levels of self-awareness, such as the meta-self-awareness introduced in Chapter 2, have been of particular interest in the artificial intelligence community. These concepts overlap significantly with meta-cognition, defined by Metcalfe and Shimamura [264] as *knowing about knowing*. Integration of AI technologies into systems that, as a result of the integration, exhibit self-awareness of this meta-cognitive form has been on DARPA's research agenda for some time [313]. Indeed, architectural issues in building such integrated systems which then exhibit self-awareness were the subject of a DARPA workshop in 2004 [12].

Cox argues that being aware of oneself is not merely about possessing information, but being able to use that information in order to generate goals, which may in turn lead to the information being modified [84]. Importantly, Cox also suggests that meta-cognition is similar to the algorithm selection problem, wherein the task is to choose the most efficient algorithm from a set of possibilities. This notion of an ability to select one's own method of collection and processing of information, according to goals which may themselves be modified by the individual, has much in common with the conceptual self discussed in Chapter 2. Cox further considers the differences between cognition and meta-cognition and argues that a meta-cognitive system is one whose domain is itself, such that it can reason about its knowledge, beliefs and reasoning process, as opposed to merely using knowledge about itself. This indeed is a capability of the conceptual self, and our corresponding notion of meta-self-awareness, discussed in Chapter 2.

3.3 Self-awareness in Collective Systems

A key concept inspiring our definition of computational self-awareness, as discussed in Chapter 2, and also our reference architecture, discussed in Chapter 4, is the notion that self-awareness can be an emergent phenomenon, arising from interactions within a collective system. Moreover, a self-aware system under consideration may be composed of a set of heterogeneous nodes which individually possess various levels of self-awareness, and it is also possible to consider self-awareness at several hierarchical levels. The notions of span and scope of self-awareness, as well as hierarchical considerations, will be elaborated on in Chapter 4.

Several directions of self-aware computing systems research consider systems with nodes at various levels of self-awareness, as well as emergent self-awareness. Examples of these are the autonomic and organic computing paradigms, and the formal modelling undertaken in the ASCENS project, which will all be covered in the following sections. One recent example of a hierarchically and collectively self-aware system is the heterogeneous swarm of autonomous underwater vehicles targeted in the CoCoRo project [354]. Here, self-awareness is considered at three hierarchical levels: the individual level, the group level, and the swarm level. At the individual level, the underwater robot needs to determine its own capacity to achieve objectives, and adapt accordingly, while at the group level awareness is present in a mechanism for locating toxic sources. Finally, at the swarm level, self-awareness allows the swarm to monitor global state information and activity, which can lead to adaptation of the swarm behaviour, or sensing when the swarm has completed a task.

3.4 Formal Models for Self-awareness

With a particular focus on distributed or collective self-awareness, recent research has also been conducted on the question of how to formally specify both knowledge and behaviour of self-aware collectives. For example, in the ASCENS[1] project, formal methods are applied to *ensembles* of components (such as robots or cloud computing services). In this project, a symbolic mathematical model for normatively describing a system of interacting components, which is referred to as the *general ensemble model (GEM)*, is proposed [177]. GEM purports to be a common integrated system model for describing components, each normally described using disparate mathematical techniques, and their interactions. In order to construct a mathematical model that can encapsulate such descriptive differences amongst components, the GEM specification works at a higher level of abstraction, namely, set theory and order theory. One then has to only consider describing an adaptive system as relations between inputs and outputs. Multiple adaptive systems can be combined using a combination operator, allowing a collective to be described as an adaptive system of adaptive subsystems. Importantly, using this combination operator, GEM is able to characterise the combination of the *environment*, the *connections* between the collective and the environment, and the *collective* itself as a system in its own right.

GEM further defines *adaptation domains*, each of which is composed of a range of environments that a collective has to be able to face, the connections between the collective and such environments, and the goals that the collective has to achieve or satisfy. Using order theory, a *preorder of adaptivity* for systems is then possible to specify. With such a formulation of systems, their adaptivity to various environments and against other systems, formal assessment of the adaptive capabilities of a system becomes possible. Indeed, a system that possesses or obtains knowledge of itself and its environment influences its own adaptivity. GEM goes on to further define the notion of *knowledge*. Knowledge that is possessed by the system at the component and at the collective level is modelled as ontologies. A meaningful data structure offers opportunities for inference, e.g., compensating for uncertainty of observations. Moreover, an ontological representation of contextual knowledge allows the component/system to identify known and unknown situations, and learn the relationships between its actions, encountered situations, and goals. Learning allows the system or its components to obtain knowledge at the time of operation in

---

[1] http://ascens-ist.eu/

the form of *patterns*, influencing how they might manage their goals in the future. Such knowledge, when employed by components to recognise changes taking place internally, at the level of the collective, and in the environment, and to learn about new situations upon detecting a change, provides *awareness* to the components. Part of this conceptualisation of awareness is *self-awareness*, which is defined as the employment of knowledge by a component to *recognise* changes taking place internally. When a component is able to recognise changes taking place at the level of the collective, such recognition abilities are termed context-awareness. Any learning which happens upon change detection forms part of the situational awareness of the component.

Vassev and Hinchey [397] discuss in some detail the current state of the art in ontology-based knowledge representation within self-aware collective systems in, and lay out both formal deterministic and probabilistic variants. They argue that this will facilitate better self-awareness through easier analysis of the states and goals of local parts of the system.

3.5 Self-awareness in Engineering

While meta-cognition or meta-self-awareness are concerned with higher reasoning abilities, and are of particular interest in artificial intelligence, efforts exist at a more fundamental level to engineer systems which explicitly consider knowledge about themselves. A case for a paradigm shift in system design practice is put forward by Agarwal and Harrod [3] and elaborated on by Agarwal et al. [4]. The idea here is to move from a procedural design methodology wherein the behaviour of the computing system is pre-programmed or considered beforehand (i.e., at design time), towards a self-aware system where this is not required and the system adapts to its context at run-time. One aim is to avoid or reduce the need to consider the availability of resources and various other constraints beforehand, instead intelligently trading-off available resources for performance at run-time.

Importantly, for a system to be self-aware it is not required that it be highly complex; indeed, the scalability of the concept means that self-awareness has also been considered in much simpler systems. An example of this is the so-called *cognitive radio devices* [133], which monitor and control their own capabilities and also communicate with other radio devices to monitor theirs. This enables them to improve the efficiency of communication by negotiating changes in parameter settings [404]. In this case, it is the system's ability to monitor and reason about its operational capabilities that is used to justify the label self-aware. Specifically, a self-aware radio device

> *"needs to understand what it does and does not know, as well as the limits of its capabilities. This is referred to as self-awareness. For instance, the radio should know its current performance, such as bit error rate (BER), signal-to-interference and noise ratio (SINR), multipath, and others. In a more advanced case, the agent might need to reflect on its previous actions and their results. For instance, for the radio to assess its travel speed a fortnight*

ago between locations A and B, it might be able to extract parameters from its log file and do the calculation. For the radio to decide whether it should search for the specific entries in the log and then perform appropriate calculations (or simply guess), it needs to know the effort required to perform such a task and the required accuracy of the estimate to its current task" [133, p. 405]

Driven by the need for satisfying heterogeneous QoS requirements across future multimedia networks, distributed monitoring and control is increasingly being seen as a viable solution for routing data as well. "Smart" networks, that use an adaptive packet routing protocol called cognitive packet network (CPN) as part of their routing architecture, are reviewed in [347]. CPN enables nodes on the network to monitor and learn at run-time the efficacy with which they can deliver packets that go through them. Node-level monitoring and learning allow the network to continuously adapt the route between a source and a destination, taking into account the potentially changing QoS requirements of nodes falling on the path between these end points. In a manner reminiscent of stigmergy, the successful delivery of a packet carves a route along the network, followed by the nodes along this route receiving feedback pertaining to this success. This feedback is key to run-time learning, enabling packet forwarding decisions to be made at each node. Nodes which are on routes leading to fewer packet losses tend to be preferred for packet forwarding, and exploration of different routes is encouraged through the allowance of a small number of random packet forwarding decisions. A network using CPN as a routing protocol is therefore seen as self-improving, and resilient to changing network conditions. The resilience of networks using CPN to denial of service attacks is shown by Gelenbe and Loukas [144], and self-awareness in the network is attributed to the use of CPN infrastructure with node-level restrictions on bandwidth. Node-level control, enabled by the CPN infrastructure, facilitates nodes' being able to detect and stifle excessive traffic passing through them.

Santambrogio et al. [349] and Hoffman et al. [170] propose the concept of *application heartbeats* as an enabling technology for self-aware computing systems. These heartbeats are intended to establish a standard way of defining application goals and evaluating the performance of a system attempting to achieve those goals. Heuristic and machine learning techniques may then be applied in order to provide adaptation and decision making behaviour. Application heartbeats form part of the SEEC approach [172] to designing self-aware systems.

When engineering a self-aware system, Agarwal argues that five design properties of the system should be considered [4]. Namely, they should be:

- Introspective, i.e., they can observe and optimise their own behaviour,
- Adaptive, i.e., they can adapt to changing needs of applications running on them,
- Self-healing, i.e., they can take corrective action if faults appear whilst monitoring resources,
- Goal-oriented, i.e., they attempt to meet user application goals, and
- Approximate, i.e., they can automatically choose the level of precision needed for a task to be accomplished.

Introspection, the ability to obtain knowledge concerning one's own behaviour, clearly forms part of self-awareness according to the definitions given in Chapter 2. However, subsequent optimisation of the system's behaviour is not directly concerned with the process of obtaining and representing that knowledge. Instead, this optimisation is concerned with the system's subsequent behaviour, thus relating to our concept of self-expression. Goal orientation, for example, requires knowledge of one's own goals, clearly a part of self-awareness. Subsequent behaviour which attempts to meet such goals, however, is not primarily concerned with obtaining or representing the knowledge, and falls under the concept of self-expression.

3.6 Self-awareness in Pervasive Computing

Self-awareness has also been of great interest to the pervasive computing community. Here, research is primarily concerned with systems that are mobile and hence their environment, performance and context changes. As such, these systems need to monitor their own state and their external environment, in order to adapt to changes in a context-specific way. Often monitoring and adaptation are studied in the context of human-computer interaction, since the interest is in how such systems self-adapt in order to be useful to humans in different situations (e.g., "going for a run"). A survey by Ye et al. [419] covers issues and challenges involved in assimilating sensor data from a myriad of sources in order for pervasive computing systems to identify situations which a human user, and hence the system itself, may be in. They show a shift in techniques over time from *logic-based* ones towards those that are *learning-based*, as obtained sensor data has become more complex, erroneous and uncertain, with sensors becoming ever more pervasive. Given currently available model building techniques, the learning of mappings between sensor data and a notion of the situation type poses several challenges. For example, the lack of training data available in rapidly changing contexts in the real world can lead to low performing models. This has been tackled by considering unsupervised learning [156, 46] and web mining [315]. Another line of research within pervasive computing [186] is concerned with constructing simulation models of contexts, in which so-called context-aware applications can learn and be tested.

3.7 Self-awareness in Robotics

One domain where self-awareness would seem an obvious property for success is the field of robotics, and in particular for autonomous and human-like robots. Indeed, architectures for autonomous robots have at least implicitly incorporated several levels of self-awareness and self-expression as defined in Chapter 2 for a long time, such as in *sense-plan-act* and other agent-based architectures [345]. Holland argues [175] that robots could benefit greatly from internal models, and identifies four major areas where such models would be advantageous: when processing novel or incomplete data, for detecting anomalies, for enabling and improving control, and for informing decisions.

Further, Winfield argues [414] that internal models in robots provide a form of self-awareness, and that this property is indeed necessary to achieve safety in unknown or unpredictable environments. By constructing and evaluating internal models of the robot itself and its surroundings, consequences of possible future actions can be evaluated. This kind of self-simulation allows for a possible moderation of the robot controller, such that future unsafe actions are not executed. Interestingly, this also opens up for a form of minimal ethical consideration in robots – robots could be programmed to avoid consequences involving humans being harmed. Another form of self-modelling in robots can be found in the Nico robot [161], where the robot's kinematic model and cameras are calibrated together. This unified approach results in a tightly calibrated self-model with low positional error on the end effector. The same approach was also used for calibrating the use of a tool, by incorporating the tool into the self-model.

Going beyond calibration, a robot building from scratch a self-model of its own morphology is presented in [41]. The model, constructed through exploratory movements, can then be used in the search of an appropriate locomotion pattern. This approach promises resilient robots—if the robot's body is damaged, it can discover this, build a new morphological self-model, and then adjust its locomotion according to the new model. The concepts of self-modelling and self-reflection in robots are further discussed in [40], where self-reflection is stated as an important aspect of self-awareness. An example of such a self-reflective robot is demonstrated in [426], where a robot with two "brains" is constructed—one neural network module monitors the actions of a lower-level neural network module. This is well in line with our concept of meta-self-awareness.

3.8 Self-awareness in Autonomic Systems

One of the key areas of computing research where the term self-awareness litters the literature is autonomic computing. Just over a decade ago, IBM proposed [180] a grand challenge to tackle *"a looming software complexity crisis"* [213]. It was argued that the complexity of computing systems, increasingly composed of distributed, interconnected components would soon be effectively unmanageable by humans, who would be unable to grasp the complexities, dynamics, heterogeneity and uncertainties associated with such systems [371]. Since then, the field of autonomic computing has developed, in an attempt to address this grand challenge. The key idea behind autonomic computing is that such complexity leaves system managers neither able to respond sufficiently quickly and effectively at run-time, nor consider and design for all possible actions of and interactions between components at design-time. Thus, in response, autonomic systems should instead manage themselves at run-time, according to high-level objectives [213].

Indeed, how to achieve effective *self-management* of complex IT systems is the fundamental aim of research in autonomic computing. Here, the concept of self-management is decomposed into four activities: self-configuration, self-optimisation, self-healing and self-protection. Unlike some of the other types of systems described in this chapter (such as meta-cognitive systems and robot swarms), these aims of autonomic systems are geared more towards achieving functional and non-functional goals associated with the management of IT systems, than achieving interesting self-adaptive or self-organising behaviour per se. However, looking deeper, we can see that such behaviours are indeed necessary for systems to self-manage. Indeed, the autonomic computing literature defines [371, 309, 105] four additional properties of autonomic systems, which are required in order to enable the activities above. These are self-awareness, environment-awareness or self-situation, self-monitoring and self-adjustment.

Self-awareness in the autonomic computing literature to date is concerned exclusively with what is described as private self-awareness in the psychological literature, discussed in Chapter 2. Similarly, our notion of public self-awareness can be compared with the notion of situated environment-awareness in autonomic computing, where the environmental knowledge concerned is built up through monitoring and observation, and constructed through modelling and learning at run-time. Finally, self-adjustment, as a form of self-expression, is concerned with behaviour of the autonomic system being based upon such learnt knowledge. Despite the prevalence of the term self-awareness in autonomic computing research, the literature offers little in the way of definition or guidance on conceptually what self-awareness might be in this context.

In order to gain some understanding of the role of self-awareness in autonomic computing, it is helpful to examine how such systems are typically built. Each component in the IT system is managed by a so-called *autonomic manager*, an autonomous agent responsible for monitoring the managed element and its environment, then constructing and executing plans based on an analysis of that information. In addition, a knowledge repository is available to the autonomic manager, containing models of the behaviour and performance of the managed component, which are typically developed and provided by an expert in advance to enable run-time planning. Together, the *monitor-analyse-plan-execute* loop and the knowledge repository are known as the MAPE-K architecture [190], which forms the basis of the design of many autonomic managers. The MAPE-K architecture is depicted in Figure 3.1; as can be seen it has much in common with other general agent architectures, such as those described by [345].

However, a decade since the original autonomic computing concepts were proposed for the management of complex IT systems, autonomic systems research now takes a broader view of applications, architectures and techniques. For example, autonomic systems are commonly now considered to include applications as diverse as pervasive computing [146], communication networks [104] and robotic swarms [427]. Indeed, any networked system which manages itself through autonomous decentralised decision making could be considered autonomic. Appropriately, a generalisation of the original MAPE control loop for autonomic systems, acknowledging

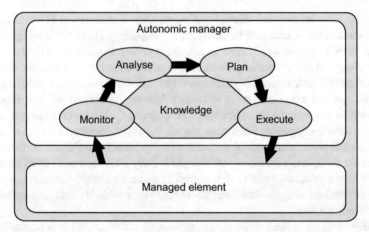

Fig. 3.1 The MAPE-K architecture for autonomic managers (inspired by [190]). The *monitor* and *analyse* components obtain and analyse sensor information. This implements functionality for what are termed self-awareness and environment-awareness in the autonomic computing literature, and can be compared with the concepts of private and public self-awareness introduced in Chapter 2.

that a multitude of techniques could be used to realise each step, is described in [105]. This more generalised autonomic control loop consists of: *collect, analyse, decide, act*. In one example of this generalisation, the analysis and planning steps from MAPE are replaced with a reinforcement learning algorithm [383]. Also in this approach, hybrid knowledge bases, consisting of available domain knowledge and that obtained during run-time, are advocated.

3.9 Self-awareness in Organic Computing

In the absence of system-wide knowledge for decision making and control, the emergence of desirable global behaviour through local interactions is one of the fundamental challenges in complex systems research today. Indeed, desirable emergence is seen as one of the key benefits that autonomous self-aware components could enable in these systems [351]. One research initiative that directly addresses the issue of *desirability* in self-organised, indeed emergent, behaviour of complex systems is organic computing [277]. Desirability is the main research focus here, and, depending on system-level requirements, takes precedence over the degree of autonomy afforded to individual components. According to Schmeck [353], *"it is not a question of whether self-organising systems will arise, but how they will be designed and controlled."*

An organic computing system is said to be strongly self-organised if its components do not rely on any central control for coordination. A weaker form of self-organisation allows for the system boundary to be relaxed by the inclusion of a central *observer* and *controller* within it, thereby making the observation of system-wide objectives carried out by some component, followed by corrective control. Indeed, lowering the degree of autonomy afforded to individual components is therefore the price paid for the achievement of desirable global behaviour. Such concession of autonomy is one major difference between organic and autonomic computing (discussed in Section 3.8), the latter favouring strong self-organisation.

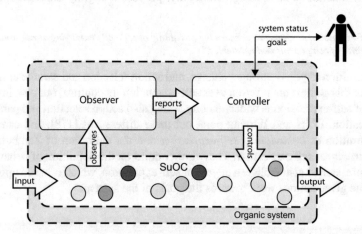

Fig. 3.2 Observer/Controller architecture (inspired by [338]). The organic system under observation and control (SuOC) consists of several self-aware heterogeneous nodes.

The observer and controller approach to governing the emerging global behaviour of a self-organising system is central in organic computing, and a generic Observer/Controller system architecture has been proposed. Here, the decentralised, self-organising system to be governed is called a *system under observation and control* (SuOC), and the observer and controller components are responsible for surveillance and feedback through a control loop [338]. An illustration of this architecture can be seen in Figure 3.2. The observer collects and aggregates behaviour information about the SuOC through sensor information. The resulting system indicators are reported to the controller which can then, based on objective functions, constraints, and overall goals, take actions to influence the SuOC. The architecture has several implementation options: both centralised and decentralised, as well as hierarchical approaches, are possible.

3.10 Self-expression in Computing

Though the notion of self-expression is not common in the approaches to self-awareness in the literature, there is little doubt that the concepts covered by self-expression are implemented and discussed in alternative ways. For example, both the MAPE-K and the Observer/Controller architectures emphasise the distinction between knowledge collection and acting based on this knowledge. Two related, but not similar definitions of self-expression were proposed at the International Conference on Self-Adaptive and Self-Organizing Systems Workshops in 2011, namely the definition of Zambonelli et al. [427] and our own interpretation [237]. Our own interpretation is presented in Chapter 2 of this book; Zambonelli et al., stemming from the autonomic computing community, propose a slightly different approach where

> *"Self-expression mechanisms concern the possibility of radically modifying at run-time the structure of components and ensembles."*

This definition implies a stronger change, and action selection and parameter modification in this context are referred to as self-adaptation. In contrast, our own interpretation of self-expression is wider, encompassing also action selection and parameter modification. Hölzl and Wirsing point out these differences [178], and categorise our definition as *behavioural self-expression* and the definition of Zambonelli et al. as *structural self-expression*. They further build on the definition of behavioural self-expression to also include a *degree* of self-expression, which is based upon how much the given actions work towards the goals of the system.

3.11 Summary

In this chapter we covered several approaches which explicitly use the term self-awareness in computer science and engineering literature, from theoretical levels to hardware implementations.

First, we saw how the artificial intelligence community took particular interest in a higher level of self-awareness—the meta-cognitive approach, similar to our notion of meta-self-awareness. Then, we saw how self-awareness is considered at agent levels, but also how it may appear collectively through interaction of simpler nodes, that is, at higher hierarchical levels in a complex system. The formal modelling approaches of the ASCENS project as well as the organic computing initiative are both explicitly emphasising this kind of collective self-awareness in systems consisting of heterogeneous self-aware nodes. On a more concrete level, the engineering approaches in cognitive radio and cognitive packet networks make use of introspection to monitor their own performance in the network. However not only private self-awareness is displayed, as the nodes also need to assess their performance in relation to the rest of the network, thus introducing public self-awareness as well. In pervasive computing and robotics we saw how self-models of the agent as well as its context were constructed, allowing for sophisticated reasoning and prediction of the outcome of future actions. These models typically comprise almost all levels of computational self-awareness according to our definitions in Chapter 2, i.e., stimulus-awareness, interaction-awareness, time-awareness, as well as goal-awareness, though not necessarily meta-self-awareness. In our final example, the autonomic computing approach tries to tackle the very concrete challenge of managing complex IT systems, through self-management. Autonomic computing distinguishes between self-awareness and situated environment-awareness, corresponding well to our concepts of private and public self-awareness.

It should be noted that there are also several other approaches within engineering and computer science which do not explicitly claim self-awareness, but still operate under similar principles. This becomes true in particular when considering the lower levels of computational self-awareness defined in Chapter 2. Often, the use of self-awareness refers to quite disparate ideas, for example, to highlight quite specific self-monitoring capabilities of a system or to indicate an awareness by the system of the user or context or that a component has a conceptual knowledge of the wider system of which it is part.

Despite the many forms of computational awareness discussed, many approaches, including the big architectural approaches like MAPE-K and Observer/Controller, do agree on the distinction between awareness and action. This key split between functionality responsible for knowledge acquisition and representation and that responsible for knowledge use and behaviour is important, and our concepts of self-awareness and self-expression follow the same line of distinction. However, though implicitly implemented, the concept of self-expression is not widely adopted in a computing context; but it has received increased attention in recent years.

There exist so far few efforts to establish a common framework for describing or benchmarking the self-awareness capabilities of computing systems, or the benefits that increased self-awareness might bring. Still, there seems to be a common understanding that studying self-awareness concepts, and implementing features of self-awareness, is a necessary path to take in order to cope with, and achieve efficiency in, increasingly complex computing systems dealing with dynamic environments. In that endeavour, common notions of what self-awareness and self-expression mean in a computational context would be of utmost importance. Common definitions and frameworks would help us not only to answer questions about the self-awareness properties of a system, such as *what kind of* self-awareness is present, and to *what extent*, but also to reason about the benefits and costs of implementing a certain form of computational self-awareness.

Chapter 4
Reference Architecture for Self-aware and Self-expressive Computing Systems

Arjun Chandra, Peter R. Lewis, Kyrre Glette, and Stephan C. Stilkerich

Abstract This chapter covers a reference architecture for describing and engineering computational self-awareness and self-expression in computing systems. The architecture provides a common language with which to engineer the capabilities exercised by a "self" at a fine resolution inspired by concepts from psychology. The "self" demarked by the reference architecture is conceptual in nature, and therefore not limited to describing single agents. Consequently, the architecture allows the engineering exercise to scale freely across systems composed of arbitrary agent collectives. Being a common language, it paves a way for identifying architectural patterns influencing the engineering of computational self-awareness and self-expression capabilities across a range of applications. The psychological basis of the architecture brings clarity to the notion of self-awareness and self-expression in computing. These foundations also serve as a rich source of ideas which can now be channelled into the computing domain and inspire the engineering of computationally self-aware and self-expressive systems of the future.

4.1 Introduction

Computational self-awareness and self-expression are processes that can realise a range of capabilities within computing systems. We introduce a reference archi-

Arjun Chandra
Studix, Norway, e-mail: arjun@studix.com

Peter R. Lewis
Aston University, UK, e-mail: p.lewis@aston.ac.uk

Kyrre Glette
University of Oslo, Norway, e-mail: kyrrehg@ifi.uio.no

Stephan C. Stilkerich
Airbus Group Innovation, Germany, e-mail: stephan.stilkerich@airbus.com

tecture for engineering such capabilities within agents and agent collectives. The architecture does not assume self-awareness to only be an add-on capability, but instead encourages methodically describing and extending the capabilities that may already be exercised by the system. The advantages to using this architecture as a design guide are threefold. First, the architecture offers tangibility over the *extent* and *scope* of the system's capabilities, irrespective of whether the system *spans* a single agent or an arbitrary collective. It does so by separating the knowledge concerns that underpin different levels of computational self-awareness, and the concerns influencing computational self-expression. This enables a high-resolution analysis and design of these capabilities, the design exercise freely scaling to include collective systems. Different implementations of the same capability can therefore be compared and evaluated. Second, it can be used as a template for identifying common ways of assembling these capabilities within systems, resulting in patterns for architecting a variety of applications. Third, the architecture provides a common and principled basis on which researchers and practitioners can structure their work. The psychological foundations of the architecture, while not strictly necessary, can serve as a rich source of inspiration that may not have occurred to engineers to have existed. From this source, a wide range of ideas could be channelled into the computing domain, thereby inspiring the design of future computationally self-aware and self-expressive computing systems.

This chapter is structured as follows. Section 4.2 questions the need for a reference architecture for engineering self-aware and self-expressive computing systems. Having established the need, Section 4.3 describes our proposed architecture in detail, enriched with example instantiations of the primitives that compose the architecture. Section 4.3.1 is mostly concerned with engineering computational self-awareness and self-expression capabilities in the context of single agents. We extend this discussion in Section 4.3.2, showing the applicability of the architecture in the context of agent collectives, and put forth the idea that these capabilities, being computational processes, can also have an emergent nature. In Section 4.4, we briefly discuss how our reference architecture is actively being used to engineer computational self-awareness and self-expression in computing systems.

4.2 Architectures for Designing Self-adaptive Systems

An agent is a computing entity *"that can be viewed as perceiving its environment through sensors and acting upon that environment through effectors"* [345], typically in order to satisfy some goals, where its actions may also depend on any relevant knowledge the agent may possess, in addition to what is immediately perceived through sensors. Russell and Norvig [345] describe a number of widely recognised architectural blueprints for realising intelligent agents. Varying degrees of knowledge acquisition and decision making capabilities, ranging from condition-action rules to the agent's modelling the environment in relation to its actions, characterise these exemplars. Computational self-awareness and self-expression, as terms, do not feature in these, yet the blueprints incorporate the foundational ideas which characterise these capabilities.

Various types of distributed systems, typically without central control, can be termed as collectives. Some examples include swarms, systems-of-systems, populations of computing entities, multi-agent systems, etc. Individual components of these systems can indeed be referred to as agents. Such collectives may or may not be composed of agents heterogeneous in their knowledge acquisition and decision making capabilities. We intend to showcase the means by which computational self-awareness and self-expression capabilities can be architected within both agents and collectives, which we generally refer to as computing systems.

Many architectures embodying (often layered) control loops have found practical use for engineering computing systems with self-awareness capabilities. Some of these do not explicitly use the term self-awareness. However, they are generally driven by operational challenges which complex computing systems face when encountering circumstances that are hard to consider at design time, e.g., faults. The space of such operational circumstances can be large, indeed unfathomable at design time, rendering run-time self-adaptation to being a fundamental architectural concern across a range of research communities. Notable amongst these include the observe-decide-act (ODA) loop and the MAPE-K [213] architecture respecting the autonomic computing paradigm [252], and the Observer/Controller [277] architecture originating in organic computing research.

All proposed ODA loop variants and their corresponding architectures are derived from the groundbreaking work on OODA (observe-orient-decide-act) loops, originally introduced by Colonel John Boyd [44]. Surprisingly, the second O-step (orient) is removed from these models. But in the OODA loops this step represents the important phase where pure observations from the first O-step are given a sense and meaning, based on a-priori knowledge and learned knowledge (experience). The reason to drop the "orient" step might be in order to simplify the model, although an important idea of OODA loops is lost in doing this. This may also be in order to not refer directly to OODA loops, which have their origin in the military domain.

One manifestation of the ODA loop is the SEEC [173] architecture. Extending the ODA loop, it decouples application and system developer concerns, with a view towards unburdening application developers from run-time operational know-how. Explicitly sitting at the interface, SEEC provides application-layer observation primitives that help monitor running applications, run-time decision making/control primitives that allow for varying degrees of deliberation on observed application data, and actuator primitives which let the system act on itself and its applications. This *self-adaptive control loop* lets the system dynamically manage both application- and system-level goals at run-time.

Another design framework characterised by a self-adaptive control loop is the RAINBOW [143] architecture. This architecture enables dynamic management of the system's components, in this case a system's computational, storage, and interface units, by *adding on* an external control layer to the system. Relying on the design specification of the system's software, this control layer monitors the run-time properties of the system, evaluating constraint violations based on the specification. Any violation is followed up by system- or component-level adaptations. Borrowing ideas from robotic system architectures inspired by the ODA loop, the three-layered reference architecture [225] for describing *self-managed* systems also relies on the use of the system's design specification for adaptations. Each layer is characterised by the degree of deliberation required for actions to follow feedback from the monitored system.

Any networked system which manages itself through autonomous decentralised decision making could be considered autonomic. A defining characteristic of the MAPE-K architecture for engineering such systems is to have an autonomic manager *added on* to each component of the system. In addition, a knowledge repository is available to the autonomic manager, containing models of the behaviour and performance of the managed component, along with goals, objectives or utility functions which describe desirable states for the managed component. The knowledge base may contain explicit system models able to predict the likely effectiveness of potential actions, which can then be used by the manager to plan appropriate actions to execute. The repository is typically developed and provided by an expert in advance. Crucially, the components of the system become self-aware by virtue of this manager, and are not so without it.

These control loop architectures have much in common with generic learning agent architectures [345], as depicted in Figure 4.1.

When a system is provided with knowledge about itself in advance, by an expert who is external to the system, we argue that the presence of this knowledge does not itself endow the system with self-awareness capabilities. We argue that the subjective nature of self-awareness requires that, in order to be considered self-aware, a system's knowledge concerning itself and its environment be obtained by the system itself, through subjective experiences. The system must exercise *processes* allowing it to learn from its own point of view. Supporting the argument that such self-awareness would be beneficial for autonomic systems, Tesauro [383] claims that the difficulties in obtaining a sufficiently accurate model of a component, especially considering the complex and dynamic nature of the environment within which the component may operate, has been a limiting factor in the adoption of this architecture and its derivatives. Instead, he argues, such models may themselves need to be adaptive, and that this is something which is very difficult to achieve following classical system modelling approaches.

However, we are not arguing for the abandonment of design-time modelling; far from it, as Tesauro [383] advocates, hybrid knowledge bases can be used, consisting of available domain knowledge and that obtained during run-time by reinforcement learning. In essence, the difficulties associated with sufficiently and accurately modelling the complexities associated with autonomic systems leads the deliberative planning process to be replaced with a more reactive reinforcement learning process.

As discussed in the previous chapter, self-awareness also finds mention in the *Organic Computing* vision [277]. The core aim of this vision has been to get a deeper understanding of the emergent dynamics of large autonomous systems. In order to have a tangible handle on the emergent behaviour, the vision prescribes *adding*

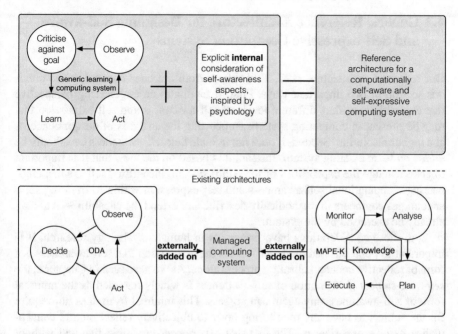

Fig. 4.1 Existing control loop architectures in the context of a generic learning agent/computing system [345]. Existing architectures prescribe the control loop be *externally* fitted on to managed components. We prescribe the explicit internal consideration of self-awareness aspects, including control loops, in the system being engineered. Any computing system that continuously learns or obtains knowledge through subjective experiences, in line with the general concept of a learning agent/computing system, including obtaining knowledge about itself, has computational self-awareness capabilities which can be described in explicit terms inspired by psychology.

on an observer and controller [353] component to the system. Architecturally, the component sits outside the system, monitoring and taking corrective actions on it when necessary. At the same time, this external component may only monitor and act on parts of the system, therefore affording varying degrees of autonomy to the emergent system.

These architectures *assume* the idea that self-awareness can be *added on* to a computing system in the form of a reflective management or control layer. Human psychology research considers self-awareness as being a more general notion, as we elaborate in Chapter 2. Self-awareness, in this broader sense, permeates all aspects of a system's behaviour. Limiting its consideration to an external feedback loop may therefore not be fully appropriate. It also limits the diversity of potential design opportunities that can be entertained by engineers. Entertaining both generality and precision with which to engineer self-awareness capabilities therefore warrants a novel architecture, which we describe next.

4.3 Generic Reference Architecture for Designing Self-aware and Self-expressive Computing Systems

Our reference architecture [127, 236] differs from the ones covered in the previous section in two important ways. First, it facilitates an engineering perspective that explicitly considers different levels of self-awareness and self-expression that may be present in computing systems, supporting the analyses of design concerns at a higher resolution. Second, it *does not assume* that self-awareness can simply be *added on* to an existing system. Instead, it is based on the view that it is important to acknowledge the capabilities of the entire system, the entire "self", when engineering computational self-awareness and self-expression within it. In doing so, it encourages engineers to methodically describe and extend the capabilities that may already be exercised by the system.

Both these considerations have firm roots in human psychology research. It is important to point out that the latter relaxes the assumption that self-awareness is a form of reflective conceptualisation process alone. As we have seen in Chapter 2, the *self-as-a-subject* [232] notion of self-awareness is widely regarded as the minimal form of self-awareness any agent can possess. This minimal form does not require for the subject to have any monitoring layer to objectively reflect on and conceptualise its own experiences. The fact that experiences are subjective and without any conceptualisation is alone sufficient to make the subject self-aware in a minimal sense. A stimulus-aware agent is therefore self-aware to some extent, and any additional conceptualisation only adds to the extent of its self-awareness capabilities. This notion has not received any attention in existing architectures. Asking engineers to consider describing the capabilities at various levels of self-awareness would therefore encourage generality and precision, indeed greater design opportunities, by letting them focus their efforts towards engineering only relevant capabilities, dictated by the wide range of challenges offered by various applications. Our reference architecture offers such design opportunities.

The architecture is based on the *three key ideas* (elaborated on in Chapter 2) underpinning computational self-awareness:

1. Computing systems can possess **public** and **private self-awareness**.
2. The *extent* of a system's self-awareness capabilities can be characterised by **levels of self-awareness**.
3. Self-awareness can be an **emergent phenomenon in collective systems**.

It is also based on the notion of an abstract computational *node*. Such a node may or may not exist as a separate physical entity in hardware or software, but more importantly represents the locality of the notion of what is considered *self* in a complex computational system. Nodes therefore represent the level(s) of abstraction at which the considered self-awareness exists. This may, in many cases, be consistent with the level at which agency is considered to exist when employing an agent-based paradigm; however, one can also think of the self as being a collective of agents, which together possesses self-awareness. A "self" is therefore an abstract boundary within which an engineer wants to give explicit consideration to realising computational self-awareness and self-expression. This "self" is the subject of experiences of its own, where its capabilities allow it to process these experiences and act accordingly. We term this abstract boundary, the *span* of a "self". The domain of the phenomena able to be sensed and modelled by the "self" in question is what we call its *scope*. As such, for a system which is only privately self-aware, the scope may be the same as the span (i.e., it has no perception of its environment). For a system which has some private and some public self-awareness, the scope would be larger than the span, and include external social or physical aspects of the environment.

This notion of a node (the *self*) being a collective is particularly relevant to the idea of distributed self-awareness, as expounded by Mitchell [272]. In this case, it is entirely possible that such self-awareness properties are present at the level of the collective, but not at the level of any individual component within that collective. In this case, we might consider that self-awareness properties have emerged from the interactions of simpler components. In summary, since a system can be a single agent or a collective, this architecture can apply equally to agents or to collectives, or both. We will discuss how it applies to agents in Section 4.3.1, and extend the discussion for its applicability within collectives in Section 4.3.2.

4.3.1 Reference Architecture for Agents

Figure 4.2 shows a schematic of our reference architecture. Our experience shows that, as a template, it brings structure to the design of self-aware systems, and helps benchmark different self-awareness capabilities. Each level of self-awareness can be studied or implemented independently or in the context of other levels. Different implementations of the same capability can be compared based on their complexities and their effects when employed by a node.

The architecture clarifies that *computational self-awareness is a process* (or set of processes). It is concerned not only with knowledge possessed by an agent at any point in time, but additionally the computational processes that enable it to continuously obtain knowledge via online learning. Such learning can result in models pertaining to the levels of self-awareness being exercised by the agent. The architecture enables reasoning about and investigating online learning in relation to an agent's self-expression capability. Driven by its goals, a self-expressive agent should be able to use the learnt models in a variety of ways so as to make decisions on how to act. Different action selection/decision making mechanisms can therefore be evaluated. Such decisions can directly or indirectly drive learning.

Given this architecture, if an agent possesses only public self-awareness then it would only be able to access knowledge of other agents or the environment the agent is operating within. Conversely, an agent which possesses only private self-awareness would have no knowledge of its social or physical environment, but would instead have knowledge about itself: perhaps its state, current behaviour or history. Possession of both public and private self-awareness allow these two sources of knowledge to be combined to provide a meaningful context for adaptation and be-

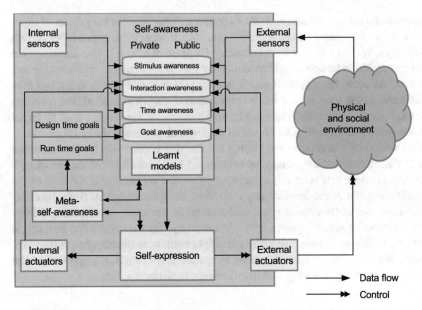

Fig. 4.2 Reference architecture [127, 236] for designing self-aware and self-expressive computing systems

havioural decisions. This knowledge will, for example, be able to support both simple reactive behaviour as well as complex learning, prediction and action selection tasks. Moreover, actions of an agent can further help it learn how and when to act towards affecting the external environment and its internal functionality, given the knowledge that forms part of its self-awareness. In other words, self-expression is a crucial companion of self-awareness. Without self-expression an agent is essentially a data sink.

An example of the benefit of considering a system's self-awareness and self-expressive properties separately can be provided in the context of the distributed smart camera system [122] described in greater detail in Chapter 13. Here, individual cameras within a decentralised network are self-aware, in that they collect and process information about their state and context, such as what they can currently see, their progress in achieving their goals (here associated with tracking seen objects), and knowledge of their interactions with neighbouring cameras in the network. However, they are also self-expressive; they make decisions about which objects to track and how to allocate tracking tasks between neighbouring cameras. Their communication behaviour determines how to balance the trade-off between overhead and performance by making use of historical knowledge. Thus, the self-awareness informs the self-expression of the camera. Clearly, the processes associated with knowledge and those with actions must both be attended to in optimising the cameras' design. Our architecture separates these concerns, thereby encouraging a focussed effort when assessing and engineering these capabilities in the context of each other. Engineers can compare and evaluate a variety of self-expression implementations for their efficacy in getting the system to achieve design and run-time goals, given the same knowledge acquisition processes.

At this stage it is helpful to visit Agarwal's [4] design properties for self-aware computing systems. More on these is detailed in Chapter 3. Each of these may be decoupled into a self-awareness component and a self-expression component. Such a proposed decoupling is presented in Table 4.1. Decoupling these design properties facilitates a thorough consideration in designing what a system knows about itself, as well as how it acts on itself and its external environment.

Original property	Self-awareness component	Self-expression component
Introspective	Knowledge based on observation and monitoring of system behaviour.	Optimisation of behaviour according to system objectives.
Adaptive	Knowledge of application or component requirements.	System adaptation appropriate to current and future application requirements.
Self-healing	Knowledge of faults in the system or utilised resources.	Appropriate corrective action.
Goal oriented	Knowledge of system level, application and user goals.	Actions taken to meet known goals.
Approximate	Knowledge of current and possible performance and capabilities, and of requirements.	Ability to select behaviours and techniques appropriate for required performance and other goals.

Table 4.1 Agarwal's [4] design properties of self-aware systems, decoupled to show self-awareness and self-expression components

4.3.1.1 Architectural Primitives

The building blocks or architectural primitives of our proposed reference architecture include:

Internal and external sensors: The private and public self-awareness of an agent rely on continuous streams of data, which are provided by the internal and external sensors respectively. Sensors are therefore the measurement apparatus of an agent, allowing it to observe phenomena on which to base its self-awareness.

Internal and external actuators: The interactions of an agent with its external environment are affected by external actuators. Similarly, the interactions of an agent with itself, or the actions of the agent that directly affect internal functionality, are exercised by internal actuators. Note that the actions taken by an agent, either external or internal, need to be observed by the agent for higher degrees of interaction awareness. The explicit flow of data directly from the actuators to the interaction awareness component depicts the knowledge of actuator status. The eventual outcome of the actions, however, may need to observed through the sensors.

Self-awareness: The computational process that realises each self-awareness capability analyses the observations provided by sensors. This results in subjective models or knowledge of the internal or external phenomena being accounted for by the agent. Additionally, the goal-awareness component helps an agent obtain and acknowledge both design and run-time goals, which are then used by various levels to construct the respective models, further affecting the actions of the agent. Meta-self-awareness plays a key role in managing the set of goals an agent works with during its lifetime. Different operational environments or internal states that an agent finds itself in can require the agent to change focus from one goal to another. The meta-self-awareness component can help an agent perceive the costs and benefits, indeed the trade-off between various goals, given the feedback from these environments and states that arise out of the agent's actions. It allows an agent to continuously monitor these goals and their relationship with its own functionality. Due to such monitoring, the meta-self-awareness component can manage the agent's functionality, specifically the degree to which its self-awareness and self-expression capabilities get realised.

Self-expression: An agent uses the knowledge and models obtained through self-awareness processes, including knowledge about goals through the goal-awareness component, when deciding upon its actions. The results of the self-expression processes are commands for the internal or external actuators. As can be expected, affecting internal functionality or the external environment can directly or indirectly influence an agent's learning, indeed self-awareness. As the self-expression component may itself involve complex decision making processes, a clear separation between this component and self-awareness can help designers and practitioners evaluate a variety of such processes explicitly.

4.3.1.2 Example Implementations of the Primitives

Below are concrete examples of the architectural primitives described above:

Internal and external sensors: Internal sensors measure aspects internal to the agent and could, for example, be temperature or battery level sensors. External sensors can include cameras or microphones.

Internal and external actuators: Internal actuators could, for example, be affecting the energy consumption of the system, like throttling the internal CPU speed, or changing properties of the sensors, such as adjusting the zoom level of a camera. External actuators, on the other hand, will affect the environment in some way, and could for instance be a radio transmitter or a loudspeaker.

Self-awareness: While at the stimulus-awareness level the agent could receive messages from neighbouring agents, at the interaction-awareness level the process could involve building a model, e.g., a spatial map, of the different agents. Advancing to the time-awareness level, one could add communications history to this map, which could be used for estimates of future communication decisions. The goal-awareness level may, for example, monitor a goal of sensory coverage in an area based on internal sensing as well as communications from other agents. The meta-self-awareness component could employ self-expression to perform algorithm selection, such as switching between sensing strategies based on knowledge about energy levels and neighbourhoods. Low energy levels or good neighbourhood coverage could activate a more power-efficient algorithm which builds a less accurate environment model based on fewer samples.

Self-expression: A self-expression process would build on knowledge from the self-aware processes, and could for example choose to rotate an on-board camera in another direction, based on knowledge about the area covered by other agents.

4.3.2 Architecting Collectives

There are a multitude of ways computational processes can be set up so as to realise various levels of self-awareness and self-expression. The capability realised is a property of some computing system. Components of this system may have autonomy, and may interact with each other following some rules of engagement, adapting to local circumstances given the costs and benefits afforded them by this autonomy. The dynamics of such a system may therefore be complex. The system may evolve through periods of instability towards exhibiting patterns of behaviour deemed desirable for it to sustain. These complex adaptive systems can in themselves be seen as computational processes which give rise to desirable systemic phenomena, making the system appear self-aware at various levels.

The emergent appearance of self-awareness can also be viewed as follows. Variedly (in terms of levels of self-awareness) self-aware agents which interact with each other only locally as part of a bigger system might not individually possess knowledge about the system as a whole (i.e., the global state). The information about the global state is distributed and statistical in nature [272], but the system is able to collectively use this information such that it appears to have a sense of its own state and thus be self-aware at one or more of the aforementioned levels.

We should emphasise that the reference architecture described in the previous section is not confined to being an agent architecture. It describes how capabilities of a computing system can be organised. It is independent of the processes that realise the capabilities, or indeed the forms of self-organisation that can be exercised by an agent collective. It describes the self, *not* how one self may interact with another. Yet, it allows the interaction between different selves to be studied and engineered in depth, by letting engineers focus on the concerned levels of self-awareness, particularly interaction-awareness, and self-expression. In doing so, it allows drawing an arbitrary boundary around a subsystem to describe its capabilities. We can therefore study capabilities of collectives, or arbitrary parts of it, under the same abstraction. This idea is depicted in Figure 4.3. As such, our reference architecture not only enables the principled engineering of self-awareness and self-expression capabilities at a fine conceptual *resolution* inspired by human psychology, it also enables this engineering exercise to freely *scale* across collectives.

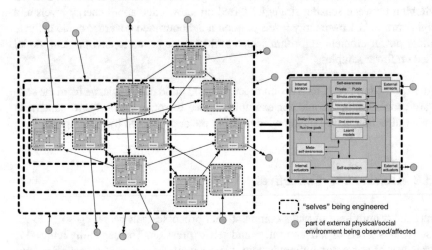

Fig. 4.3 Reference architecture demarking the "self", composed of either individual agents or an arbitrary agent collective. The boundary of each "self" defines the *span* of the subject exhibiting computational self-awareness and self-expression capabilities. Given the span, a principled study and engineering of the extent and scope of self-awareness capabilities and self-expression, realised by the "self", can be carried out.

We now give some examples of collectives exhibiting emergent behaviour reminiscent of the characteristics attributed to a "self" with different levels of computational self-awareness and self-expression. Consider ant colony optimisation; an artificial ant senses pheromone levels to act (the system is stimulus-aware) and it interacts with others via stigmergy, where the system maintains and updates a memory (the system is time-aware); and this system appears to be able to find the shortest path even if there are disruptions (by way of dynamics in the optimisation problem in question), making it goal-aware as well.

Significant research effort has been expended in recent years towards agent-based modelling of collectives where individuals within the collective have competing goals. One particularly active direction has been in terms of modelling economic interactions [60] and how to engineer market-based agent collectives [240, 321]. Amongst other things, these efforts have shown that desirable systemic characteristics can emerge through decentralised agent interactions, specially when these agents can adapt their behaviours through online learning. One of these systemic characteristics is that of the system resolving individual conflicts of interest and reaching equilibrium states. Individual agents do not share any knowledge of, nor a means to cooperate towards, equilibrium states. These systems appear to be goal-aware without their components having any notion of these goals. Having obtained the knowledge that their actions affect their social and/or physical environment, by way of sensing the changes to the costs and benefits their agency affords them, the components continuously adapt to meet their individual goals. The components therefore exhibit interaction awareness.

4.4 Reference Architecture in Practice

In recent years, we made use of this reference architecture to aid the engineering of computational self-awareness and self-expression across a wide variety of applications. We have found its use advantageous, helping advance the state of the art across these applications. Some of these include:

- Decentralised service selection in cloud-based collectives (Chapter 5).
- Run-time hardware reconfiguration (Chapter 8).
- Run-time reconfiguration of the Internet protocol stack (Chapter 10).
- Acceleration of financial market computations on heterogeneous compute clusters (Chapter 12).
- Object tracking with smart cameras (Chapter 13).
- Encouragement of human participation in single and multi-user active music environments (Chapter 14).

The above applications are covered in detail in the remainder of the book. One benefit of using the architecture is that it provides a common language with which to describe the range of capabilities each "self" in these applications possesses, be it an FPGA, a smart camera, an interactive musical device, a host in the cloud, or indeed a collective of these computing entities.

Any common description language carries within it the potential for exposing similarities across phenomena it tries to explain. Our reference architecture, having been used as such a language across the applications mentioned above, has exposed common ways of assembling computational self-awareness and self-expression in order for the "selves" to meet various quantitative and qualitative requirements, be they functional, non-functional, or constraints, posed by these applications. As such, we have formulated a wide range of architectural patterns, each characterising the effects of assembling and realising one or more levels of computational self-awareness and self-expression within computing systems. These patterns can be referred to by engineers and practitioners when challenged by achieving these effects. Chapter 5 describes these patterns and a systematic pattern selection method which uses a set of questions to help the designer identify application-specific requirements in relation to each level of computational self-awareness.

Part II
Patterns and Techniques

Part II outlines some common architectural primitives and guidelines for engineering self-aware systems using design patterns and different knowledge representation techniques, respectively. Chapter 5 provides design patterns and primitives for designing self-aware and self-expressive computing systems in a principled way. It discusses how the proposed patterns and primitives can be used in real software system projects. Chapter 6 explains issues which may be present in self-aware and self-expressive systems, such as adaptivity, robustness, multi-objectivity and decentralization, and discusses their implications in terms of knowledge representation and modelling choices. Finally, Chapter 7 concludes the second part by introducing common techniques that could be used in self-aware and self-expressive systems, including classical online learning, nature-inspired learning and socially-inspired learning in collective systems.

Chapter 5
Design Patterns and Primitives: Introduction of Components and Patterns for SACS

Tao Chen, Funmilade Faniyi, and Rami Bahsoon

Abstract When faced with the task of designing and implementing self-aware and self-expressive computing systems, researchers and practitioners need guidelines on how to use the concepts and foundations of self-awareness. This chapter provides such guidelines on how to design self-aware and self-expressive computing systems in a principled way. We have documented different levels of self-awareness and proposed architectural patterns. We have also discussed common architectural primitives and attributes for architecting self-aware and self-expressive systems. Drawing on the knowledge obtained from the previous investigations, we discuss how the proposed patterns and primitives can be used in real software system projects.

5.1 Introduction and Motivation

We have developed the notions of self-expression and different levels of computational self-awareness, inspired by psychology and cognition. In the context of architecture, we refer to different levels of computational self-awareness, the self-expression, the sensors and the actuators as the **capabilities** of systems to obtain and react upon certain knowledge. The levels of self-awareness describe different types of self-knowledge which a system may possess and learn. Subsequently, the presence of the different types of knowledge may lead to different classes of behaviours being possible. The categorisation of different types of self-knowledge has the possibility to ensure that, when designing self-aware systems, only relevant types of knowledge are included, and their inclusion is justified by the identified benefits.

Tao Chen
University of Birmingham, UK, e-mail: t.chen@cs.bham.ac.uk

Funmilade Faniyi
University of Birmingham, UK, e-mail: f.faniyi@gmail.com

Rami Bahsoon
University of Birmingham, UK, e-mail: r.bahsoon@cs.bham.ac.uk

This is extremely important as there is no need for a system to become unnecessarily complex, learning and maintaining knowledge which does not advance its outcomes, generating only overheads. Consequently, design processes for self-aware systems will need to take account of the necessity or otherwise of different levels of self-awareness, and hence different types of capability. In Section 5.2, we document different categories of self-awareness and self-expression levels using architectural patterns. In Section 5.3, we also document common architectural primitives and attributes for architecting self-aware and self-expressive systems. The potential way of using the proposed patterns in real software system projects is discussed in Section 5.4.

5.2 Patterns for Self-aware Architecture Style

In this section, we present the categorisation of different capabilities from the architecture perspectives. Particularly, we have codified the knowledge about how to design self-aware applications in the form of architecture patterns where each pattern contains different capabilities. Here, an architecture pattern refers to an architectural problem-solution pair, which uses the capabilities in a given context. We have elicited some patterns, where each pattern is decentralised by design. That is, structurally our self-aware patterns resemble a peer-to-peer network of interconnecting self-aware nodes, varying only in the number of the capabilities and the type of connection between them.

5.2.1 Basic Notations

Until recently, architecture patterns for self-adaptive systems have received little attention [412]. Many existing patterns target specific application domains [263], limiting their reuse outside the domains for which they were originally conceived. Weyns et al. [412] argued that UML notations are limited in their ability to characterise self-adaptive architecture patterns; hence they proposed a simple, generic notation for describing patterns for the Monitor-Analyse-Plan-Execute (MAPE) architecture style. Our patterns are distinct in focus from Weyns' in the sense that while we model self-aware capability and knowledge concerns in the architecture, their attention was about MAPE component interaction.

To achieve such, we adopt a pattern notation, similar to the one in [412] for describing our self-aware patterns. First, Weyns's notation [412] is simple and easy to comprehend. Second, we believe describing our self-aware patterns using existing notation in the self-adaptive community makes our work accessible to other researchers and paves the way for others to build on our work. Existing works on architecture patterns focus on modelling the **components** and **connectors** of architecture; in such contexts, components are specialisations of modules in the architecture and therefore have attributes and operations, but are also associated with the *provide* and *required* interfaces; and a connector could be the assembly that connects the *required* interface of one component to the *provided* interface of the second one, or could be the delegation that links the ports of a component to its internal parts. In our self-aware patterns, instead of modelling components, we model the capabilities of self-awareness and self-expression (e.g., stimulus awareness) in the architecture. In this way, our patterns preserve flexibility for the concrete architectural implementation; since whether two or more capabilities are combined and realised in one component, or one capability is implemented in separate components could be based on the context. On the other hand, the connectors in our patterns are based on the standard definition but are associated with capability rather than component. Although the capabilities of patterns are designed in a flexible manner, it is important that the interactions amongst these capabilities not be violated when realising the pattern. For instance, one should not realise a direct interaction between a sensor and an actuator if it is not presented in a pattern. The pattern notation is depicted in Figure 5.1.

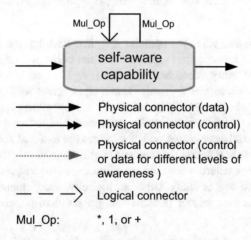

Fig. 5.1 Notation for describing self-aware architecture pattern

Two types of connectors are used to express the logical and physical interactions. A physical connector means there is a direct interaction between two or more capabilities (from the same node or different nodes), and each capability is required to directly interact with the others. Notably, a physical connector (between different levels of awareness), or the red arrow, particularly refers to the interactions for the self-awareness of different types (e.g., goal and time awareness); in contrast, the other, solid black arrows represent the interactions for the self-awareness of the same type (e.g., the interaction-awareness from different nodes). On the other hand, the logical connector does not require direct interaction, but rather the data or control in

the interaction is sent/received through the other capabilities (e.g., Sensors and Actuators), which have the physical connector. For instance, self-expression might be
logically required to reach consensus amongst different nodes, but such interaction
is physically realised through Sensors and Actuators. The benefit of additionally
introducing the logical connector is that, when designing a self-aware capability
where the communication protocol (e.g., local/remote function call, multi-cast and
broadcast) is not needed, the pattern can still show that such capability needs to interact with the others. This provides the designers with a more precise view about
the architecture.

We have used a multiplicity operator to represent how many capabilities and their
components (a capability can be realised in one or more components), including
those from different nodes, are involved in the interaction. There are three types of
multiplicity operators (mul_op):

- **+** expresses that the number of capabilities of the same type in the interaction is
  restricted to at least one.
- **1** indicates that one and only one capability of the same type is permitted.
- ***** indicates that zero, one or many of the type specified are permitted in the
  interaction.

It is worth noting that when the operator is *, it means that the associated interaction may not exist but does not represent that the corresponding capability can be
eliminated. A capability interacting with itself, e.g., a + on both sides of the intra-
capability arrow of a capability, means that it can interact with the same capability
implemented in other nodes. To better clarify the operators, suppose that there is
a physical interaction between stimulus awareness and external sensors where the
stimulus awareness is associated with 1 whereas the external sensors is associated
with +. This means that within the interaction, the stimulus awareness can only have
one external sensor whereas the number of external sensors presented in the interaction needs to be one or many. Other multiplicity arrangements can be similarly
interpreted. We document our patterns using a standard pattern template [54] as
follows.

- Problem/Motivation: A scenario where the pattern is applicable
- Solution: A representation of the said pattern in a graphical form
- Consequences: A narration of the outcome of applying the pattern
- Example: Instance of the pattern in real applications or systems

5.2.2 The Self-aware Patterns

We present eight self-aware patterns using the template described above. For our
purposes, the state of the node comprises the self-aware capability and knowledge
captured in its self-awareness processes. Thus, self-expression can be thought of as
behaviour based on self-awareness.

5.2.2.1 Basic Pattern

Problem/Motivation. In some cases, a system may need to trigger some actions in order to cope with emergent events and stimuli. Such capacity could greatly help to manage the system at runtime. As a result, there is an increasing need for systems to react upon stimuli, based on either static or dynamic rules.

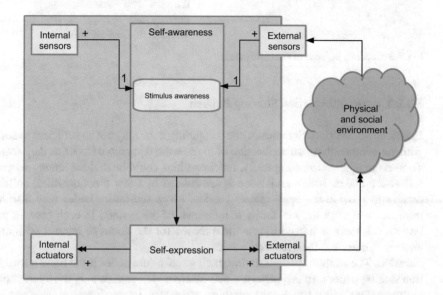

Fig. 5.2 Basic pattern

Solution. The simplest pattern to enable a self-aware node is what we call Basic Pattern, as shown in Figure 5.2. This pattern contains only stimulus awareness which receives data flow from sensors and actuators. Proper actions of self-expression could be triggered based on the type of stimulus detected. A concrete example has been shown in Figure 5.3, where each node aware only of its own stimuli.

Consequences. A major limitation of this pattern is that no information is shared amongst nodes; therefore the node is not aware of the environment and the other node. This could become a major problem in some cases (e.g., the smart camera case study described in Chapter 13), where there are intensive interactions and/or interferences amongst nodes.

Examples. Consider the case of server farm or private cloud where the numbers of deployed applications/services are limited. The basic pattern could be realised in such a context by defining *if-condition-then-action rules*, in which case the conditions could be various stimuli (e.g., QoS is low and utilisation is low); the action could be changing software configuration and/or resource provisioning.

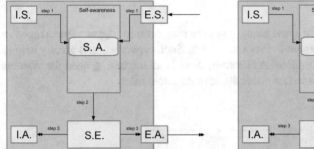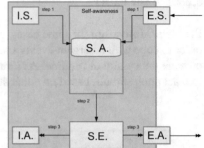

Fig. 5.3 Concrete instance of the basic pattern

5.2.2.2 Basic Information Sharing Pattern

Problem/Motivation. Sometimes one computing node may not be sufficient to cope with the complexity of an application or to meet the demands of users as they scale. To manage application complexity, functionalities could be divided among several self-aware nodes, where each node is specialised in a few functionalities, collaborating to provide the application's service. More self-aware nodes may also be introduced to meet the scalability requirement of the system. In each case, at the basic level, there is a need to provide a means for the nodes to interact with one another to carry out their respective roles.

Solution. The simplest pattern for interacting self-aware nodes is the basic information sharing pattern. In this pattern, a self-aware node contains only the interaction-awareness capability besides the stimulus-awareness. Interaction-awareness can be connected to one or more self-aware nodes as shown in Figure 5.4. Each self-aware node may have one or more sensors (internal/external) and actuators (internal/external). The underlying characteristic of this pattern is that peers are linked only at the level of interaction-awareness. It is important to note that nodes can interact not only with their neighbours but also with their environment. For example, in the financial modelling application described in Chapter 12, interaction is all about communication between nodes and the market rather than amongst nodes themselves.

An example of the basic pattern where two nodes are connected via their interaction-awareness capabilities is shown in Figure 5.5. Although only two nodes are shown in Figure 5.5, the number of connected nodes is not limited to two. The number of nodes is limited by the scalability of the interaction mechanism. For instance, a broadcast mechanism may limit the number of interconnected nodes when compared to a gossip protocol. In practice, a node may be connected to either all or a subset of nodes in the system depending on its role in the system.

Consequences. Self-aware nodes could use the interconnection between them to negotiate the protocol to use for communicating in a network. As observed in the smart camera case study, this pattern can be used to facilitate sharing information among nodes about neighbourhood relations in a network of smart cameras.

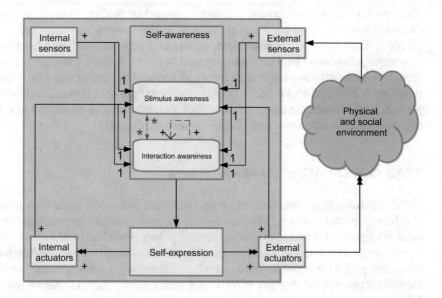

Fig. 5.4 Basic information sharing pattern

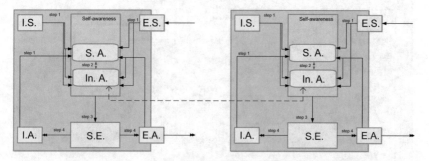

Fig. 5.5 Concrete instance of the basic information sharing pattern

Crucially, in this pattern each self-aware node maintains its autonomy with regard to how to make adaptation decisions via its self-expression capability. This means that each node is responsible for its interpretation and reaction to the information shared via interaction-awareness. Therefore, this pattern is not suitable for cooperative problem-solving scenarios, where nodes need to reach an agreement among themselves about the best course of action for the problem. This limitation is addressed in the *coordinated decision-making pattern* (see the next section). The basic information sharing pattern assumes the system's goal is predefined at design time, consequently constraining the system's adaptation.

Examples. Federated datacentres and clouds, owned by distinct entities, are good candidate applications of the basic information sharing pattern. The owners of such clouds or datacentres may choose only to share status information about availability of resources or current load and not cooperate beyond this level. Thus, each cloud provider maintains autonomy over its resources while collaborating with other cloud providers in a limited way to facilitate outsourcing of resources, if required. Participants in a grid computing set-up utilise a similar communication model and rely on incentive-based mechanisms to facilitate resource sharing [417].

5.2.2.3 Coordinated Decision-Making Pattern

Problem/Motivation. Decisions made by individual self-aware nodes in a group may be suboptimal due to their limited view of the system and its operating environment. As noted in the basic information sharing pattern, individual self-aware nodes do not cooperate when making decisions. In applications requiring near-optimal and consistent global decision making in a cooperative setting, a more advanced architectural pattern may be required. In particular, such a pattern should make it possible for nodes to synchronise their self-expressive actions.

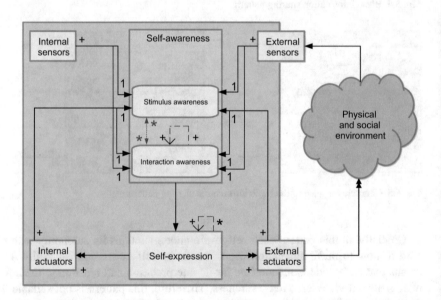

Fig. 5.6 Coordinated decision-making pattern

Solution. The coordinated decision-making pattern provides a means of coordinating actions of multiple, interconnected self-aware nodes. Figure 5.6 shows this pattern. It differs from the basic pattern in that self-expressive nodes are linked to one another, such that they are able to agree on *what* action to take. It is clear that the coordinated decision-making pattern is a pattern related to the basic information sharing pattern as they only differ on the self-expression capability. However, they are designed for different problems and forces; therefore such separation of concepts paves a better way in pattern selection.

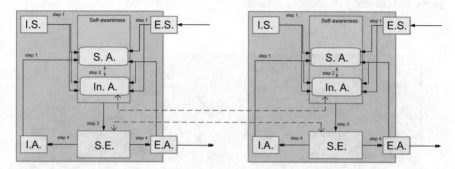

Fig. 5.7 Concrete instance of coordinated decision-making pattern

Consequences. Unlike the basic pattern, given the * to 0 multiplicity on the self-expression capability in Figure 5.6, it is not mandatory for nodes to link their self-expression capabilities to each other. This makes it possible for nodes to form clusters, where nodes in a cluster cooperate to solve problems in one part of a system, while nodes in other clusters cooperate to solve problems in other parts. Figure 5.7 shows an example where two self-aware nodes instantiate this pattern. As argued in the case of the basic pattern, using two nodes to illustrate the pattern as shown in Figure 5.7 does not limit the number of nodes that can realise the pattern in a real system.

The downside of this pattern is that although nodes are able to form clusters and cooperate on *what* action to take, they are unable to decide the timing of such actions, i.e., *when* to act. This notion of time insensitivity is addressed in the *Temporal Knowledge Sharing Pattern* (see next section). The temporal knowledge sharing pattern incorporates time-awareness capabilities into the coordinated decision making pattern.

Examples. Large-scale cloud federations where providers agree to implement unified resource allocation policies, irrespective of how such policies are enforced at individual cloud levels, are a candidate application of this pattern. In such federated clouds, policy changes are negotiated via interaction-awareness capabilities; upon agreement, the self-expression capability of each cloud enforces the agreed policy within its (local) cloud.

5.2.2.4 Temporal Knowledge Sharing Pattern

Problem/Motivation. As stated in the previous section, the coordinated decision-making pattern does not provide a means of coordinating the *timing* of actions agreed upon by cooperating nodes. This limitation may not be tolerated in applications where timing of actions has an impact on the integrity of the application. Also historic knowledge may be required to forecast future actions, in order to improve the accuracy of adaptive actions.

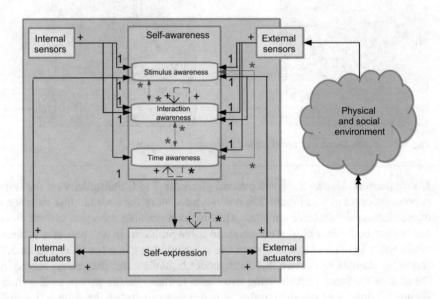

Fig. 5.8 Temporal knowledge sharing pattern

Solution. The temporal knowledge sharing pattern solves this problem by incorporating time-awareness capabilities into the coordinated decision-making pattern. As shown in Figure 5.8, each self-aware node has a time-aware capability which is, optionally (as denoted by its multiplicity), linked to other self-aware nodes to represent timing information. An example where two nodes are connected using this pattern is shown in Figure 5.9. This timing information can be exploited by the self-expression capability to manage the timing of adaptation actions across multiple nodes.

Consequences. The knowledge of timing information provides a rich basis to enrich the power of the adaptation action that is possible. However, there are a lot of design considerations left to the application designer who instantiates the style. For example, how often should timing information be recorded? In storage constrained systems, how long should acquired knowledge be stored for before forgetting (removing) it? Should the forgetting process be total, i.e., delete all knowledge acquired

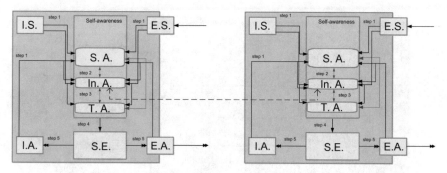

Fig. 5.9 Concrete instance of temporal knowledge sharing pattern

within a period at once, or selective? Depending on the concerns of the application at hand, these questions will have different answers.

Examples. Clusters in cloud datacentres, where the servers in the cluster cooperate to execute tasks assigned to the cluster head, are able to exploit this pattern. For example, a parallel scientific application assigned to the cluster, requiring coordination across different time-steps of the application, could utilise the pattern to ensure actions taken in each time-step are coordinated to avoid compromising the integrity of the result.

5.2.2.5 Temporal Knowledge-aware Pattern

Problem/Motivation. The knowledge of timing enables the capability of proactive adaptation and potentially better adaptation quality. Within the previously mentioned pattern, the Temporal Knowledge Sharing pattern is the only one that applies time awareness capability. However, a drawback of such a pattern is that the interaction awareness capability might not be a necessity; therefore it could affect the self-aware system as it is suffering unnecessary overhead.

Solution. As shown in Figure 5.10, the temporal knowledge-aware pattern solves this problem by incorporating only time awareness working in conjunction with stimulus awareness. Again, the time awareness capabilities of different nodes are logically linked together (optionally). This pattern allows the knowledge of timing to assist the self-expression capability and offset the extra overhead produced by unneeded level of awareness. A concrete example has been shown in Figure 5.11.

Consequences. There are scenarios where the software designer is uncertain about whether the lack of environmental information could affect the modelling of timing knowledge. This is highly dependent on the concrete time-series prediction technique in the time awareness capability. An inappropriate use of certain time-series prediction technique could result in low accuracy, which eventually affects the quality of adaptation. As a result, the decision of which time-series prediction technique

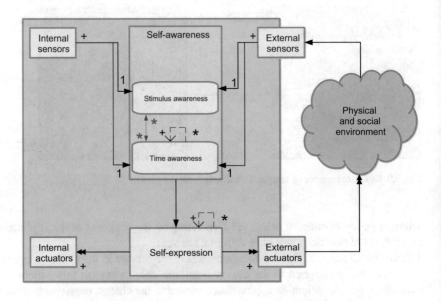

Fig. 5.10 Temporal knowledge-aware pattern

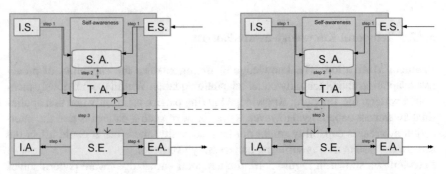

Fig. 5.11 Concrete instance of temporal knowledge-aware pattern

is to be used is critical and designers are advised to consult experts of time-series modelling when applying this pattern.

It should be noted that until now non of the patterns discussed cater to changing goals. That is, they assume the goal of the self-adaptive system is known at design-time and statically encoded in the system, without an opportunity to modify it at run-time. The pattern discussed in the next section—the*Goal Sharing Pattern*—will address the challenge of modifying or changing goals at run-time.

Examples. A cloud is an environment where resource is shared via a Virtual Machine (VM) on each node. In this context, by leveraging the historical usage of resources, time-series prediction would be able to predict the demand of VMs on a node for the nearfuture, which assists proactive provisioning of resource and potentially, prevents SLA violation and/or resource exhaustion on a node.

5.2.2.6 Goal Sharing Pattern

Problem/Motivation. User preferences are mostly dynamic, i.e., users want different things at different times. As an example, a user who is satisfied operating a computing system using a touch screen at one time may prefer voice interaction at another time. These changes in user preferences may range from simple changes, such as mode of user interaction, to more advanced ones. Furthermore, a computing system may itself decide to change its goal, depending on the amount of resources available to it. In the smart camera case [125], a camera running low on battery may choose to bid for only the most valued objects within its field of view instead of aiming to track all objects in its vicinity. A specialised pattern that allows explicit representation of run-time goals, and facilitates changes to these goals as the system evolves, is needed.

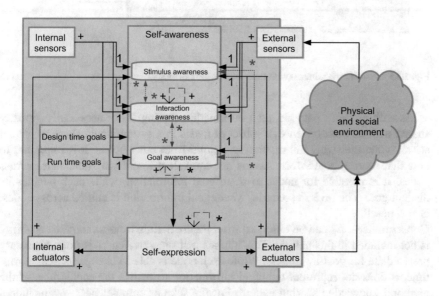

Fig. 5.12 Goal sharing pattern

Solution. Figure 5.12 shows the goal sharing pattern that addresses the concern of representing run-time goals. A goal-awareness capability represents knowledge about run-time goals, which can be changed as the system evolves. The goal-

awareness capability in a self-aware node can, optionally, share its state information
with goal-awareness capabilities in other self-aware nodes.

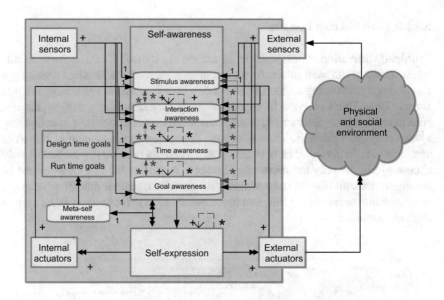

Fig. 5.13 Goal sharing pattern (with time-awareness capability)

As with previous patterns, goal information sharing is not necessarily globally
shared with all nodes. Hence, a subset of nodes in a system could share their goal
states, while their goal information is disjoint from other nodes. It is important to
note that sharing goal information is not equivalent to unifying goal states across
nodes. It is possible for nodes to share goal information, while each pursues its
distinct goal. The reverse scenario, where goal information is unified across nodes,
is also possible.

Consequences. As can be observed from Figure 5.12, a time-awareness capability
is not included in this pattern. This implies that time-awareness is not a necessary
prerequisite for goal-awareness. While each node is able to change its goal at run-
time, it does not represent temporal information to realise the capabilities of the
temporal knowledge sharing pattern. For the sake of completeness, we include a
different pattern that addresses this limitation (see Figure 5.13). The pattern in Figure 5.13 makes the inclusion of temporal knowledge explicit, making it suitable for
application domains where changing goals and forecasting are required.

In both patterns, the self-expression capability makes use of the goal-awareness
capability to make strategic decisions in line with the system's current goal. Figure 5.14 shows an instance of the pattern (without time-awareness), while Figure 5.15 an instance (with time-awareness).

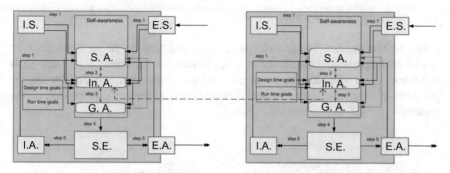

Fig. 5.14 Concrete instance of goal sharing pattern (without time-awareness capability)

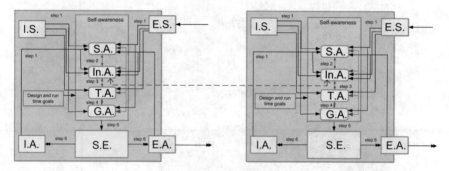

Fig. 5.15 Concrete instance of goal sharing pattern (with time-awareness capability)

Examples. Service-based applications operating in dynamic, open cloud environment are possible candidates of this pattern. Here, applications are composed from cloud services which are selected based on QoS and cost considerations. A service that is highly performing at one time may degrade in quality at later times due to overloading of the service. Each application has service-level agreements (SLAs), to which it must adhere. Application goals encoded in SLAs may themselves change as users demand different levels of service from time to time.

Using the goal-sharing pattern with time-awareness capability (see Figure 5.13) in this scenario has the benefit of making each application capable of representing temporal knowledge about service performance and forecasting which service(s) are likely to be more dependable and long-lasting. Also, the goal-awareness capability makes it possible to represent SLA terms of users and adapt such goals as they change. Lastly, by sharing temporal knowledge, applications can exchange knowledge of service performance among themselves. It should be noted that this introduces opportunities to falsely downplay or inflate performance of services. Considerations for filtering out good knowledge are left to the computational models used to implement time- and goal-awareness.

5.2.2.7 Temporal Goal-aware Pattern

Problem/Motivation. The knowledge of goals and time might not necessarily be shared amongst nodes, especially in cases where the optimisation of local goals could lead to an acceptable approximation of the global optimum. As a result, the presence of an interaction awareness capability could cause extra overhead on the system.

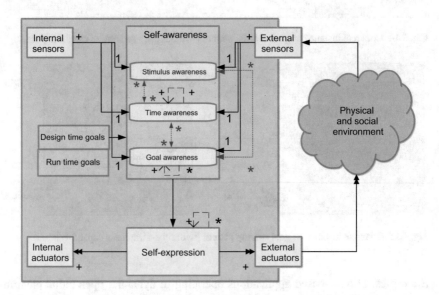

Fig. 5.16 Temporal goal-aware pattern

Solution. As shown in Figure 5.16, the temporal goal-aware pattern solves this problem by removing the interaction awareness capability. In this pattern, there is no notion of "sharing" as the nodes are not aware of any interactions and therefore not aware of the presence of the other nodes. It is worth noting that the absence of interaction awareness does not mean there is no interaction—nodes and the environment could still interact with each other, but the nodes are not aware of it. A concrete example has been shown in Figure 5.17.

Consequences. The removal of interaction awareness implies that the nodes could be in an inconsistent state. The designer should carefully verify that such a situation will not result in violations of system requirements. In addition, the self-expression capability should not use any information from other nodes when making decisions.

Examples. Orchestrating a fully decentralised harmonic synchronization amongst different mobile devices requires each node to be aware of stimulus, time and goal but not necessarily for interaction. In such a case, each node receives phase and fre-

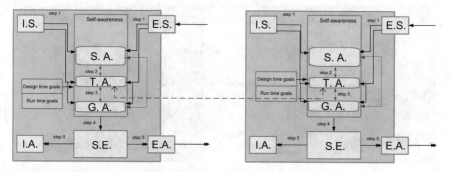

Fig. 5.17 Concrete instance of temporal goal-aware pattern

quency updates from the other nodes or the environment, and reacts based on its own time and goal information. This is a typical example where there are occurrences of interaction, but no occurrences of interaction awareness; because the nodes are only aware of the incoming phase and frequency updates but have no knowledge of where they come from, the sources could be other nodes, the environment or even some unexpected noise.

5.2.2.8 Meta-self-awareness and Self-aware Patterns

Meta-self-awareness is useful for managing the trade-off between various levels of self-awareness and for modifying goals at run-time. Since reasoning at the meta level is considered an advanced form of awareness, it may be beneficial or necessary in some contexts and not beneficial in others. This section especially treats the relation between meta-self-awareness and the patterns discussed in previous sections.

We reiterate that one of the distinct benefits of the self-aware style is its ability to reduce the complexity of modelling adaptive behaviour when compared to non-self-aware approaches. For the sake of illustration, consider the problem of modelling and tuning an online learning algorithm, e.g., a neural network, for deciding actions of an application in different scenarios. It is known that this task is time-consuming and requires expertise in mathematical skills, which may not be readily available [116]. Additionally, in some use cases, small changes in the application scenarios may render the solution proffered by the algorithm invalid or incorrect—another cycle of algorithm tuning may be needed to cater to these changes. An alternative approach is to provide families of algorithms for different contexts and dynamically select the appropriate algorithm at run-time using online learning capabilities of the meta-self-aware capability.

While the first approach offers faster adaptation, if application scenarios are relatively stable, the second approach is able to better cope with complexity in highly perturbed environments, where one algorithm is insufficient to cover the scope of

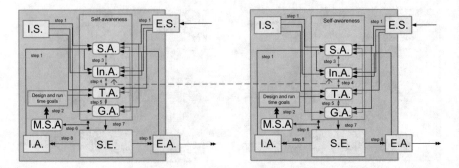

Fig. 5.18 Concrete instance of goal sharing pattern (including meta-self-awareness capability)

adaptive behaviour. Accordingly, we recommend that every pattern optionally incorporate the meta-self-aware capability depending on the complexity to be managed and the expertise of the application designer. Figure 5.18 shows the goal-sharing pattern with time-awareness capability where a meta-self-aware pattern is present to manage trade-off between goal-, interaction-, and time-awareness capabilities. Presumably, the presence of meta-self-awareness capability could help to switch between different patterns at runtime, which could be a very interesting direction for future work.

There are also other examples of meta-self-awareness capability in real applications. For instance, in the smart camera network [125], the meta-self-awareness is used to switch between different behavioural strategies in the interaction awareness and self-expression capabilities. In the active music application [61], the meta-self-awareness can help to control the degree of stimulus awareness based on confidence measure. In the cloud autoscaling system [67], [68], [69], the meta-self-awareness can be used to select the best learning algorithm online, and thus provide better modelling of QoS in a dynamic environment.

5.3 Architectural Primitives and Attributes for Self-aware Systems

The aforementioned architectural patterns provide proven solutions to recurring design problems that arise in the context of self-aware and self-expressive systems. However, the pattern itself is an abstract form of the architecture. Often, the abstract patterns can be used as the first step for architecting and engineering self-aware systems by mapping them with an existing/candidate architecture instance which is application-specific. We have observed that the same pattern can be used in many different ways depending on the problem context and the concrete architecture instances, e.g., whether to use meta-self-awareness; whether to use a particular technique to realise a capability; whether to realise two capabilities in the same component. However, the concrete applications of patterns in different contexts still share many underlying concepts. These shared concepts lead to the notion of **architectural primitive**, which refers to the common concepts that can be constructively used to form a concrete architecture instance of a pattern. The architecture instances of a pattern differ in terms of the candidate techniques and/or attributes (i.e., a particular form of an architectural primitive) used to realise each architectural primitive. These differences amongst architecture instances of a pattern are referred to as **pattern variability**; similarly, the alternative architecture instances of a pattern are the **variants** of this pattern.

Architectural primitives and pattern variability are long-studied problems [262]. However, unlike traditional work that aim for generic system architectures, we are specifically interested in how the common concepts can be used in self-aware systems. In particular, we aim to link these well-studied concepts to the self-awareness principles [32], which are all about different levels of knowledge awareness. In this section, we have documented the architectural primitives in four categories (i.e., capability, behaviour, interaction and topology), each of which covers one aspect of a self-aware system. We anticipate that the benefit from this will be reduction in the chances of introducing faults, and easier fault detection, when using the proposed patterns during design. The architecture primitives and candidate techniques and/or attributes have been used in many real applications (e.g., [125], [61] and [67]) to form different variants of a pattern.

5.3.1 Taxonomy of Primitives

To better understand the design of self-aware systems, we have codified four categories of architectural primitives. This is motivated by our investigation of the critical aspects that could affect the functional and/or non-functional requirements of a self-aware system. The categorisation is as shown below:

- **Capability**: The capability of the system to obtain certain knowledge or to react based upon the knowledge, e.g., the levels of awareness and expression.
- **Behaviour**: The process of a capability regarding how input data is consumed and output data is produced.
- **Interaction**: The relationship between the capabilities or within a capability itself as expressed by the multiplicity operators.
- **Topology**: The deployment of capabilities and their distribution to the components in the architecture instance.

5.3.2 List of Architectural Primitives and Attributes

We have structured the architectural primitives to express (i) the characteristic of a component or connector and (ii) the function of a component or connector. The architectural primitives and their attributes with respect to the four categories are as shown below:

- **Capability**:
  - *stimulus-awareness*: the capability to obtain knowledge of stimuli
  - *interaction-awareness*: the capability to obtain knowledge of stimuli and its own actions with other nodes and the environment
  - *time-awareness*: the capability to obtain knowledge of historical and/or likely future phenomena
  - *goal-awareness*: the capability to obtain knowledge of current goals, objectives, preferences and constraints
  - *meta-self-awareness*: the capability to obtain knowledge of its own level(s) of awareness and the degree of complexity with which the level(s) are exercised
  - *self-expression*: the capability to react based upon the node's state, context, goals, values, objectives and constraints

- **Behaviour**:
  - *send*: the output process of a sender (e.g., capability, sensors and actuators); it has two attributes
    - · *synchronous*: the scenario where after the sender sends a request; it needs to wait for a reply from the receiver before transitioning to the next actions
    - · *asynchronous*: the case where the sender does not require such blocked communication
  - *handle*: the input process of a receiver (e.g., capability, sensors and actuators); it has two attributes
    - · *sequential*: the case where incoming data is processed by a specific sequence using a queue
    - · *parallel*: the scenario where the incoming data is simply processed in parallel upon its arrival
  - *state*: the behaviour of a capability; it depends on the candidate techniques that realise such capability; mainly, it has two attributes
    - · *proactive*: the behaviour that predicts and reasons about when the change is going to occur, which it then acts upon
    - · *reactive*: the capability responds when a change has already happened
  - *transit*: a specific behaviour of meta-self-awareness capability; it aims to reason about and switch on/off one or more other capabilities

- **Interaction**:
  - *link*: the physical and logical relationship between the capabilities; it has four attributes
    - *one-to-many*: the capability of a type interacts with many sensors or actuators
    - *many-to-many*: the capability of the same type from different nodes interact with each other
    - *one-to-one*: one capability of a type interacts with one capability of another type
    - *none*: there is no interactive relationship

- **Topology**:
  - *structure*: how the capabilities are mapped to the components of an architectural instance; it has three attributes
    - *combine*: the case where a capability is realised in conjunction with other capabilities within a single component
    - *separate*: a single capability is realised using separate components
    - *compact*: a capability is realised using exactly one component
  - *existence*: the existence of an optional capability of the pattern; it has two attributes
    - *exist*: the optional capability is included
    - *non-exist*: the optional capability is not included

The use of these primitives is application-specific and depends on whether they would affect the functional and/or non-functional requirements of the self-aware systems.

5.4 Discussion

In this section, we discuss the method regarding how the proposed patterns can be used in a given application context by leveraging on the Architecture Trade-off Analysis Method (ATAM) [312]. In the following, we briefly explain the major phases contained in the method; more details and comprehensive examples can be found in our handbook [70].

5.4.1 Phase 1: Collect Requirements and Constraints

The first phase involves collecting requirements and constraints from the stakeholders and environment for engineering self-aware and self-expressive systems. As in the step in ATAM [312], the purpose here is to gain deep knowledge about the problem context; to operationalise both functional and non-functional requirements, to facilitate communication between stakeholders, and to develop a common vision of the important features the system should support.

The requirements can be either functional or non-functional; for instance, *the system should be able to record historical data for analysing its behaviour* is clearly a functional requirement. On the other hand, *the delay of adaptation in the system should be no more than 1 minute* is an example of non-functional requirement. Constraints are also another important factor to be considered; in particular, they could come from the stakeholders, e.g., *the hourly cost of infrastructure for running the system should not be more than 10 dollars*, or from the environment, e.g., *the topology of nodes in the context should be dynamic so that nodes can join/leave on demand*.

5.4.2 Phase 2: Propose Candidate Architecture

We assume that there is at least one candidate architecture of the system. This is because our self-aware patterns are generic and do not rely on any assumptions about the application domain. Therefore, in order to apply the pattern in practice, it is essential to obtain certain knowledge about the architecture related to the context of given application domain (often, this knowledge is represented as a collection of components and connectors, which we call existing candidate architecture). This provides two major benefits: (i) It provides a clear view about what is really needed by the problem domain from an architecture perspective, and thus assists the engineers in reasoning about and selecting the right pattern. (ii) It makes it possible to refine the existing candidate architecture when mapping the components and connectors to the capabilities of chosen pattern(s). We believe this is a rational assumption as once the important requirements and constraints have been determined, the engineer should be able to propose a candidate architecture based on the obtained information. In addition, design almost never starts from a clean slate: legacy systems, interoperability, and the successes/failures of previous projects all constrain the space of architectures.

The candidate architecture must be described in terms of architectural components/elements. In particular, the architecture should express the module/component view [311] of the system, which is usually used to reason about work assignments and information hiding. In this work, we do not assume any prerequisite of the modelling notations; therefore the engineer could use either the standard ones (e.g., UML) or create his or her own notations.

5.4.3 Phase 3: Select the Best Pattern(s)

This phase is concerned with selecting the suitable pattern(s) based on the functional and non-functional requirements as well as the constraints. The pattern helps the engineer to rethink the domain-specific problem in a self-aware computing sense. The patterns differ mainly in terms of the self-awareness capabilities, which play an integral role in satisfying the functional requirements; therefore the selection of pattern is equivalent to the selection of the right set of self-awareness capabilities.

To promote a systematic approach for pattern selection, we design certain questions and ask the software engineer to consider these questions. The suitable pattern(s) could be determined based on these answers. Specifically, one should answer the following questions for each self-awareness and self-expression capability:

- What does the capability mean in your problem context?
- What are the functional requirements that are affected by this capability?
- What are the non-functional requirements that are affected by this capability?
- What are the constraints that could affect this capability?
- Is this capability necessary or beneficial?

If the problem domain requires two patterns, then there can be two sets of answers for each self-aware capability. Finally, the answers of these questions provide an insight about what level of awareness is necessary, which can be used to select pattern(s).

5.4.4 Phase 4: Fit the Selected Pattern(s)

Once the best pattern(s) has been determined, we can now fit the components/elements from the candidate architecture to the capabilities described in the pattern(s). It is worth noting that our architectural patterns preserve the flexibility for the concrete architecture, since whether two or more capabilities are combined and fit in one component or one capability is fit in separate components can be based on the requirements and constraints. It is also an opportunity to refine/improve the candidate architecture during the mappings.

5.4.5 Step 5: Determine the Important Primitives and the Possible Alternatives for Non-functional Requirements

The next step is to determine the important architectural primitives and the relevant alternatives (i.e., certain forms of techniques) for the given problem context. Particularly, we should consider the non-functional requirements here and link them to the architectural primitives, including their attributes and the relevant techniques. Those primitives and their attributes and techniques that could influence the non-functional requirements would be extensively modelled and examined for justifying their benefits during the next steps. In addition, this is also an opportunity to eliminate the primitives and techniques, which could be easily selected or are too trivial to be considered. For example, if the performance requirement is much more important than the other quality attributes and the environment conditions are dynamic, then the behaviour primitives (e.g., the primitives send and handle) can be eliminated as it is almost certain that only multicast and parallel interactions are feasible here.

From this step forward, our method would start focusing on the non-functional requirements, which are more difficult to satisfy because their compliances are often related to runtime uncertainty. In addition, the non-functional attributes are usually highly sensitive to the variants of patterns. The problem becomes even more complex when the non-functional requirements are conflicted, e.g., accuracy vs. overhead. In extreme cases, it is possible to go back to Step 2 if one feels that the chosen pattern(s) is (are) inappropriate. This iterative process can continue till the final suitable pattern(s) is (are) determined.

5.4.6 Step 6: Create Scenarios

At this stage, we create the likely runtime scenarios that can significantly influence the non-functional attributes. There is no restriction on the number or the granularity of scenarios per non-functional attribute. Consider, for example, that one could have a faulty scenario for a system running in stable state in order to examine the influence to the availability attribute. In addition, he or she could also create a scenario under the unstable state (e.g., when new nodes join in or existing nodes leave). It is worth noting that the predefined scenarios do not need to be exhaustive as these scenarios are not intended to limit the developed system, but rather to provide assurance and confidence in system development by justifying the benefits of possible techniques, especially under the likely scenarios.

5.4.7 Step 7: Score the Alternative of Primitives Against Each Non-functional Attribute Using Analytical or Simulation Models

In this step, we need to determine the score for each alternative of a primitive against each non-functional attribute under the considered scenarios. The purpose is to assess each alternative and justify its benefits with respect to the non-functional attributes. We assume that there is no or limited dependency between the architectural primitives in terms of how they affect the non-functional attributes; therefore we assess each primitive in isolation. In particular, we aim to gain a relative score for every alternative in the context of each architectural primitive. Scoring an alternative can be achieved by using either an analytical model or a simulation: the former both refer to the empirical or statistical analysis of a particular alternative against a non-functional attribute, e.g., the complexity analysis; on the other hand, the latter uses the quantitative results from some simulations on an alternative under the considered scenarios. In particular, if the techniques of a primitive are difficult assess both ways, then the software engineer could score the alternatives using empirical knowledge by assigning weights based on experience [9]. It is worth noting that in cases where the non-functional attributes can only be assessed using multiple primitives, when scoring the alternatives of a primitive it is important to ensure that the used alternatives for other primitives are equivalent. For example, when deciding the alternatives of stimulus-awareness in terms of a system's overall adaptation quality, the chosen alternative for self-expression should be consistent. Precisely, the scoring process has three phases:

First, weight the relative importance of different non-functional attributes after negotiation amongst the stakeholders. In particular, this can be achieved using the pair-wise comparison in AHP [346]. This alternative can be used to measure how much importance of attributes Q_a is over the attribute Q_b based on some scale.

Second, score each alternative against each non-functional attribute under every defined scenario. Having this done, we then obtain a matrix of n times m for the hth primitive under a given non-functional attribute k, as shown below:

$$P_h = \begin{matrix} & SC_1, ..., SC_m \\ \begin{matrix} A_1 \\ ... \\ A_n \end{matrix} & \begin{pmatrix} S_{1,1}^k & ... & S_{1,m}^k \\ ... & ... & ... \\ S_{n,1}^k & ... & S_{n,m}^k \end{pmatrix} \end{matrix} \tag{5.1}$$

where n is the total number alternatives of the hth primitive P_h and m denotes the total number of considered scenarios for non-functional attribute k. SC_m denotes the mth scenario. A_n is the nth alternative and $S_{n,m}^k$ denotes the score of the nth alternative under the mth scenario. In cases where there is more than one scenario for a non-functional attribute, the score of an alternative would be the aggregative result of the scores under all scenarios. Therefore, we calculate the total score of an alternative for the hth primitive under non-functional attribute k by aggregating its score of all scenarios. The matrix can be then reduced to a vector as shown below:

$$P_h = \begin{matrix} A_1 \\ ... \\ A_n \end{matrix} \begin{pmatrix} S_1^k \\ ... \\ S_n^k \end{pmatrix} \quad s.t. \quad S_n^k = \sum_{x=1}^{m} S_{n,x}^k \tag{5.2}$$

where S_n^k is the aggregative score for the nth alternative of the hth primitive under non-functional attribute k.

Finally, we normalise the raw score for the alternatives in each architectural primitive against a non-functional attribute k. To achieve this, we use the following formula:

$$\text{the } A_a \text{ of primitive } P_h = \begin{cases} NS_{a,h}^k & \text{if } max\, Q_k \\ 1 - NS_{a,h}^k & \text{if } min\, Q_k \end{cases} \quad s.t. \quad NS_{a,h}^k = \frac{S_a^k}{\sum_{x=1}^{p} S_x^k} \tag{5.3}$$

where $NS_{a,h}^k$ is the normalised score for the ath alternative of the hth primitive; p denotes the total number of alternatives for the hth primitive. By doing so, we can normalise the score scaling from 0 to 1. If the non-functional attribute is to be minimised, the final normalised score would be calculated as $1 - NS^k_{a, h}$.

5.4.8 Step 8: Find the Best Alternatives for the Final Architecture View

Once all the scores of alternatives have been obtained, the final task is to identify the best combination that produces the highest total score. Specifically, we need to maximise the formula below:

$$argmax \sum_{k=1}^{x} (w_k \times \sum_{h=1}^{y} NS^k_{s, h}) \tag{5.4}$$

where $NS^k_{s, h}$ is the score of the sth selected alternatives for the hth primitive; w_k is the weight for the kth non-functional attribute. x and y are the total number of non-functional attributes and architectural primitives respectively. It is easy to see that Equation 5.4 can be solved by any optimisation algorithm, and finally the method produces a pattern-based architecture with the best selected alternatives for all architectural primitives.

5.5 Conclusion

Engineering self-aware and self-expressive systems is a widely important and complex activity. This is because the process involves making many decisions on the possible alternatives, even in the early stage of development. In this chapter, we defined patterns and architectural primitives for engineering self-aware and self-expressive systems. Given a problem domain, these principles serve as the guideline to justify the benefits of different self-awareness capabilities, and their possible implementation techniques. The patterns and primitives are discussed with a step-by-step method that demonstrates how they can be used in a concrete and practical way. We believe that the proposed patterns and primitives can better assist engineers to make principled design decisions concerning whether, how, and to what extent to include self-awareness and self-expression capabilities in the system.

Chapter 6
Knowledge Representation and Modelling: Structures and Trade-Offs

Leandro L. Minku, Lukas Esterle, Georg Nebehay, and Renzhi Chen

Abstract As explained in Chapter 5, self-aware and self-expressive systems can be designed based on a number of patterns and primitives. In this chapter, we discuss issues to be considered when developing such systems, especially when going through phases 3 (selecting the best pattern) and 5 (determining primitives and alternatives), and possibly also phase 7 (score alternative primitives) of the methodology for designing and implementing self-aware and self-expressive systems described in Section 5.4. Specifically, we explain several features which may be present in self-aware and self-expressive systems, namely adaptivity, robustness, multi-objectivity and decentralisation. We discuss their implications in terms of knowledge representation and modelling choices, including potential trade-offs among different choices. Knowledge representation is interpreted loosely, referring to any structure used to store knowledge, whereas knowledge modelling is considered to be the process used to create and update such knowledge structures. The discussion raises awareness of general issues to be considered and carefully reflected upon when developing self-aware and self-expressive systems.

Leandro L. Minku
University of Leicester, UK, e-mail: leandro.minku@leicester.ac.uk

Lukas Esterle
Alpen-Adria-Universität Klagenfurt, Austria, e-mail: lukas.esterle@aau.at

Georg Nebehay
Austrian Institute of Technology, Austria, e-mail: gnebehay@gmail.com

Renzhi Chen
University of Birmingham, UK, e-mail: rxc332@cs.bham.ac.uk

6.1 Introduction

As explained in Chapter 5, self-aware and self-expressive systems can be designed based on a number of patterns and primitives. In this chapter, we discuss issues to be considered when developing such systems. Specifically, we explain different features which may be present in self-aware and self-expressive systems, and their implications in terms of knowledge representation and modelling choices. The following features are considered: adaptivity (Section 6.2), robustness (Section 6.3), multi-objectivity (Section 6.4), and decentralisation (Section 6.5).

These features and their implications should be considered specially when going through phases 3 (selecting the best pattern) and 5 (determining primitives and alternatives) of the methodology for designing and implementing self-aware and self-expressive systems described in Section 5.4, and possibly also phase 7 (score alternative primitives).

In the context of self-aware and self-expressive systems, the terms knowledge representation and modelling are interpreted in a loose way. Knowledge representation is not restricted to formalisms such as semantic nets, frames, rules or ontologies. Instead, it refers to any structure used to store knowledge. Knowledge modelling can then be seen as the process used to create and update such knowledge structures/-models.

The discussion provided in this chapter highlights general choices and their trade-offs in the specific context of self-aware and self-expressive systems, and is not intended to be exhaustive and comprehensive or empirically demonstrated through experiments. In particular, different applications may be affected by different choices in different ways, and certain choices may be more relevant for some applications than for others. It would thus be infeasible to list all possible choices and to empirically analyse how these choices affect all possible applications of self-aware and self-expressive systems. Rather than that, the discussions provided in this chapter aim at raising awareness of general points to be considered and carefully reflected upon when developing self-aware and self-expressive systems.

The rest of this chapter is organised as follows: Sections 6.2 to 6.5 explain the features of self-aware and self-expressive systems and discuss their implications to knowledge representation and modelling, and Section 6.6 provides a summary of the chapter.

6.2 Adaptivity

In order to perform well in a given environment, systems usually need to be adapted not only to their goals, but also to the environments where they are embedded. In standard computing tasks, the environment where a system operates typically exhibits a static behaviour. When the type of expected input is well known, it is relatively easy to devise a program to map the input to the desired output. As computing tasks get more and more complex, this static property can no longer be relied upon, because the environments where these tasks must be performed are frequently dynamic and present uncertainties. Therefore, self-aware and self-expressive systems usually need the ability not only to self-adapt to a given environment condition, but also to self-adapt to (possibly unexpected) changes in the environment. In some applications, the goals of the system may also change with time, resulting in the need for self-adaptation to new goals.

The need for adaptivity in self-aware and self-expressive systems leads to several general knowledge representation and modelling choices, such as whether and how to model time and what structures to use for dealing with changes. Each individual application of self-aware and self-expressive systems also involves choices related to whether and how to model the environment; whether and how to represent the suitability of a solution to the environment; and what components of the system to adapt. Section 6.2.1 explains adaptivity in self-aware and self-expressive systems and gives some examples of applications involving adaptivity. Section 6.2.2 explains some general implications of adaptivity on the development of such systems.

6.2.1 Definition and Examples

Two definitions of the verb *adapt* given by the Oxford dictionary are as follows [303]:

- (With object) Make (something) suitable for a new use or purpose; modify.
- (No object) Become adjusted to new conditions.

In biology, adaptation can be seen as the process whereby an organism becomes better suited to its habitat(s) [106].[1] Similarly, in the context of self-aware and self-expressive systems, *adaptation* can be seen as the process whereby a system becomes better suited to its environment, given its purpose. If its environment or purpose changes, the system must adjust to the new conditions. The term *adaptivity* refers to the ability of a system to perform such adaptation automatically, i.e., the ability of a system to self-adapt. It differs from the term *adaptability*, which refers to systems that can be substantially customised by users through tailoring activities by themselves [413].

Adaptivity can be advantageous for many self-aware and self-expressive systems. For example, let's consider a credit card approval machine learning system whose aim is to accurately predict whether a customer should be given credit or not [407, 268]. Learning consists in creating a model of the relationship between customers' features (e.g., age, salary, gender, etc.) and a dependent variable describing whether or not they have paid all their bills in, say, the last twelve months. This model represents the system's knowledge of its environment, and is created based

---

[1] Please note that we do not intend to provide a comprehensive set of possible definitions of adaptivity, but focus on a couple of definitions that can help us understand the concept of adaptivity in self-aware and self-expressive systems.

on examples of existing customers and their payments (training examples). Predictions on whether a new customer should be given credit or not are performed based on this model. The environment in which credit card approval systems operate may change due to several factors. For example, customers who usually paid all their bills may become less likely to pay their bills due to some economic crisis. A credit card approval system should thus be able to adapt to such changes. This can be achieved by monitoring how well existing models of the environment perform on new incoming training examples, and updating such models or creating new models to reflect new environment conditions [407, 24, 268, 141, 223, 325].

Another example of self-aware and self-expressive system where adaptivity can be advantageous is that of distributed smart camera networks [125] (Section 7.4.1). Such networks can be used, for example, to track objects. In order to accomplish this goal, each smart camera must implement a handover mechanism, which refers to finding the next camera to observe the target object once it leaves the field of view of the current camera [121]. This mechanism could be based on a static known vision graph whose neighbourhood structure is encoded a priori in the cameras. However, it would require extensive manual design-time work to determine this graph. Moreover, if a change such as the failure of a given camera or a camera being moved to a different position occurs, the cameras would need to be manually updated to reflect the new situation. A system where cameras can automatically learn their vision graph and adapt their handover mechanism would be desirable to avoid heavy manual (re-)tuning of the system. This can be achieved, for example, by using a market-based mechanism where each camera "bids" for objects with the goal of maximising its own utility [125]. Each camera's utility is updated with time, and increases based on the objects it tries to track, their visibility in the camera's field of view, the performance of the camera in tracking these objects, and the payments it receives from other cameras. The utility reduces based on the amount it pays to sell objects to other cameras. Each camera's relative utility is then used to automatically determine which camera the object should be handed over to. A model of the environment (vision graph) can be automatically created and updated based on the cameras that trade objects with each other.

An additional example of a self-aware and self-expressive system where adaptivity is used is the decentralised system for synchronisation of music agents described by Nymoen et al. [295] (Section 7.3.1). This system is designed for use in collaborative active music mobile apps, where a group of non-musicians use their mobile phones (agents) to interact with music together. This interaction could have either the purpose of creating music or playing back pre-recorded music. The goal of the synchronisation system is to synchronise the timing of the several agents who are participating in the musical interaction. In order to achieve this goal, each agent must be able to adapt its timing to the timing of other agents. For that, each agent has knowledge of its own timing, which is represented by a state that oscillates in cycles with a certain phase and frequency. Whenever the cycle of an agent is complete, it emits a short sound that can be heard by other agents. This sound is used by agents to update their states so that their timing becomes more similar to the timing of the agent who emitted the sound.

In general, the advantage of adaptivity is to enable automated reactive modification of behaviour at run-time in order to suit (i) a given environment condition and (ii) a set of goals. Adaptivity not only avoids the need for pre-defining the system's behaviour for a particular environment, but also allows it to operate in possibly unforeseen situations. Such advantage does not come without risks. For instance, in some systems, it may be difficult to ensure that adaptation to changing conditions is successfully achieved in time. Depending on the application, the result of unsuccessful self-adaptation could range from slight under-performance to overall system failure.

Adaptivity is achieved based on the capabilities possessed by an agent and/or collective. In order to be able to adapt, a system must be able to monitor its own state and/or the state of its external environment. For example, the credit card approval system explained above is able to monitor its own state through a performance measure which is updated whenever new training examples are made available from the environment, besides being able to maintain and update a model of its external environment by learning such incoming training examples. In the distributed smart camera example, cameras can monitor their own utility, besides being able to detect currently visible objects given their position and acquire knowledge of their relationship to neighbouring cameras in the network. In the music synchronisation system, each agent has knowledge of its own timing and can listen to sounds emitted by other agents in the environment. As explained in Chapter 2, a self-aware system could be defined as a system able to obtain knowledge about its internal state and/or knowledge about its environment. Such knowledge should be obtained by the system on an ongoing basis throughout the system's lifetime, rather than being programmed by a domain expert. Therefore, we could see self-awareness as an enabler for adaptivity.

6.2.2 Implications

This section describes the implications of adaptivity on knowledge representation and modelling. Adaptivity is usually associated to a cost; a limited amount of adaptivity may have lower cost than a more thorough level of adaptivity [387]. Therefore, separating different types of adaptivity into different levels of self-awareness can help to avoid the cost associated to unnecessary adaptation. This section first discusses implications of adaptivity in terms of choices of levels of self-awareness (Section 6.2.2.1). Then, it discusses implications in terms of modelling states and environments (Section 6.2.2.2), modelling time (Section 6.2.2.3) and dealing with changing environments (Section 6.2.2.4).

6.2.2.1 Choice of Levels of Self-awareness

Stimulus-awareness: This type of self-awareness allows a node to adapt to *static* environmental conditions. When the environment is dynamic, stimulus-awareness is also capable of adapting to changes, even though this might be rather limited. The efficiency and effectiveness of the adaptation would highly depend on the type and severity of the change in the environment, given that this level of self-awareness alone would not be able to distinguish between past, present and future stimuli.

Interaction-awareness: When a system is composed of several agents, a given agent will frequently need to adapt not only to its external environment, but also to other agents in this environment. Interaction-awareness allows multi-agent self-aware systems to achieve that. For example, the distributed smart camera system described in Section 7.4.1 [125] is interaction-aware, because each smart camera can interact with other smart cameras in the environment in order to hand objects over to them. It is interesting to note that, when a system is composed of several agents, adaptivity of the collective system as a whole may be achieved through the adaptivity of each of its agents.

Time-awareness: This type of self-awareness can be very helpful for achieving adaptivity to dynamic environments, which are environments that may suffer changes with time. Time-awareness allows a system to distinguish between past and current events, helping it to acquire knowledge about the current environment condition and to adapt to any changes suffered by the environment. For example, a credit card system able to adapt to dynamic environments such as the one mentioned in Section 6.2.1 [407, 268] must be able to distinguish between past and current examples while monitoring the performance of existing models, so that the monitored performance reflects the current situation of the environment rather than being outdated.

Goal-awareness: Given that adaptation is the process whereby a system becomes better suited to its environment *given its purpose*, the concept of adaptation is closely linked to the goal(s) of the system. Therefore, many self-aware systems will use explicit knowledge of their goal(s) in order to achieve adaptivity. For example, a credit card system such as the one explained in Section 6.2.1 could explicitly try to minimise the error of its predictions by using a certain error measure. Some systems are composed of several agents who are aware of their own goals, even though each agent may not be aware of the goals of the system as a whole. For example, each smart camera in the system explained in Section 7.4.1 [125] is unaware of the goals of the system as a whole. Instead, each smart camera tries to maximise its own goal, which is represented by its utility value. This allows each camera to self-adapt to its environment and to other cameras, leading to the achievement of the goals of the system without explicitly considering the goals of the system as a whole. A system able to adapt to changing goals would also normally be expected to be goal-aware. However, a system does not necessarily need to be aware of its goal(s) in order to achieve adaptivity to changes in the environment, as long as its adjustments, which can possibly be made via implicit goals, make it more suitable to the environment given its purpose.

Meta-self-awareness: Self-adaptive systems do not need to be meta-self-aware. However, meta-self-aware systems allow a more thorough level of adaptivity, where systems are able to adapt the way in which the levels of self-awareness are realised (e.g., by changing algorithms), or adapt the levels of self-awareness themselves (e.g., by activating/deactivating certain levels). For instance, the active music system described in Chapter 14 can automatically decide which strategy to adopt for mapping between a gestural controller and sound engine in order to adapt the music to the current user intention. Another example of meta-self-awareness would be to to use hyperheuristic optimisation algorithms [53] to automatically decide which meta-heuristic optimisation algorithms to use in goal-aware systems.

While the different levels of self-awareness may help a system to achieve better adaptivity, they also require additional computation time. This may not be relevant to standard computing systems, but is crucial in applications requiring near-real-time performance or in embedded devices where resources are very limited. Especially with meta-self-awareness, the system may need to rely on multiple learning techniques simultaneously. A trade-off between the level of self-awareness that a system possesses and the amount of learning that it is able to perform arises.

6.2.2.2 Modelling States and Environments

In order to achieve adaptivity, self-aware and self-expressive systems will frequently rely on learning models of the state of their nodes, or models of the environment where they are embedded. However, the nature of the data available for learning must be considered when deciding what type of model to build and which type of learning algorithm to use. With respect to the nature of the data, models can be built based on the following types of learning algorithms:

Supervised learning algorithms: Data come in the form of examples with inputs and desired outputs (labelled data). Desired outputs frequently need to be provided by a "teacher" from outside the computing system. While knowledge of desired outputs may facilitate learning, providing desired outputs can be a time-consuming and difficult task. An example of supervised learning system is a credit card system fed with data describing the features of customers and explicit information on whether these customers have paid their bills or not [407, 268].

Unsupervised learning algorithms: Data come in the form of input attributes with no desired outputs provided (unlabelled data). This avoids the need for a possibly costly external "teacher" to provide desired outputs. However, in self-aware and self-expressive systems, this frequently means that nodes have to learn in a completely autonomous fashion, employing only the information present within the node itself or within nodes in the immediate environment. An example of unsupervised learning algorithm would be a clustering algorithm to identify the characteristics of different groups of customers, but without pointing out whether these customers are good or bad payers.

Semi-supervised learning algorithms: These algorithms can be used to achieve a compromise between the two extremes represented by supervised and unsupervised learning. In semi-supervised learning, part of the data come labelled and part come unlabelled. Semi-supervised algorithms can be used to find hidden structures in the data by using unlabelled data and infer outputs based on the labelled data.

Reinforcement learning algorithms: Data come in the form of feedback from the environment representing how close the system is to achieving its goal(s), or how well it is performing its task(s). This type of learning algorithm is adequate when it is not possible to obtain desired outputs, but feedback from the environment is available. For instance, a robot learning how to walk may receive feedback from the environment in the form of how far it moved forward. An example of a reinforcement learning algorithm is the algorithm for scheduling tasks in wireless sensor networks presented by Khan and Rinner [215], where each node in the network uses a reward function to decide which task to perform next.

6.2.2.3 Modelling Time

Stimuli from the environment will frequently come in the form of data and be used to learn and update a model representing the knowledge that the system has of the environment. Therefore, the choice of how to model time can be seen as the choice of whether and how to distinguish the time stamps in which different data points were produced or presented to the system. There are different ways to process data with respect to time, and the relevancy of their advantages and disadvantages can vary considerably depending on the intended application. In general, the following options are available:

Offline data processing algorithms: These algorithms create models with pre-existing data sets describing the environment, and there is no distinction between the time stamp of different data points within pre-existing data sets. Therefore, models of the environment are built before the system is put into operation. The advantage of that is that the designer of the system can frequently have a good idea of how well adapted the model is to the environment before it starts to be used. The disadvantage is that it is not possible to update the model with additional data incoming with time. Therefore, the model cannot be adapted to any changes in the environment. Traditional machine learning algorithms typically process data in offline mode, e.g., backpropagation using several epochs for learning [38].

Online data processing algorithms: These algorithms are aware of the chronological order of the data points. Each data point is processed separately and then discarded. The advantage of that is that models can be improved over time with incoming data. As old data points are not re-processed, these algorithms can also be memory and time efficient, being suitable for applications with strict memory and time constraints or for applications where large amounts of data need to be processed. These algorithms can also be combined with strategies to adapt models to changes in the environment (see Section 6.2.2.4). The disadvantage is that it may be difficult to ensure beforehand how well the model will behave over time. An example of an algorithm that processes data online is the multi-objective ensemble method for class imbalance learning [407], which has been applied to fault detection, credit card approval and network intrusion detection. Another example is the learning algorithm proposed by Fern and Givan [132], which has been applied to prediction of conditional branch outcomes in microprocessors in order to cope with the strict memory and time constraints of this application.

Chunk-based data processing algorithms: These algorithms process data in chunks. There is no distinction between the time stamp of different data points within a chunk, but the relative chronological order of the chunks themselves is known. Each data chunk is processed and then discarded, but data points within a chunk can be re-processed several times before the chunk is discarded. The advantage of chunk-based algorithms is that models can be improved over time with incoming data. Chunk-based algorithms can also be combined to form strategies to adapt models to changes in the environment (see Section 6.2.2.4). As data points belonging to a given chunk can be re-processed several times before a chunk is discarded, chunk-based algorithms may be able to achieve higher accuracy on each chunk than online algorithms. The disadvantage is that this is less memory- and time-efficient than online learning. In addition, chunk-based algorithms need to wait for a whole chunk of data to arrive before the model can be updated, being unable to update the model with continuously incoming examples. Choosing the best chunk size may also be difficult. A too large chunk would lead to slow adaptation, whereas a too small chunk may cause the system to perform poorly, depending on the algorithm being used. An example of an algorithm that processes data in chunks is the the music synchronisation algorithm [295] explained in Section 6.2.1. Rather than immediately updating the frequency of an agent once it hears a short sound from another agent, each agent collects the short sounds received from other agents within a whole oscillation cycle before updating its frequency. In this application, this can prevent premature convergence of the system to an undesirable state [295], i.e., waiting for chunks can help the system to adapt better.

6.2.2.4 Dealing with Changing Environments

There are different possible strategies for dealing with changing environments, and each strategy incurs different decisions in terms of knowledge representation and modelling. For instance, a system may or may not use knowledge models able to detect changes, i.e., change detection methods.

Change detection method: In this case, explicit mechanisms are used to monitor the environment or the suitability of the system to the current environment condition, in order to detect changes. When a change is detected, a mechanism to adapt the system to the change is triggered. For example, in the credit card system explained in Section 6.2.1, the accuracy of the predictive model being used could be monitored and updated based on new incoming training examples. When the accuracy suffers a significant drop, a change is detected [141]. The predictive model could then be deleted and a new model could be created to start learning the new situation of the environment. More advanced strategies could also incorporate mechanisms to accelerate the learning of the new environment condition rather than having to learn it from scratch [268]. Another example of an approach using a change detection method is the multi-objective ensemble for online class imbalance learning [407]. Class imbalance learning refers to learning algorithms able to deal with classification problems where at least one class is under-represented in comparison to other classes (see Section 7.2.3.2). The approach presented by Wang et al. [407] uses an explicit method to detect whether a certain learning problem is a class imbalance learning problem and which classes should be currently considered as minority and majority classes. The advantage of using explicit change detection methods is that a system can be designed to swiftly react to a change once it is detected (e.g., by deleting old models as in the example above). The disadvantage of using change detection methods is that they may trigger false positive drift detections, which may hinder the system's performance if not carefully catered for. For more details on algorithms based on change detection, refer to Section 7.2.3.3.

No change detection method: In this case, mechanisms to adapt the system to changes are continuously active, without the need for being triggered by explicit change detection methods. For example, an ensemble of learning algorithms can be used to create a credit card system such as the one explained in Section 6.2.1. This ensemble could maintain different models and assign a weight to each model based on its accuracy (i.e., a weight can be used to monitor the suitability of each model). This weight could be updated with new training examples by using a decay function that would give higher emphasis to newer examples [223]. When the system performs wrong predictions, new models could be created and added to the ensemble. Therefore, if some change has happened and is the cause for the wrong predictions, then the new models may be able to learn the new situation from scratch. When the weight of a given model is below a certain threshold, this model could be deleted for not representing the current environment well [223].

Another example of a mechanism that does not use an explicit change detection is the mechanism used by the smart camera system in Esterle et al.'s work [125]. Cameras exchanging an object to be tracked should not be communicating with other cameras that are too far away, in order to reduce the communication overhead of the system. By keeping track of the exchanged objects, each individual camera is able to learn about its neighbouring cameras. Inspired by the ant foraging mechanism, the cameras use artificial pheromones to depict a graph of their local neighbourhood. With every exchanged object, additional pheromones are deposited on the respective link. However, if there are no exchanges of objects between two cameras the pheromone evaporates, allowing the system to overcome changes in topology or the cameras' field of view over time.

The advantage of not using an explicit drift detection method is that it is not necessary to decide when exactly a change occurred. Deciding when exactly a change detection should be triggered may be particularly difficult when changes are slow and take some time to complete. The disadvantage is that the lack of a change detection method may in some cases lead to a difficult trade-off between stability and plasticity. If the system is set to forget old knowledge quickly, it may be too unstable to perform well. If the system is set to forget old knowledge slowly, it may take too long to adapt to changes. For more details on algorithms without change detection, please refer to Section 7.2.3.3.

6.3 Robustness

As described in Section 6.2, adaptivity is a property necessary for a computing system to be able to *react* to changes. In contrast, robustness is necessary in order to address changes in a *proactive* way, so that mechanisms can be adapted beforehand to reduce negative effects that future events could incur to the system. The development of robust systems involves making suitable knowledge representation and modelling choices such as how to model anticipation mechanisms and how to maintain different solutions so that they can be quickly recovered if necessary. Section 6.3.1 further explains and exemplifies robustness, and Section 6.3.2 explains the knowledge representation and modelling choices related to robustness.

6.3.1 Definitions and Examples

We define robustness as every kind of proactive behaviour with the aim of avoiding or diminishing the detrimental effects caused by changes, dynamic events or certain choices. This aspect of self-aware and self-expressive systems is important in many different systems, for instance in the work of Nymoen et al. [295], where multiple nodes strive towards the common goal of being in synchrony in order to engage in the creation and/or playback of collaborative music, as briefly described in Section 6.2.1. In order to achieve this goal, the phase and the frequency of the oscillators of the participating nodes need to be adjusted accordingly. Adaptivity is achieved by each node's reacting to short sounds emitted by other nodes, reflecting their timing/state. It is however not clear how this adaptation should be performed in an optimal way. If the adaptation is too slow, it will not reach the goal of synchrony in time. If the adaptation is too fast, multiple nodes will try to achieve synchrony at the same time and will not find a stable state. In order to make the system more robust, each node can vary its degree of self-adaptation based on its anticipated value (self-confidence) of being in synchrony.

A second example is the work of Nebehay et al. [282], which was performed in the context of visual object tracking. We present this work in more depth here, because it implements robustness in different ways, anticipating not only certain changes, but also the effect of certain choices. The goal of visual object tracking is to follow one or more objects of interest in a video stream as long as possible. However, several different types of change can (and frequently do) occur, making tracking difficult. For example, there may be changes in illumination, object or camera pose, as well as substantial changes to the object of interest itself. Achieving robustness to such changes is thus arguably one of the most desirable properties of visual object tracking systems.

Nebehay et al. [282] propose the keypoint-based method Consensus-based Matching and Tracking (CMT) for long-term object tracking, where the idea of robustness takes a central role. Here, a model of the object being tracked consists of multiple keypoints on the object of interest, which are nothing more than individual nodes working towards the common goal of tracking an object. In each new frame of the video sequence, each keypoint in the object model aims at finding its correct location on the object in the image, as defined by two measures of visual similarity [235, 251] that correspond to both matching and tracking of keypoints.

In fact one single keypoint would be sufficient to localise the object of interest. As visual information is highly ambiguous, it is however rather unlikely that each keypoint will position itself correctly on the object of interest. This is caused for instance by similar objects appearing in the background or by changes on the object that disallow a correct re-identification. Inevitably, some keypoints will end up on wrong parts of the object or even on different objects. In CMT, it is anticipated that these kinds of errors will occur and *redundancy* is introduced to address them as a form of robustness. Instead of creating a single hypothesis, each keypoint provides its own hypothesis for locating the object of interest by voting for its centre, as shown in Figure 6.1. These votes are combined robustly by means of clustering them directly in the image space. A basic assumption here is that the relative majority of the keypoints is able to correctly identify the object centre. In Figure 6.1, the majority cluster is shown in blue, while all minor clusters are shown in red, representing keypoints that failed to establish the correct position on the object of interest. By removing all keypoints in minority clusters, a much better object tracking result can be obtained, for instance by averaging all votes in the majority cluster as an estimate for the object centre.

An interesting aspect about CMT is that the object model is never updated, which is in strong contrast to the prevalent paradigm in object tracking, where the object model is updated continuously. The problem of updating the object model continuously is that it always bears the danger of introducing new errors. For instance, when the localisation is not exact, an update might lead to the incorporation of background information. From the experimental evaluation presented in [282], it appears that the robustness of the overall system in CMT is strong enough to make adaptation of individual nodes unnecessary. While beneficial to the tracking performance, one drawback of the proposed approach of achieving robustness lies in its relatively high computational cost, where the communication between the individual nodes contributes the largest share.

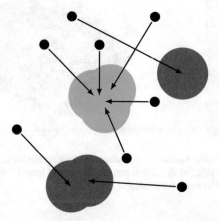

Fig. 6.1 Finding consensus in voting behaviour. Cast votes are clustered based on their proximity. The consensus cluster (shown in blue/light grey) is identified based on the highest number of votes. The minority clusters (shown in dark red/dark grey) are not used for estimating the actual location of the object.

Another aspect in CMT that was born out of considerations for improving robustness is the question how voting is performed. As shown in Figure 6.2 on the left side, the voting vectors are initialised in the first frame, pointing to the object centre. In later frames, these votes have to undergo a certain transformation in order to reflect changes in the location, pose and appearance of the object of interest. If this transformation is modelled with a degree of freedom as high as possible, this introduces a series of problems. First, it is very difficult to estimate the correct parameters for a complex transformation as the amount of training examples is strongly limited. Second, the chances that multiple votes agree for one location get lower for more complex transformations. Third, overly complex transformations might further increase the computational load. In CMT, it is anticipated that the changes in the object of interest will be composed of translation, scaling and rotation, and the transformation of votes are restricted to these transformations, as shown on the right side of Figure 6.2. This class of transformations nicely balances the trade-off between robustness and expressiveness that is present in this kind of problem. In summary, robustness is achieved by a *restriction* of the output space. Obviously, one drawback of this kind of robustness is a reduction of the class of problems that can be handled, as already, for instance, perspective transformations are out of scope.

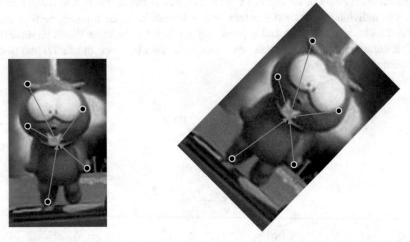

Fig. 6.2 Left: In CMT, voting vectors of keypoints are initialised in the first frame. Right: In frame t, each keypoint casts a vote for the object centre. The object transformation is anticipated to be composed of translation, scaling and rotation.

6.3.2 Implications

6.3.2.1 Choice of Levels of Self-awareness

Stimulus-awareness: Stimulus-awareness can be used to obtain data to learn how to behave pro-actively. However, stimulus-awareness on its own may not be enough to learn how to anticipate changes, because it would not enable the system to distinguish between past, present and future stimuli. It is worth noting that stimulus-aware systems do not necessarily present robustness, but could certainly benefit from mechanisms to achieve certain types of robustness. For instance, the functional correctness of a computing system is highly dependent on its input. Therefore, it is important to ensure the correct functioning of its direct connection to the environment. For example, it may be desirable to achieve robustness to failure in the sensors of the individual nodes that allow them to learn about and interact with their immediate environment. Section 6.3.2.2 discusses some strategies that could be used to achieve that.

Interaction-awareness: As shown in the object tracking examples explained in Section 6.3.1, interaction-awareness can be used to provide a great degree of robustness in cases where the correctness of the output of the individual nodes is not ensured. Another example is the multi-camera object tracking application that will be presented in Chapter 13, where the use of multiple nodes makes the system robust to failures of individual nodes, for instance when a camera loses connection to the network.

Time-awareness: Time-awareness is an enabling component for designing robust self-aware computing systems. It allows the system to learn how to anticipate changes by making predictions about possible changes in the future based on how changes occurred in past events. For instance, if errors made in previous actions or decisions are recognised, then behaviour in the future can be adapted accordingly. An example application in this respect can be found in Chapter 14, where time-awareness enables extracting a rhythmic pattern from users shaking input devices in the SoloJam application, a process that has to be robust to changing input patterns.

Goal-awareness: In general, similar to adaptivity, robustness will frequently be achieved with respect to certain goals, and being aware of such goals could thus be helpful. Moreover, a goal-aware system could assess the importance of different goals as an interesting way of establishing robustness to changes. For instance, a self-aware node monitoring its battery status might decide that now is the time to pause striving towards its current goal of high priority and rather find a way of recharging itself in order to avoid running out of battery in the future. Note that this way of reasoning is different from adaptivity—in this case, anticipation about one's own state and the consequences of (not performing) actions is of utmost importance. Goal-aware systems could also establish certain objectives to evaluate how well a system is likely to perform in the future and in the event of changes, rather than only evaluating its current suitability to the present environment [138]. It is clear that, even though achieving robustness through this level of self-awareness can be beneficial, it requires a substantial amount of knowledge about oneself and its immediate environment, which might be difficult to obtain for certain systems. In SoloJam (Chapter 14), goal-awareness is also present. The concrete goal of a node in this application is represented by the Hamming distance between the leader's rhythmic patterns and its own pattern. A change in the activity of a user then directly has an influence on its goal.

Meta-self-awareness: Achieving robustness on a meta-self-awareness level can be considered the holy grail of achieving robustness in a computing system. A self-aware computing node might, based on the outcome of its self-awareness components, decide what kind of robustness is actually important, and then take action to establish it. On the positive side, the computing system might discover aspects of its task during run-time that have not been under consideration at design-time and that lack a sufficient degree of robustness. On the negative side, this kind of robustness is very difficult to achieve as it requires a very abstract understanding of the environment consequences of actions, goals and oneself. In the multi-camera object tracking application (see Chapter 13), one form of meta-self-awareness is achieved by dynamically deciding on the strategy for handing over objects from one camera to another camera. This increases the robustness of the system compared to a static strategy selection mechanism.

Similarly to what has been discussed in Section 6.2.2.1, while the different levels of self-awareness may help a system to achieve better robustness, they also require additional computation time, which can be particularly concerning in applications requiring near-real-time performance or in embedded devices where resources are very limited.

6.3.2.2 Robustness Through Redundancy of Components

Redundancy of components can be a very useful way to achieve robustness in self-aware and self-expressive systems. In general, redundancy could be employed in two different ways:

Deployment of multiple identical components: This could be used, e.g., to achieve robustness to failure of a given type of component. For example, it might be reasonable to employ multiple *identical* sensors to account for malfunctioning hardware components. This way, correctness of the input is ensured even during changes to the proper functioning of individual system components. Another example is the multiple keypoints used by the object tracking system explained in Section 6.3.1 in order to achieve robustness to wrong votes given by single keypoints. In [409] a memetic algorithm is presented that finds a solution to the Redundancy Allocation Problem (RAP), aiming at optimally allocating redundancy to components under some resource constraints.

Deployment of multiple different components: In [355] a system is presented that improves the fault-tolerance of electronic circuits. Here, robustness is achieved by creating multiple circuits exhibiting different error patterns that are then combined into a single strong circuit. However, failure of a component to function well may be caused not only by an error in the component, but also by changes in the environment which may cause a given component to be unsuitable. For example, one type of sensor might work well during daytime (such as colour cameras), while other sensors work well only at night time (such as infra-red cameras). It therefore makes sense to combine multiple types of sensors to account for changes in the environment. Another example is the use of a set of different models representing different environment conditions in anticipation of possible changes in the environment [268, 181]. Even though a certain model may not be appropriate for the current environment condition, it may be appropriate for future conditions.

When employing multiple components, there are also different choices in terms of how to use the outputs produced by these components. For example, one could use one or more of the following:

Combining nodes' outputs: The output of multiple nodes is combined into a single output. In this case, there is the question of how to combine the outputs. If one node has to be designated as a coordinating node, this introduces a potential single point of failure. This is related to the issue of decentralisation, further explained in Section 6.5. Alternatively, one may wish to have no central node. In this case, nodes can either all contribute equally to the combined output, or they could be given different emphases, depending on how suitable they are to the environment.

Selecting nodes: Erroneous nodes might be detected and actions might be taken to prevent these nodes from participating in future actions. Care has to be taken about the quality of the failure: a hardware error might be easy to recognise (for instance by measuring the component temperature), while it is difficult to assess the correctness of the high-level self-expression of individual nodes. For instance, in the multi-camera object tracking application (Chapter 13), a node is given full control when its confidence about having localised the object correctly is high enough, as it can be assumed that it is well adapted to its environment.

Node communication: Multiple nodes might be used in parallel and communicate with each other to account for potential defects in hardware or software.

The positive side of redundancy of components is that it could enable robustness to certain events or changes. However, it is clear that employing multiple nodes can also lead to unwanted side-effects, such as increased power consumption, increased processing time, and conflicting nodes data.

6.3.2.3 Learning How to Anticipate Events or Changes

Anticipation can be either incorporated in the system at design time (e.g., the object tracking example given in Section 6.3.1) or learnt during the system's lifetime. Learning models able to anticipate events or changes can provide a much higher level of autonomy to a self-aware and self-expressive system. However, the nature and availability of the data that can be used for learning must be considered. This issue is also relevant to adaptivity of models of the environment, and has been discussed in Sections 6.2.2.2 and 6.2.2.3.

6.4 Multi-objectivity

Self-aware and self-expressive systems are built with one or more goals/objectives to be accomplished. These goals/objectives can be either explicit or implicitly implemented in the system. The process of finding a best possible solution given a set of objectives and a set of limitations is referred to as optimisation process. The traditional optimisation problem aims to minimise/maximise a specific objective (usually in the form of a function). This type of problem is called the Single-objective Optimisation Problem (SOP). However, only optimising one objective cannot satisfy actual demands in several cases, because most of the real-world optimisation problems need to achieve a balance among multiple objectives, which are frequently in conflict with each other. Hence, optimising one objective with respect to an SOP often results in unacceptable results in the other objectives. For example, when buying a new laptop, we want it as cheap as possible but with the best quality. Or, when scheduling tasks on a chip [65], we want to improve the performance while keeping the chip cool. Obviously these objectives conflict with their counterparts. We may not be able to buy a laptop which is both cheap and powerful. Similarly, we may not be able to obtain very high performance when scheduling tasks at the same time as keeping the chip cool. SOPs cannot handle this type of balance well. In this case, we extend the SOPs to Multi-objective Optimisation Problems (MOPs). Instead of minimising/maximising one objective, MOPs aim to achieve a balance among several objectives. As mentioned above, it is difficult to achieve a perfect multi-objective solution to minimise/maximise every objective. Therefore, we need to use some other method to find a reasonable solution which satisfies the objectives at an acceptable level.

Besides being relevant for optimisation tasks, multi-objectivity can also be relevant for machine learning tasks. For example, when learning how to detect faults in some machinery, one may wish to minimise both false positive and false negative fault detections. The issue of multi-objectivity has been investigated more in depth in the optimisation than in the machine learning literature. This section will thus concentrate mainly on multi-objectivity in optimisation. However, works on multi-objectivity in machine learning can also exist [407, 269, 62].

Section 6.4.1 further explains multi-objectivity, whereas Section 6.4.2 explains the implications of multi-objectivity on knowledge representation and modelling choices.

6.4.1 Definition and Examples

The general multi-objective optimisation problem can be defined as:

$$
\begin{aligned}
\textit{minimise } \mathbf{F}(\mathbf{x}) &= (F_1(\mathbf{x}), F_2(\mathbf{x}), ..., F_k(\mathbf{x}))^T \\
\textit{s.t.} \qquad g_i(\mathbf{x}) &= 0, i = 1, 2, ..., a \\
h_j(\mathbf{x}) &< 0, j = 1, 2, ..., b
\end{aligned}
\tag{6.1}
$$

where k is the number of objectives, a is the number of equality constraints, b is the number of inequality constraints, $\mathbf{x} \in X$ is a vector of decision variables, X is the search space (set of all candidate solutions), and $\mathbf{F}(\mathbf{x})$ is the vector of objective functions $F_i(\mathbf{x})$, $\mathbf{F}(\mathbf{x}): X \to R^k$. Objective functions can also be called goal functions, payoff functions, cost functions or fitness functions. Please note that functions to be maximised can be easily converted to functions to be minimised. Therefore, without loss of generality, the MOP can be defined as a minimisation problem as above.

In the example of buying new laptops, we have two conflicting objectives: lower price and higher performance. Consider that there are n alternative laptops $X = \{\mathbf{x_1}, \mathbf{x_2}, ..., \mathbf{x_n}\}$ available for purchase, and that their costs are $c_1, c_2, ..., c_n$ and performances are $p_1, p_2, ..., p_n$. A solution $\mathbf{x} \in X$ to this problem is a vector of size 1 representing a single laptop to be purchased. The objectives of this MOP can be described as: minimise c_i for lower prices and minimise $1/(p_k + \varepsilon)$ for higher performance, where ε is a small constant to avoid division by 0.

In the chip scheduling example, X represents all the possible schedules of tasks on the chip. If we have n cores and m tasks, then $|X| = C_m^n$. We can use a vector to stand for a given scheduling of tasks to cores: $\mathbf{x} = (u_1, u_2, ..., u_n)$, where $u_i \in (1, m) \cup \{-1\}$. If $u_i \in (1, m)$, then that means that core i runs task u_i; otherwise it means that nothing is running on core i. It is worth noting that other structures could be used to represent a solution for the same problem. For example, having a vector with $n + m$ entries instead of n could be used to allow certain tasks to be assigned to no core, while still enforcing that all tasks are mentioned at least once in the vector. This could be desirable, for example, to allow tasks that are unscheduled in two solutions to be scheduled in a new solution created by crossing over these two solutions in evolutionary algorithms [65]. The objectives can be described as:

- Maximise the overall performance of the chip:

$$F_1 = \min(-\sum_{j=1}^{m} \overline{p_j}(\mathbf{x}))$$

- The average temperature of the chip:

$$F_2 = \min(\sum_{j=1}^{m} \overline{\theta_j}(\mathbf{x}))/m$$

where $\overline{p_j}(x)$ and $\overline{\theta_j}(x)$ are the performance and temperature of core j under scheduling $\mathbf{x}$, respectively.

As the objectives in an MOP are frequently conflicting, it is impossible to find a single solution that optimises all objectives simultaneously. Instead, we are usually interested in obtaining a set containing the best trade-off solutions, which are called Pareto optimal solutions. Pareto optimal solutions are solutions non-dominated by any other solution. The definition of a solution $\mathbf{x}$ being dominated by another solution $\mathbf{y}$ ($\mathbf{x} \prec \mathbf{y}$) is as follows:

$$\mathbf{x} \prec \mathbf{y} \iff \forall i, F_i(\mathbf{x}) \le F_i(\mathbf{y}) \land \exists j, F_j(\mathbf{x}) < F_j(\mathbf{y}) \qquad (6.2)$$

Given the set of Pareto optimal solutions (Pareto set) for a given MOP, the decision maker would be able to select his or her desired solution with his or her preferred trade-off among all optimal solutions in the Pareto set.

In order to find the Pareto set, we need optimisation algorithms specifically designed for MOPs. Alternatively, it is not uncommon to convert MOPs to SOPs and then use existing SOP algorithms to solve them. For instance, most current research on temperature control of multi-cores converts this MOP to an SOP by using linear weight methods. Nevertheless, SOP algorithms aim at finding a single optimal solution. Converting MOPs to SOPs leads to some additional problems to be tackled, e.g.:

1. How to set the relative importance of different objectives

   Given that SOPs will attempt to find a single optimal solution, approaches to convert MOP to SOP usually require some method to set the relative importance of the different objectives beforehand. The algorithm will then look for a single best solution considering this particular relative importance. However, it may not be straightforward to chose values to represent relative importance, and a wrong choice will lead the algorithm to miss useful trade-offs among the objectives. Moreover, in certain problems, the relative importance of different objectives should ideally not be constant, making it difficult to choose ideal values manually. This is the case, for example, of the chip task scheduling problem. When the current temperature of the chip is already too high, we should give higher importance to temperature in order to cool down the chip more quickly. When the current temperature is low, we do not need to pay too much notice to increasing temperature. So, the relative importance of these objectives can change with time.

2. How to normalise the objective functions

   The objective functions F_i, $1 \leq i \leq k$, are the values which will stand for the quality of the solution. However, these values can be incomparable. For instance, it is hard to say which unit is larger in the chip scheduling problem—one unit of performance or one unit of temperature. We need methods to normalise these values into comparable ones. For instance, we can normalise them by:

$$f_i(\mathbf{x}) = \frac{\max_{\mathbf{x_1} \in X} \mathbf{F_i(x_1)} - \mathbf{F_i(x)}}{\max_{\mathbf{x_1} \in X} \mathbf{F_i(x_1)} - \min_{\mathbf{x_2} \in X} \mathbf{F_i(x_2)}}, \mathbf{1 \leq i \leq k}$$

where k is the number of objective functions. Or we can normalise them by:

$$f_i(\mathbf{x}) = \frac{\mathbf{F_i(x)}}{\max_{\mathbf{x_1} \in X} \mathbf{F_i(x_1)}}, \mathbf{1 \leq i \leq k}$$

The two methods above can normalise the objective values but will lead to different results, and it is unclear which of them to adopt.

Given the problems of converting MOPs to SOPs, methods specifically designed for solving MOPs are desirable. For more details on MOPs, we recommend Deb's book [94].

6.4.2 Implications

6.4.2.1 Choice of Levels of Self-awareness

The issue of multi-objectivity can be relevant to systems with any level of self-awareness. This is because any self-aware system will be designed with a purpose in mind. This purpose can only be achieved if there is some "force" pushing the system towards eventually achieving this goal. In systems that are not goal-aware, this "force" can be viewed as one or more implicit goals. If the system is goal-aware, then the goal(s) will be explicit in such a way that the system can reason about it (them). Therefore, multi-objectivity could be tackled by any level of self-awareness.

Stimulus-awareness: The ability to perceive stimuli is essential to evaluate how well a system performs in a given environment, or to decide which actions to take so as to perform well in a given environment. Stimuli can be used directly or indirectly to compute objective functions, no matter whether these functions are explicit or implicit.

Interaction-awareness: Systems to deal with multiple objectives do not necessarily need to be interaction-aware. However, interaction among nodes could be beneficial if different nodes are responsible for different objectives. Interaction-awareness could also be used by a node to acquire information for evaluating its objectives. For instance, each node in the smart camera system [125] explained in Section 7.4.1 uses payments made to and received from other cameras in order to compute its utility function, as mentioned in Section 6.2.1.

Time-awareness: Given that this type of self-awareness allows a system to distinguish between past, current and future events, it can help to compute different objectives for reflecting the suitability of the system to situations encountered in the past, to the current situation, and to possible future situations. The multi-objective ensemble approach presented by Wang et al. [407] makes use of time-awareness by computing each objective based on a time-decayed function, reflecting the suitability of the system to the current situation of the environment.

Goal-awareness: This type of self-awareness is the most relevant for multi-objectivity. Goal-awareness would allow a system to explicitly consider its objectives and reason about them. This could facilitate dealing with multiple objectives by making it easier to consider different trade-offs among objectives, and by allowing objectives to be added/removed over time. Therefore, goal-awareness could make it easier to design a system to cope with multiple conflicting objectives. An example of system that benefits from goal-awareness to deal with multiple objectives is the multi-objective ensemble approach for class imbalance learning presented by Wang et al. [407]. The system is designed to perform classification tasks and its learning procedure explicitly considers classification performance in terms of recall on each existing class separately.

Meta-self-awareness: Meta-self-awareness is not essential for a system to be able to deal with multi-objectivity. However, multiple objectives could be designed in order to represent how beneficial different levels of self-awareness are in order to develop a meta-self-aware system.

Similarly to what has been discussed in Section 6.2.2.1, while the different levels of self-awareness may help a system to better handle multi-objectivity, they also require additional computation time, which can be particularly concerning in applications requiring near-real-time performance or in embedded devices where resources are very limited. These issues must be considered when deciding what level of self-awareness to adopt.

6.4.2.2 Searching for Solutions with a Particular Trade-Off Among Objectives

Weighted methods: Classical methods for solving MOPs usually convert MOPs into SOPs, and then use a single objective optimisation algorithm to find a best solution. The simplest method to do that is to multiply objectives by a manually set weight and then sum them up [425, 224, 19, 254]. Therefore, the MOP is redefined as an SOP as follows:

$$
\begin{aligned}
minimise\ & F(x) = \sum_{i=1}^{k} w_i F_i(\mathbf{x}) \\
s.t.\quad & g_i(\mathbf{x}) = 0, i = 1, 2, ..., a \\
& h_j(\mathbf{x}) < 0, j = 1, 2, ..., b
\end{aligned}
\tag{6.3}
$$

Another method is to use the weighted distance between the value of each objective for a given solution $\mathbf{x}$ to the best achievable value $\mathbf{z}^*$:

$$
\begin{aligned}
minimise\ & F(x) = \left(\sum_{i=1}^{k} w_i(|F_i(\mathbf{x}) - z_i^*|^p)\right)^{1/p} \\
s.t.\quad & g_i(\mathbf{x}) = 0, i = 1, 2, ..., a \\
& h_j(\mathbf{x}) < 0, j = 1, 2, ..., b
\end{aligned}
\tag{6.4}
$$

where p can be any value from 1 to ∞. This method will be the same as the previous one when $p = 1$.

A similar method [112] uses the objective values $\mathbf{z}^0$ of a feasible non-Pareto-optimal solution rather than the ideal $\mathbf{z}^*$:

$$
\begin{aligned}
maximise\ & F(x) = \left(\sum_{i=1}^{k} w_i(|F_i(\mathbf{x}) - z_i^0|^p)\right)^{1/p} \\
s.t.\quad & z_i^0 < F_i(\mathbf{x}), i = 1, 2, ..., k \\
& g_j(\mathbf{x}) = 0, j = 1, 2, ..., a \\
& h_l(\mathbf{x}) < 0, l = 1, 2, ..., b
\end{aligned}
\tag{6.5}
$$

Likewise, we can also multiply all the objectives instead of adding them up, as follows:

$$minimise\ F(x) = \prod_{i=1}^{k}(F_i(\mathbf{x}))^{w_i}$$
$$s.t.\quad g_i(\mathbf{x}) = 0, i = 1, 2, ..., a \quad\quad (6.6)$$
$$h_j(\mathbf{x}) < 0, j = 1, 2, ..., b$$

The advantage of this type of method is that, once the weights are set, SOP algorithms can be used to find a solution with the trade-off represented by these weights. However, finding a suitable weight vector may not be easy, especially when the number of objectives is larger than three. The weight vector will strongly affect the results obtained by the SOP algorithms.

Interactive methods: The main idea of these methods is to guide the search direction according to the information provided by the decision maker during the runtime of the optimisation algorithm, rather than having all information on preferences set beforehand. The main process is as follows:

1. Initialise the algorithm and generate a starting point for the search.
2. Ask the decision maker for preference information, e.g., desirable objective function values, or number of new solutions to be generated.
3. Generate new solutions according to the preference and show them to the decision maker.
4. Ask the decision maker to select the best solution so far and adjust the preference according to the selection.
5. Stop if decision maker wishes to; otherwise return to step 2.

Popular methods include the Interactive Tchebycheff Metric approach [372], NIM-BUS [266] and Guess Methods [50]. The advantage of interactive methods is that they require less prior information than the weighted methods. However, they may cause user fatigue, as the decision maker may need to provide inputs several times throughout the optimisation process.

6.4.2.3 Searching for a Set of Solutions with Different Trade-Offs

There are different frameworks for optimisation algorithms specifically designed to search for the Pareto optimal set of solutions for MOPs [430]. One of the main choices in terms of knowledge representation and modelling in these frameworks is how to model the relative quality of the solutions. This section will briefly explain some different strategies that can be used to compare different solutions in MOP algorithms.

Pareto dominance-based comparisons: Most optimisation algorithms for solving MOPs are evolutionary algorithms based directly on the concept of Pareto dominance explained in Section 6.4.1 [369, 179, 435, 434, 95]. These algorithms consider that a solution that dominates another solution is better than this other solution. They also frequently use some strategy to maintain the diversity of the search, given that concentrating on dominance on its own could lead to lack of diversity and thus early convergence to local optima. For instance, mechanisms such as fitness sharing and crowding distance have been proposed [179, 434, 95] to prioritise solutions that will lead to higher diversity. A drawback of these algorithms is that they may struggle to cope with large numbers of objective functions (e.g., larger than three).

Indicator-based comparisons: Another way to compare solutions is to use a scalar quality indicator/metric [433, 29, 49]. The indicator can be used to compare pairs of solutions [433]. However, a very popular approach is to use scalar quality indicators to evaluate sets of solutions to guide the search. For instance, several algorithms based on the hypervolume have been proposed [35, 49, 23]. These algorithms can perform better than algorithms such as the ones based on Pareto dominance, especially when the number of objectives is large [23, 35]. Depending on how the indicator is defined, it is also possible to avoid the need for separate mechanisms to preserve diversity. However, care must be taken when choosing an indicator or designing an algorithm to use indicators, as some indicators can be expensive to compute [35]. Section 6.4.2.4 further explains indicators/metrics.

MOP decomposition: The strategy adopted by the multi-objective evolutionary algorithms based on decomposition (MOEA/D) [430] is to decompose the MOP into a number of SOPs, where each SOP is a weighted aggregation of the individual objectives. A neighbourhood between SOPs is modelled based on the distances between their weight vectors [430] or some other neighbourhood structure [244]. Each SOP is then optimised simultaneously by using information from its neighbours. The advantage of MOP decomposition is that solutions for a given SOP can be compared against each other based on the scalar corresponding to their single objective, eliminating the need to decide how to compare solutions based on multiple objectives. However, care must be taken to design effective neighbourhood structures.

Algorithms for MOPs usually work efficiently when dealing with two and three objectives. When the number of objectives increases, i.e., when we have many-objective problems, Pareto dominance-based algorithms' performance deteriorates [216]. One of the possible reasons is that nearly all solutions are non-dominated when the number of objectives increases. For instance, Ishibuchi et al. [196] showed that 200 random solutions can be all non-dominated when the number of objectives is over 16. Indicator-based algorithms take the indicator function as the only objective, avoiding this problem. However, calculating indicators such as hypervolume is usually a significantly time-consuming process. For methods based on MOP decomposition, it is difficult to decide on a good balance among all the objectives when dealing with many-objective problems. Currently, there is plenty of work trying to solve the problems caused by many objectives. For example, Narukawa et al. [279] incorporated the preference of a decision maker with a dominance-based algorithm, Ishibuchi et al. [195] proposed an iterative indicator-based approach which attempts to decrease the cost of computing the indicator, and Wang et al. [242] proposed an improved version of the Two-Archive Algorithm to try to incorporate a ranking mechanism for updating the convergence of the archive.

It is worth noting that it is possible to hybridise algorithms in order to combine their advantages. For instance, the memetic multi-objective evolutionary algorithms proposed by Ishibuchi and Murata [194] randomly draw a scalar function to evaluate solutions in the evolutionary search process and then use a local search method to further improve solutions.

6.4.2.4 Metrics for Comparing Sets of Solutions

Algorithms for SOPs generally aim at finding a single best solution to the problem. However, as explained in Section 6.4.1, we are frequently interested in finding the Pareto set. The Pareto set is frequently too large and it could be even infeasible to find the whole set. As a result, many algorithms will find a set of non-dominated solutions expected to be a good approximation of the Pareto set. Therefore, we may need to compare sets of non-dominated solutions, especially when comparing the effectiveness of different algorithms. There are mainly two types of quality metrics:

Unary quality metrics: Unary quality metrics give a quality value to each set of solutions. The advantage of such measures is that they provide a scalar indication of the quality of a set of solutions with respect to a reference point, where the reference point should ideally be the quality of the best or worst possible solution. However, the disadvantage of this type of metric is that it is not always possible in real world applications to determine the ideal reference point. Moreover, these measures need to consider not only the quality of the solutions in terms of convergence with respect to the reference point, but also how diverse the solutions in the set are. This is because it is desirable to have a diverse set so that the decision maker can choose his or her desired trade-off among objectives. However, it has been argued that both aspects of convergence and diversity cannot be properly measured by a single unary metric [436]. Examples of unary metrics are [283]:

- Generational distance. This measure represents how far a set of solutions is from the Pareto set on the objective space:

$$GD = \frac{\sqrt{\sum_{i=1}^{n} d_i^2}}{n} \qquad (6.7)$$

  where n is is the number of solutions in the set of non-dominated solutions found and d_i is the Euclidean distance (measured in objective space) between each of these solutions and the nearest member of the Pareto set. It is clear that $GD = 0$ indicates that all the generated solutions are in the Pareto set. In order to get reliable results, objective values are normalised before calculating this metric.
- Hypervolume. This measure calculates the volume (in the objective space) covered by the solutions in the non-dominated set M being evaluated with respect to a reference point x_{ref} representing the worst possible point. This is done by

calculating the volume Λ of the union of the hypercubes a_i defined by each non-dominated solution $m_i \in M$ and the reference point x_{ref}:

$$S(M) = \Lambda(\{\cup_i a_i | m_i \in M\}) = \Lambda(\{x | m \prec x \prec x_{ref}\}) \qquad (6.8)$$

Larger hypervolume values represent better values. As this measure is also dependent of the scale of the objectives, objectives should be normalised before computing it.

Binary quality metrics: Binary quality measures assign a quality value not to a single set of solutions, but to the relation between two sets. It can overcome the drawbacks of unary quality measures mentioned above, as they do not consider convergence and do not need an ideal reference point to be set. However, its disadvantage is obvious: for k solutions, we have $k(k-1)$ values for every pair of solutions, while in unary quality measures, there are only k values. This makes it harder to analyse the results. Examples of metrics are:

- Coverage of two sets [432]. Given sets of solutions A and B, this measure is defined as:

$$CS(A,B) = \frac{|\{\mathbf{b} \in B | \exists \mathbf{a} \in A, \mathbf{a} \succeq \mathbf{b}\}|}{|B|} \qquad (6.9)$$

The value $C(A,B) = 1$ means that all solutions in B are dominated by or equal to solutions in A. The value $C(A,B) = 0$ means that none of the solutions in B are covered by the set A. Both $C(A,B)$ and $C(B,A)$ should be considered, as $C(B,A)$ is not necessarily the same as $1 - C(A,B)$.

- I_ε metric [436]. In order to define this metric, we first need to define a new relation called ε dominance ($\prec_\varepsilon$). Given two solutions $\mathbf{a}$ and $\mathbf{b}$, this relation is defined as follows:

$$\mathbf{a} \prec_\varepsilon \mathbf{b} \iff F_i(\mathbf{a}) \leq \varepsilon F_i(\mathbf{b}), i \in 1, ..., k \qquad (6.10)$$

The metric I_ε for two sets of solutions A and B is then defined as:

$$I_\varepsilon(A,B) = \inf_{\varepsilon \in R} \{\forall \mathbf{b} \in B \, \exists \mathbf{a} \in A : \mathbf{a} \prec_\varepsilon \mathbf{b}\} \qquad (6.11)$$

Therefore, $I_\varepsilon(A,B)$ equals the minimum factor ε such that any objective vector in B is ε-dominated by at least one objective vector in A. In the single-objective case, $I_\varepsilon(A,B)$ is simply the ratio between the two objective values represented by A and B.

6.5 Decentralisation

Self-aware and self-expressive applications are frequently composed of multiple agents without a central unit to control them. While this might not be true for all applications, decentralised approaches generally have various advantages over centralised applications. For instance, distributed agents may interact with each other to acquire knowledge and perform certain tasks. As each agent performs its task independently, tasks can be processed concurrently. Due to this autonomy of each agent, the robustness of the entire system is increased. While acquired knowledge can be exchanged among the independent agents, this exchange is not mandatory. This reduces the communication overhead in comparison to centralised systems. Section 6.5.1 explains what is meant by decentralisation and provides examples of applications using decentralisation, and Section 6.5.2 explains the implications of decentralisation in terms of the capability primitives introduced in Section 5.3.2, knowledge representation and modelling choices.

6.5.1 Definitions and Examples

In a centralised computing system, all information is collected at and all resources are coordinated by a central node. While central/networked nodes may be able to make their own computations, only the central node will be able to make the decisions for the entire network. Furthermore, besides being responsible for making decisions for the entire network, this central node is usually also responsible for coordinating tasks among all participating nodes. A typical example of centralised systems are Automated Teller Machines (ATMs) where the ATM receives information about the clients' current balance from a central component. The key fact here is that the relevant information about the current balance is not defined by the individual teller machine, but provided by the central component. There are various advantages and disadvantages of a centralised system.

- **Advantages:**
  - + *Coordination:* All information is gathered at a single entity. This single computing node only submits processed information to the requesting components. A coordination of which node does what at what time is not necessary.
  - + *Maintenance:* As there is only a single entity responsible for performing the different tasks, there is only one single entity to maintained. Also, localisation of problems is much easier in a centralised system.

- **Disadvantages:**
  - − *Bottlenecks:* In a large system, the centralised node might receive a lot of different requests simultaneously. This creates bottlenecks on two different fronts: the processor and the network. Having the different entities sending requests for processing their raw data to the server might lead to congestion when it comes to the server receiving this data. Furthermore, the server might be overwhelmed by the amount of work. Finally, storing the raw data on site for further processing might become problematic.

- *Single point of failures:* Even though a single centralised node responsible for processing all data can have advantages when it comes to maintenance and localisation of errors, it is highly disadvantageous when it comes to robustness, reliability and availability. If the central node fails, the other nodes in the system are paralysed and cannot perform any further operations.

In comparison to centralised computing systems, decentralised or distributed computing systems exploit a set of autonomous computers. In the absence of a central node, the computers in the network have to coordinate their required tasks autonomously. This means that each node in a decentralised computing system has to be able to make its own decisions. These decisions can be based on its very own information, on interactions with or on information from other nodes. While in such a distributed approach, the autonomous computing nodes share their information about their own state, desired goals and required tasks among each other, a complete exchange of information might be infeasible. Especially larger systems containing nodes which change their state frequently will inevitably congest their own network by transmitting such updates. Hence, a single machine has virtually no chance of having complete information about the state of all nodes in the network. This also means that having a single machine fail does not bring the entire system to a halt. In a centralised system, if the central server crashes the other nodes of the system will not be able to perform any further tasks. An example in the area of decentralised systems is the telephone network or the well known Peer-to-Peer networks, where upon connection of two or more participants no central component is required for exchanging information between the participants. The advantages and disadvantages of decentralised systems are as follows:

- **Advantages:**
    + *Concurrency:* A single node is not responsible for processing all tasks. This allows the system to distribute the workload and process multiple tasks concurrently on different nodes.
    + *Reliability:* Having multiple entities capable of performing the same tasks makes failing of single entities have less impact than in centralised systems. While failing of single entities might slow down the system or require it to repeat tasks, it does not paralyse it completely. This is related to robustness by redundancy of components, explained in Section 6.3.2.2.
    + *Scalability:* In a decentralised system, each node is independent. New nodes can be added to the system without need for them to be registered at a central component. This allows new nodes to join at any time without prior notice to prepare the system. Additionally, new nodes will be integrated autonomously and they can take over tasks immediately.
    + *Shared resources:* Resources such as processing power or memory can be shared among the different nodes in the network. This allows nodes with less hardware capabilities to transfer unneeded data and workload to other nodes with sufficient resources available.

- **Disadvantages:**

  - *Coordination:* In order to ensure tasks are not unnecessarily performed twice, the system has to coordinate tasks among the nodes. This coordination requires additional effort from the system.
  - *Security:* Having multiple nodes interacting with each other opens up various security risks. In sensitive applications, transmissions between the different nodes have to be protected, requiring a secure network communication. Furthermore, malicious entities have to be detected and neutralised in order to protect the system's original purpose.
  - *Error localisation:* In case of failures in the system, the error has to be localised among all participating nodes. This can be very hard, especially when interdependent tasks are distributed among multiple nodes.
  - *Replicated data:* To allow for reliability but also robustness of a decentralised system, often data has to be replicated multiple times on different nodes. This replication consumes unnecessary resources from a system-level point of view, but allows for concurrent processing of information, increased reliability and robustness.

In the following, two examples of self-aware and self-expressive systems that can benefit from decentralisation will be explained. The first one is a multi-camera system, where a set of deployed cameras pursue a common goal. This goal could be tracking of objects, as in the example given in Section 6.2.1, or monitoring a specific area or detecting and identifying certain behaviour. Using standard cameras requires a centralised approach, where all cameras send their information to a central server for further analysis. Not only will the large amount of video data congest the network, but also video analysis is a resource intensive task. A central server will require a lot of computational power in order to deal with the video streams of multiple cameras consecutively and in near-real-time. Alternatively, standard cameras could be replaced with the so-called smart cameras [342, 341, 336]. Smart cameras combine the image sensor with a processing unit and a network interface. Here, the data can be preprocessed and only relevant or even aggregated data is transmitted to a central control. This reduces the workload for the central server tremendously and distributes the work among the nodes of the network. However, in such a scenario, the server still collects and coordinates information and assigns tasks to the cameras as necessary. While this approach is in general much more scalable than a completely centralised approach, in the case of a server failure, the system is still halted. Only if the decision making capability is transferred to the individual cameras a central coordination can be omitted and a completely decentralised system developed. In this case, each camera still may have to collect various pieces of information from other cameras. However, the whole system becomes more robust against camera failures, i.e., if a camera fails, the entire system will not be paralysed [124].

The second example of a self-aware and self-expressive system that can benefit from decentralisation is the active music system. Active music allows non-musicians to participate in a music-oriented activity using hand-held devices (e.g., a mobile phone) for interaction. A synchronisation mechanism that can be used in this system has been briefly discussed in Sections 6.2.1 and 6.3.1, but the system itself will be briefly discussed in this section from the perspective of decentralisation. The self-aware and self-expressive active music system recognises the interactions of the person with a device and interprets it as music. In a group, multiple persons can join together in a bigger ensemble to create music. A central server would be able to collect all information from the different nodes and synthesise music, but might face limitations on the network as well as with respect to its own processing power. Simple coordination based on data aggregated by a server poses a risk to robustness in case the server fails. Only when decisions are being made by the individual nodes and music generated by the collective behaviour of the nodes, can the workload can be distributed accordingly among the network and failure of individual nodes does not stop the system from producing music.

6.5.2 Implications

When designing a decentralised or distributed system, a system designer has to make various choices. In this section we focus on important choices to be made for self-aware and self-expressive systems, focusing on choices of levels of self-awareness and structures to be used.

6.5.2.1 Choice of Levels of Self-awareness

Stimulus-awareness: The individual nodes of a stimulus-aware, decentralised system are only able to interact with each other by reacting to stimuli and triggering new external stimuli in return. An excellent example in nature is the frequency synchronisation of fireflies. Only by perceiving the stimulus of the blink of a light, unaware of its origin or exact time of blink, does the firefly react by blinking itself. An example in the context of self-aware, self-expressive systems is the synchronisation mechanism [295] that can be used in systems such as active music. It uses a stimulus-aware approach in order to synchronise musical patches, as explained in Section 6.2.1. Section 7.3.1.3 gives a more detailed explanation of a computational self-aware approach for frequency synchronisation.

Interaction-awareness: Typical decentralised systems are expected to implement interaction-aware behaviour. Here the individual nodes can not only distinguish between the different stimuli but are also able to differentiate between the same stimuli from different nodes. Furthermore, the individual nodes can select a specific subset of nodes in the system for interaction. This allows them to not only improve the performance of the entire system, but also make collective decisions and exploit local behaviour. For example, the smart camera system [125] uses an economy-inspired interaction-aware approach in order to coordinate tracking responsibilities.

Time-awareness: Time-awareness allows a system to adapt to and anticipate changes not only in the environment, as discussed in Sections 6.2 and 6.3, but also in the behaviour of other nodes. Additionally, the system is enabled to 'forget' previously learnt information which has become obsolete. An example for such time-awareness is the pheromone-based foraging process of ants. Pheromones are strengthened while the food source lasts, but evaporates over time, when the food source is depleted and ants do not deploy pheromones anymore. Artificial pheromones can be seen as a way for nodes in decentralised systems to interact with each other in a time-aware fashion. Two different computational time-aware approaches employing artificial pheromones are used in the smart camera system and the active music system. They are discussed in Section 7.4.1.1 and Section 14.3.1

Goal-awareness: In the absence of a central component, goals are defined and represented by the individual nodes and are based on their individual capabilities. This abstraction enables the system to deal with heterogeneous nodes and scale better. Additionally, each node is focussed on its own capabilities and goals, allowing the system to distribute load more equally among all nodes and execute tasks concurrently. In the smart camera system the individual cameras are aware of their own goals without having a central control steering them.

Meta-self-awareness: While meta-self-awareness is not essential to a decentralised system, it allows the system to improve its own performance based on the environment it has been deployed in. Additionally, each individual node can reason about its own behaviour in the dynamic environment and the actions and reactions of its neighbouring nodes. An example of computational meta-self-awareness in the smart camera system is given in Section 7.4.1.3 and by Lewis et al. [239], where each node learns about the performance of a set of actions and over time selects the best for its given situation.

Similarly to what has been discussed in Section 6.2.2.1, while the different levels of self-awareness may help a system to better handle decentralisation, they also require additional computation time, which can be particularly concerning in applications requiring near-real-time performance or in embedded devices where resources are very limited.

6.5.2.2 Choice of Neighbourhood

In a distributed system, each node is able to build up its social neighbourhood, i.e., those nodes it is interacting with during its lifetime. Essentially, a designer of a self-aware and self-expressive system has to make a choice regarding the neighbourhood being limited or unbound in terms of the number of neighbours.

Limited neighbourhood: In a limited neighbourhood, each node is only able to interact with a limited number of other nodes from the network. This limit can be either fixed or relative to the system size, but is defined by a system designer before deployment. A limit to the neighbourhood size allows the system designer to control the resource consumption to some extent. On the other hand, it also limits the possible capabilities of the system when it comes to requesting new resources or processing capabilities from other nodes.

Unbound neighbourhood: In contrast to a limited neighbourhood, in an unbound neighbourhood the node itself decides upon the number of neighbouring nodes it wants to interact with. An unbound neighbourhood allows each individual node to adapt its own neighbourhood during runtime by adding new nodes or removing unneeded ones. This allows the individual node to access more resources through other nodes or reduce its communication overhead by removing nodes from its network. Nevertheless, even unbound neighbourhoods are limited by the communication and sensing capabilities of the respective node. For example, a node will not be able to communicate with all nodes in a wide-area network with single-hop communication over a wireless network interface (e.g., in the active music system). In the smart camera system, the field of view of the individual cameras naturally limits the size of the neighbourhood.

6.5.2.3 Choice of Information Accessibility

In a distributed system, resources as well as information is shared among all nodes in the network. This means a node can access resources and information locally or remotely. Local information is based on the experiences a node makes by itself. In contrast, with remote information nodes explicitly exchange knowledge about their own experiences (e.g., sensor data or available resources). While accessing remote resources allows us to distribute workload, information from remote nodes may have benefits as well drawbacks for a node in comparison to using only local information. We will briefly discuss the benefits and drawbacks of limiting the access of a node.

Local information: A node only relies on its very own experiments. Only local information is used in order to make decisions and continue towards a global, system-wide goal. While this may limit the capabilities of the entire network in terms of possible performance, nodes may not rely on possibly incorrect information from other nodes. Additionally, it prevents nodes from exchanging irrelevant information among each other, which would otherwise increase the network traffic unnecessarily. For example, the smart camera networks, where each node only values objects it is able to "see", are based on local information. This means a camera ignores objects in neighbouring fields of view, even if they might enter its own field of view in the near future.

Remote information: In contrast to limiting a node to only local information, remote information enables each entity to draw from experiences other nodes have already had. This allows faster transition of knowledge but requires clear structures on how knowledge is represented. If facts in a given situation are omitted, a seemingly similar situation for one node might result in worse performance when drawing from another node's experiences. Taking the smart camera system as an example, when one camera omits its available resources when representing their knowledge and experience, another node might use a similar behaviour but may perform worse if it has less resources at its disposal.

Mixed information: While information from remote nodes might be insufficient to be applied at a given node, it can also be very helpful. In a mixed approach each node could individually learn how much remote information it wants in addition to local information. Furthermore, nodes could even learn which remote node provided what kind of information to what degree of usefulness. This would allow them to distinguish between information from different nodes, filter noise from useful information and learn faster in their environment. Various learning methods will be discussed in the following Chapter 7.

6.6 Summary

This chapter discussed different features that may be present in a self-aware and self-expressive system from the perspective of knowledge representation and modelling. The features discussed were adaptivity, robustness, multi-objectivity and decentralisation. A common theme in the discussion of these features was that different levels of self-awareness can be helpful for successfully implementing these features. However, trade-offs in terms of the benefit that such levels of self-awareness can provide and computation time must be carefully considered, especially in systems that must operate in near real-time or in embedded systems. Besides the choice of the level of self-awareness, several other knowledge representation and modelling choices must also be made to decide how exactly each of these levels of self-awareness should be implemented. These choices include, but are not limited to, how and whether to model states and environments, model time, deal with changing environments, use redundancy of components, anticipate changes, formulate multiple objectives, compare solutions with different objectives, implement nodes' neighbourhoods, and exchange information among nodes. Different options must be carefully considered when developing a self-aware and self-expressive system. The best choice will depend on the application to be developed and the type of environment in which it will be embedded.

Chapter 7
Common Techniques for Self-awareness and Self-expression

Shuo Wang, Georg Nebehay, Lukas Esterle, Kristian Nymoen, and Leandro L. Minku

Abstract Chapter 5 has provided step-by-step guidelines on how to design self-aware and self-expressive systems, including several architectural patterns with different levels of self-awareness. Chapter 6 has explained important features in self-aware and self-expressive systems, including adaptivity, robustness, multi-objectivity and decentralisation. To allow such self-aware capabilities in each design pattern and enable those system features, this chapter introduces the common techniques that have been used and can be used in self-aware (SA) and self-expressive (SE) systems, including online learning, nature-inspired learning and socially-inspired learning in collective systems. Online learning allows learning in real time and thus has great flexibility and adaptivity. Nature-inspired learning provides tools to optimise SA/SE systems that can be used to reduce system complexity and costs. Socially-inspired learning is inspired by common social behaviours to facilitate learning, particularly in multi-agent systems that are commonly seen in SA/SE systems. How these techniques contribute to SA/SE systems is explained through several case studies. Their potentials and limitations are widely discussed at different self-awareness levels.

Shuo Wang
University of Birmingham, UK, e-mail: s.wang@cs.bham.ac.uk

Georg Nebehay
Austrian Institute of Technology, Austria, e-mail: gnebehay@gmail.com

Lukas Esterle
Alpen-Adria-Universität Klagenfurt, Austria, e-mail: lukas.esterle@aau.at

Kristian Nymoen
University of Oslo, Norway, e-mail: kristian.nymoen@imv.uio.no

Leandro Minku
University of Leicester, UK, e-mail: leandro.minku@leicester.ac.uk

7.1 Introduction

Chapter 5 has provided step-by-step guidelines on how to design self-aware and self-expressive systems, including several architectural patterns with different levels of self-awareness. Chapter 6 has explained important features in self-aware and self-expressive systems, including adaptivity, robustness, multi-objectivity and decentralisation. To allow such self-aware capabilities in each design pattern and enable those system features, this chapter introduces the learning methods that are widely applied and considered in SA/SE systems. First, online learning (Section 7.2) contains various techniques that can potentially encourage time awareness by updating learning models with time. Some online learning methods are designed to anticipate changes in learning data and environments, which may contribute to goal awareness and stimulus awareness. They also maximise system adaptivity, in order to handle changing environments. Second, nature-inspired learning (Section 7.3) provides important ideas for designing large and complex systems. Meanwhile, it includes great techniques to realise multi-objectivity. Third, socially-inspired learning (Section 7.4) includes a set of techniques for multi-agent systems, which can decentralise the system and reduce system complexity.

The above learning techniques are defined and explained in the following sections. In particular, example case studies are given to illustrate how they have been used in SA/SE systems at the levels of self-awareness. For each type of learning technique, related learning methods that may potentially contribute to SA/SE systems are also introduced to widen the application for future use.

7.2 Online Learning

Online learning has been extensively studied and applied in the field of machine learning in recent years. Most machine learning methods operate in offline mode. They first learn how to perform a particular task and then are used to perform this task. No task can be performed during the learning phase and, after the learning phase is completed, the system cannot further improve or change [267]. However, a large number of real-world problems do not allow us to see all the data in advance, and they are not intrinsically static. For example, consider an information filtering system which predicts the user's reading preferences. It is not possible to collect all users' information beforehand for learning. New users can join in and old users may leave at some point. Users can also change their preferences with time and a system which learnt how to predict them in the past will fail if it cannot update according to the new ones [356]. In a recommender or advertising system, customers' behaviour may change depending on the time of the year, on economic inflation and on new products made available. Other examples include systems for computer security [142], market-basket analysis [14], spam detection [288] and web pages classification [10].

Differently from offline learning algorithms, online learning is brought up to perform a particular task at the same time as the learning occurs. Online learners are updated whenever a new training example is available, being able to perform lifelong learning in the sense that they do not ever need to stop learning. So, they can attempt to adapt to possible changes in the environment, if properly designed [267]. In a more formal definition, online learning algorithms process each training example once "on arrival" without the need for storage and reprocessing, and maintain a model that reflects the current concept to make a prediction at each time step [305] [408]. The online learner is not given any process statistics for the observation sequence, and thus no statistical assumptions can be made in advance. A closely related concept to online learning is incremental learning. Incremental learning algorithms can also operate in changing environments, but training data are processed in chunks [267]. From another point of view, online learning can be viewed as a strict case of incremental learning, which processes data iteratively. It can be used to solve both online and incremental problems. Due to its low time and memory requirements, besides being useful for applications in which training data arrive continuously (streams of data), online learning is also useful for applications with tight time and space restrictions, such as prediction of conditional branch outcomes in microprocessors [132].

Considering the nature and advantages of online learning, it contains useful techniques for SA/SE systems, as many SA/SE applications produce data continuously and the data may not be in a static state. For example, one of our applications is to perform object tracking and object detection in single-camera and multi-camera scenarios. In order to locate the object, the object detector must keep monitoring the location of targets in consecutive video frames. Meanwhile, the object tracking/detection can suffer from clutter and occlusion difficulties, as well as the execution time issue. Therefore, a fast online object detection method is adopted, to learn the appearance of objects in image streams and to re-detect the objects of interest after occlusions. More details on the application and the adopted techniques will be given in the following subsections.

7.2.1 Example Application

In object tracking, the general goal is to find the location of one or more objects of interest in consecutive frames of a video sequence, as depicted in Figure 7.1. Under

Fig. 7.1 Tracking a single object of interest (in red)

some specific scenarios, training data may be available to aid the process of recognizing objects. A learning algorithm can then be used in order to better understand the variability in the appearance of the objects of interest. In many cases, the object of interest is unknown beforehand and is indicated only by a single user-defined label that is typically given in the first frame of a sequence. Although no supervised training data is available, it is still often desirable to perform an update of the object model in order to reflect changes in the appearance of the object of interest. Many approaches model the problem of object tracking as binary classification in the online learning scenario. The two classes are the object of interest and background. We will now give a brief overview of existing techniques that fall into this category.

Collins et al. [80] were the first to employ binary classification in a tracking context. They employ feature selection in order to switch to the most discriminative colour space from a set of candidates and use mean-shift for finding the mode of a likelihood surface, thereby locating the object. In a similar spirit, Grabner et al. [148] perform online boosting and Babenko et al. [22] use multiple instance learning in order to find the location of the object. All of these methods use a form of reinforcement learning, meaning that the prediction of the classifier is directly used to update the classifier. While this approach enables the use of unlabelled data for training, it typically amplifies errors made in the prediction phase, thus leading to a degradation of tracking performance. In [149], this problem is addressed by casting object tracking as a semi-supervised learning problem, where only the first appearance of the object is used for updating. Both Kalal et al. [206] and Santner et al. [350] employ an optical-flow-based mechanism for labelling the available data in order to reduce the errors made in the prediction phase and demonstrate superior results. All of these methods employ online learning to incorporate new appearances of the object of interest into the object model with the aim of improving the general tracking performance.

In the following sections we will now explain the TLD approach of Kalal et al. in more detail. Section 7.2.1.1 gives a high-level overview of the approach. Section 7.2.1.2 introduces the learning component of TLD, a Random Fern classifier. Finally, Section 7.2.1.3 explains how online learning is performed in TLD.

7.2.1.1 Tracking-Learning-Detection

Kalal et al. [206] propose a solution to the tracking problem which they call *Tracking-Learning-Detection* (TLD). TLD consists of two separate components. The first component is a frame-to-frame tracker that predicts the location L_j of the object in frame I_j by calculating the optical flow between frames I_{j-1} and I_j and transforming L_{j-1} accordingly. Clearly, this approach is only feasible when the object is visible in the scene and fails otherwise. When the object is presumably tracked correctly (according to certain criteria) the location L_j is used in order to update the second component, which is a Random Fern classifier [306]. The Random Fern classifier is updated with positive training data from patches close to L_j and negative data from patches that exceed a certain distance. This classifier is then applied in a sliding-window manner (see Figure 7.2) in order to re-initialise the frame-to-frame-tracker after failure.

Fig. 7.2 In TLD, a binary ensemble classifier is used to locate the object of interest by applying it in a sliding-window manner. The ability for multi-scale detection is achieved by scaling the size of the detection window. Image is from [281].

7.2.1.2 Random Fern Classifier

The Random Fern classifier [306] operates on binary features $f_1 \ldots f_n$ calculated on the raw image data. These features are randomly partitioned into groups of so-called *ferns* $F_1 \ldots F_m$ of size s

$$\overbrace{f_1 \ldots f_s}^{F_1}, \overbrace{f_{s+1} \ldots f_{2s}}^{F_2} \ldots \overbrace{f_{(m-1)s+1} \ldots f_{ms}}^{F_m}. \tag{7.1}$$

Ferns essentially are non-hierarchical trees, meaning that the outcome of each fern is independent of the order in which features are evaluated. The main reason for favouring ferns over trees is that they can be implemented extremely efficiently, an important property for real applications.

Suitable features for random ferns are proposed in [306], where a feature vector of size s consists of s binary tests performed on greyscale image patches. Each test compares the brightness values of two random pixels (See Figure 7.3). The locations of the tests are generated once at startup and remain constant throughout the rest of the processing. The same set of tests is used with appropriate scaling for all subwindows. Input images are smoothed with a Gaussian kernel to reduce the effect of noise.

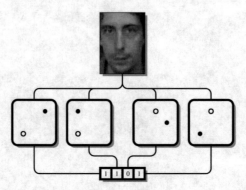

Fig. 7.3 Feature values depend on the brightness values of pairs of two random pixels. In this case, the outcome is the binary string 1101. Image is from [281].

7.2.1.3 Online Learning of the Random Fern Classifier

An important property of a classifier is the amount of time it takes to re-train the classifier as soon as new training data becomes available. In this respect, Random Ferns are superior to other classifiers such as Random Forests, as their formulation allows for a straightforward application in an online learning scenario. The posterior probability for each fern is

$$P(y=1|F_k) = \frac{P(y=1)P(F_k|y=1)}{\sum_{i=0}^{1} P(y=i)P(F_k|y=i)}, \qquad (7.2)$$

where $y = 1$ refers to the event that the object of interest is present in the subwindow. In TLD, the prior is assumed to be uniform, and the $P(F_k|y=i)$ are modelled as the absolute number of occurrences $\#p_{F_k}$ for positive training data and $\#n_{F_k}$ for negative training data. Therefore, the posterior probability becomes

$$P(y=1|F_k) = \frac{\#p_{F_k}}{\#p_{F_k} + \#n_{F_k}}. \qquad (7.3)$$

When $\#p_{F_k} = \#n_{F_k} = 0$, then $P(y=1|F_k)$ is assumed to be 0 as well. The update procedure in TLD employs all training instances (i.e., all subwindows) for updating the classifiers that are currently misclassified. A decision is obtained by employing a threshold θ on the posterior probabilities combined using the mean rule

$$\frac{1}{m} \sum_{i=1}^{m} P(y = 1|F_i) \geq \theta. \tag{7.4}$$

7.2.2 Benefits and Challenges at Levels of Self-awareness

In this section we examine TLD running on a single node with respect to the levels of self-awareness, pointing out benefits and drawbacks. It has to be noted that TLD runs on a single node.

7.2.2.1 Stimulus-awareness

A stimulus causing the node to react refers to self-awareness on a very basic level. In TLD, this level of self-awareness is present in the computation of the optical flow that provides a predefined reaction to a certain optical stimulus provided by an image sensor. The reaction is in the form of computing the appropriate displacement vectors for the sparse motion field of the target. A very important aspect in this respect is the time that the node takes to react to the stimulus.

Benefits: In TLD, the employed method for optic-flow allows us to compute the response tremendously fast.

Drawbacks: The optical flow component is unable to adapt itself.

7.2.2.2 Interaction-awareness

While the nodes of TLD in the form of ensemble members of the random fern classifier might be seen as virtual nodes, they act completely independently, not performing any kind of interaction with the exception that their output is combined into a single response.

Benefits: By not modelling interactions between virtual nodes, computational efforts are kept low.

Drawbacks: Wrong decisions by individual virtual nodes are not identified.

7.2.2.3 Time-awareness

Time-awareness is an enabling component for learning from experience, as errors made in the past are remembered and individual virtual nodes are adapted in order to prevent similar errors in the future. In TLD, time-awareness is present in the individual virtual nodes that each remember a history of misclassified feature values. It is, however, clear that this level of awareness comes at a certain computational cost.

Benefits: Time-awareness enables adaptation in order to avoid future errors.
Drawbacks: Time-awareness increases the complexity of a node considerably.

7.2.2.4 Goal-awareness

The single goal in TLD is to track an object of interest as long as possible. While this goal is present implicitly in the adaptation of the posterior probabilities of the individual ferns, there is no explicit modelling of this goal.
Benefits: N/A
Drawbacks: N/A

7.2.2.5 Meta-self-awareness

There is an interesting feedback loop in TLD that we consider to render TLD meta-self-aware. This feedback loop refers to how the training examples in TLD are obtained, namely by extracting them by making use of its very own stimulus-aware component, the optical-flow-based tracker. The output of this optic flow-based tracker serves as a means to distinguish between positive and negative training examples in the environment and to adapt the behaviour of the node to improve in the future.
Benefits: Meta-self-awareness allows for the automatic extraction of training data.
Drawbacks: Absolute correctness of the extracted data is hard to verify.

7.2.3 Other Related Techniques

This section discusses other related techniques, namely ensemble learning, class imbalance learning, techniques for dealing with concept drift, and reinforcement learning. It does not provide a comprehensive overview of learning methods, but focuses on techniques that we believe to be more relevant to self-awareness and self-expression.

7.2.3.1 Ensemble Learning

Ensemble learning techniques have been given particular attention in online learning. The idea of ensemble learning is to employ multiple learners on a given problem and combine their outputs as a "committee" to make a final decision for better accuracy. The individual committee member is sometimes called base learner. In classification, ensemble learning is also referred to as multiple classifier system [166], classifier fusion [228], committee of classifiers, classifier combination, etc. The members' prediction might be real-valued numbers, class labels, posterior probabilities, or any other quality. To make the best use of the strengths of the individuals and make up their weaknesses, how to combine the predictions is important and has been studied extensively in the literature [359] [219].

Bagging [47] and AdaBoost [136] are the most popular offline ensemble algorithms in the literature, based on which numerous variations have appeared for different learning scenarios. There also exist other popular ones, such as Random Subspace [165], Random Forests [48] and Negative Correlation Learning [248].

Ensemble learning methods have some desirable features, which encourage the rapid growth of related research. For theoretical reasons [101], every single learning model has limitations and may perform differently due to insufficient data. Averaging many of them can reduce the overall risk of making a poor prediction [317]. It is innate behaviour to consult others before making a decision. Second, certain learning algorithms confront the local optima problem, such as neural networks and decision trees. Ensembles may avoid it by running local search from different views, where a better approximation to the true function is expected. Besides, some problems are just too difficult and complex, beyond the learning ability of the chosen models. Ensembles allow partition of the data space, where each individual only learns from one of the smaller and simpler sub-problems. Their combination can then represent the whole problem better. For practical reasons [324], real-world problems can be very large or small. A large data set can be divided into several smaller subsets, which will be processed by multiple learners in parallel. In the case of too little data, resampling techniques can be used in ensembles for drawing overlapping subsets to emphasise the available data.

In addition to the above advantages, in the context of online learning, ensemble learning has become a preferable solution, because it has a more flexible and robust training framework than other single-model training methods, especially for non-stationary online scenarios. On one hand, building and maintaining multiple learners allows model updating without forgetting previously learnt information. Online Bagging [304] and Online Boosting [132] [304] are two successful online extensions of the well-established offline Bagging and Boosting. They have been widely used for processing static data streams. On the other hand, an ensemble model can be expanded for future data coming from a new data concept. For example, when processing image data streams, the image result can be affected by the machine capturing images and the environment where the machine is placed. If any change occurs, for a single-model learner, it has to be re-trained to learn the new data concept, whereas for an ensemble learner, a new learner can be simply trained based on the new data, which can then be added to the ensemble model, considering that the existing learners in the ensemble may still contain some useful information. Section 7.2.3.3 further explains how ensembles can be used to deal with changes in data concept.

The aforementioned techniques can intrinsically facilitate learning tasks in online SA/SE systems. With regard to the five levels of self-awareness, it is apparent that the characteristic of real-time processing of online ensemble learning methods allows time-awareness. The time information in data streams can be utilised for time series modelling and/or anticipation. Furthermore, due to the flexibility and robustness in dynamic environments, online ensemble learning methods may encourage goal-awareness. If the learning objective is changing over time, appropriate ensemble techniques can be applied to sense the change and/or adjust the learning process correspondingly and adaptively.

7.2.3.2 Class Imbalance Learning

It is worth mentioning that there is a specific class of online ensemble learning methods aiming to tackle imbalanced data streams. "Class imbalance" is a type of classification problems where some classes of data are heavily under-represented compared to other classes. It is commonly seen in real-world applications, such as intrusion detection in computer networks and fault diagnosis of control monitoring systems [406]. This type of data suffers from learning issues caused by the relatively or absolutely under-represented class (minority) that cannot draw equal attention to the learning algorithm compared to the majority class. It often leads to very specific classification rules or missing rules for the minority class without much generalization ability for future prediction. This problem is aggravated when data arrive in an online fashion. Therefore, special treatments are required to overcome class imbalance.

When the received data become imbalanced, it is necessary for the relevant nodes in the SA/SE systems to have the knowledge of data status (seen as a type of stimulus), so that corresponding events can be triggered to maintain the system performance. Among very limited research that explores this type of stimulus-awareness, Wang et al. [405] developed a class imbalance detector that reports the real-time class imbalance status online. Once class imbalance occurs, it will inform the running online ensemble learner, which can then adopt some state-of-the-art class imbalance techniques to adjust the learning bias. Based on the traditional Online Bagging, several resampling-based ensemble variations have been proposed, which apply oversampling or undersampling techniques to boost the role of the minority class and give online predictions, such as OOB, UOB [405], MOSOB [407] and WEOB [408]. Generally speaking, we expect the training framework of online class imbalance learning to introduce a certain degree of stimulus-awareness into the ensemble learning framework with time-awareness and goal-awareness.

7.2.3.3 Concept Drift Techniques

As mentioned in the beginning of Section 7.2, many SA/SE applications produce data continuously and the data may not be in a static state, i.e., the joint probability distribution of the data may change with time. Such changes are referred to as *concept drifts* [141]. Online learning algorithms can implement strategies to deal with concept drifts, so that the detrimental effect of changes can be diminished and the learning models can recover from changes. As explained in Section 6.2.2.4, such algorithms can be based either on explicit change detection methods or on mechanisms that do not use explicit change detection. That section concentrated on the advantages/disadvantages of using change detection methods. The current section concentrates on explaining existing online supervised learning algorithms themselves.[1]

Algorithms with explicit change detection methods:

Change (a.k.a. drift) detection methods are usually based on the idea that the performance of a model improves over time when the data concept is stable. Therefore, such methods detect a concept drift when a considerable drop in the performance is observed. For example, the Drift Detection Method (DDM) algorithm [141] monitors the error rate and detects a concept drift if the error rate increases above the confidence interval of the minimum error rate so far. Early Drift Detection Method (EDDM) [24] monitors the average distance between any two misclassifications and detects a change if the current average distance is larger than the minimum distance so far by a pre-defined threshold. Statistical Test of Equal Proportions (STEPD) [289] monitors the difference in accuracy on older and more recent training examples and uses a statistical test to detect changes. These algorithms maintain a learning model which is reset upon change detection [141, 24, 289].

In order to reduce the problems caused by false positive change detections (Section 6.2.2.4), the algorithm Two Online Classifiers for Learning and Detecting Concept Drift (Todi) [288] maintains two classifiers in the learning system instead of one. Only one of these classifiers is reset upon change detection, so that the model representing the old data concept can still be used in the case of a false positive drift detection. Another approach able to avoid problems with false positives is Diversity for Dealing with Drifts (DDD) [268]. It maintains different ensembles to achieve better robustness to different types of changes. Among them, an ensemble representing the old data concept well is kept both for dealing with slow changes and false positive change detections. There are also algorithms to cope with recurring concepts, i.e., with concept drifts that take the current data distribution to a previously seen state. For instance, Just-in-Time Classifiers for Recurrent Concepts [11] keeps a representation for each concept that has been identified so far. Whenever the concept observed after a change detection matches an existing concept representation, a model created for that representation is retrieved so that this concept does not have to be learnt from scratch.

Algorithms with no explicit change detection method:

Most algorithms with no explicit change detection method are ensembles of learning machines specifically designed to deal with concept drift. These algorithms usually maintain a set of models which possibly represent different data concepts. New models are created to represent new concepts, whereas existing models deemed to be out of date can be deleted to gradually forget old concepts. Then, by emphasis-

---

[1] It is worth noting that there are also several other algorithms for dealing with concept drift in chunk-based learning (i.e., algorithms that are not online algorithms in the strict sense) [330, 76, 375, 119].

ing the predictions given by the models more likely to perform well on the current data concept, these ensembles can deal with concept drifts.

For example, Concept Drift Committee (CDC) [370] maintains an ensemble whose base models are weighted based on their accuracies. A new base model is created whenever a new training example is made available. When the maximum ensemble size is reached, a new model is added only if an existing model can be deleted. Models are deleted based on their current accuracy. Different from CDC, Dynamic Weight Majority (DWM) [223] creates new base models only at every p training examples, if the ensemble misclassifies the training example. This helps DWM to maintain a more diverse set of base models representing different concepts. Rather than assigning weights to base models according to their overall recent accuracy, Dynamic Classifier Selection [396] assigns weights to base models based on their local accuracy. Local accuracy is the accuracy on a subset of validation examples most similar to the example being predicted. Even though this algorithm may achieve better performance by tailoring weights to the specific example being predicted, a set of validation examples must the stored.

7.2.3.4 Reinforcement Learning

In self-aware systems reinforcement learning deals with nodes that are allowed to take certain actions in order to maximise an objective function. For instance, in the context of people tracking, one action for a camera node might consist of "hand over the current object to the next camera in the environment". The objective function in such a camera network might consider the balance between accurate tracking and the overall computational load of the camera nodes. A camera node might therefore decide to stop tracking the object in order to reduce the penalty imposed by its computational load when other cameras have a better view of the object. In order to be able to perform reinforcement learning, a node must be able to analyse its environment using its sensors as well as to interact with the environment using its self-expression capabilities.

The most striking difference between reinforcement learning and classical machine learning is the absence of labels. Instead of looking at examples (as in supervised machine learning), reinforcement learning algorithms perform a form of trial-error in order to explore its environment and find out which actions lead to a long-term increase of the objective functions. An intriguing problem is that the environment is constantly changing, possibly also due to the action of the node. For instance, handing over an object from one camera might be a good idea if the other camera is able to continue tracking the object. However, if the other camera quickly loses the object, the benefit of reducing the computational load is thwarted by the loss of tracking accuracy. Reinforcement learning algorithms usually employ some form of probabilistic model that are commonly based on Markov Decision Processes.

7.3 Nature-Inspired Learning

The field of nature-inspired learning is an inter-disciplinary area of research concerned with the problem-solving and computational capabilities of natural systems. Nature provides great sources of inspiration to both develop intelligent systems and find solutions to complicated problems. For example, the analogy between the human nervous system and computational devices has been studied and exploited comprehensively from theory to application. It leads to the development of mathematical models of computation, such as neural networks [28]. The resulting techniques have contributed to substantial real-world applications successfully, especially in pattern recognition. Taking animals for example, evolutionary pressure forces them to develop highly optimised organs and skills to survive. Those organs and skills can be well refined as optimisation algorithms, and the evolution can be described as the process of fine-tuning the parameter settings in the algorithms. Genetic algorithms (GAs) and ant colony optimisation (ACO) are popular nature-inspired learning algorithms. Stated simply, nature-inspired learning, namely natural computing, is the computational version of the process of extracting ideas from nature to develop 'artificial' (computational) systems, or using natural media (e.g., molecules) to perform computation [59].

The nature-inspired learning approaches are of great use in the following particular situations: (i) the problem to be solved is very complex; for example, there are too many variables, or the problem is highly dynamic and non-linear; (ii) a potential optimal solution is not guaranteed; (iii) the problem to be solved cannot be modelled properly; (iv) a single solution is not good enough and more than one solution is needed [59].

With the behaviour of natural systems, nature-inspired learning approaches are also shown and expected to possess greater adaptivity and resilience to applied environments than other learning systems. This adaptivity and resilience arise from several aspects[255]: interchangeable elements in the system, loose and flexible interconnections between elements, the differences between elements, and the diverse responses to a changing environment.

The above aspects of nature-inspired learning approaches provide good reasons to solve complex problems in SA/SE systems that are usually formed of multiple SA/SE modules and software agents.

7.3.1 Example Application

We consider an example of a nature-inspired self-aware and self-expressive computing system within the field of *active music*. In an active music system computational nodes play malleable music that may be changed by a user or by the computational node itself (see further description and more examples of active music systems in Chapter 14). When a group of nodes play music together, the most important challenge faced is timing. By sharing some common notion of time, the nodes may create various controlled musical patterns, such as rhythms or melodies.

In a setup where a collaborative active music system is implemented on mobile phones, the distribution of the system on separate units of the system calls for a mechanism for *decentralised* synchronisation. That is, there should not be a central unit to which the remaining nodes synchronise. A decentralised approach would allow nodes to leave or enter a musical performance at any time.

The example algorithm given in this section is inspired by fireflies who synchronise their flashes in a decentralised manner. The phenomenon has been studied for almost a century, and has more recently been proven useful when synchronising devices in machine-to-machine systems [39]. The algorithm allows pulse-based communication, and thus individual musical tones may be used for communication between nodes. The algorithm presented was introduced in [295], and is based on Mirollo and Strogatz' firefly algorithm [271] with various changes to compensate for delays in the system and faulty nodes, as will be explained in the coming section. More examples of how nature-inspired learning may be used to obtain self-awareness in active music systems is given in Chapter 14.

7.3.1.1 Fireflies as Inspiration for Pulse-Coupled Oscillators

The phenomenon of certain species of firefly on riverbanks in Asia that synchronise their flashes has been observed and studied for more than a decade [151]. Throughout the twentieth century John Buck proposed several explanations to the phenomenon [51, 52], although a satisfactory mathematical model was not proposed until the last part of the century. Inspired by Peskin's work on heart physiology [318], Mirollo and Strogatz succeeded in proving that the phenomenon could be described as *pulse-coupled oscillators* [271]. Their proof, however, required the fire events to be infinitely short impulses, with no delay in the communication, and also required that oscillator frequencies be equal.

In our example system, the synchronising units are musical nodes, each with an internal oscillator with a certain frequency (ω) and phase (ϕ). To enable interaction between such nodes, they must be *coupled*, either continuously by having each node monitor the phase of the oscillators of the other nodes (phase-coupling), or discretely though pulse-like communications (fire events) whenever an oscillator node reaches a phase threshold in its cycle (pulse-coupling). While phase-coupled systems provide continuous updates, and theoretical models show how they can be programmed to obtain synchronisation between oscillators, these systems are difficult to implement in real-world scenarios [39]. Pulse-coupled systems, on the other hand, are less difficult to implement in terms of inter-node communication, and as such Mirollo and Strogatz' firefly model is well suited for enabling synchronisation of active music nodes.

The fundamental operation of the algorithm described by Mirollo and Strogatz is that the phase of an oscillator is increased by a small amount whenever a fire event is received from another oscillator. The amount by which the phase is increased depends on the internal phase of the receiving oscillator. It is imperative to their model that the larger the phase of the receiving node, the larger should be the phase jumps that occur. The process is illustrated for two nodes in Figure 7.4.

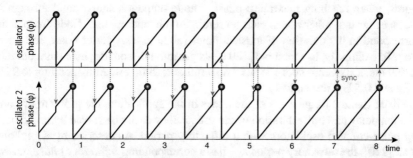

Fig. 7.4 Synchronisation of two firefly nodes by the Mirollo/Strogatz algorithm. The two nodes have identical frequencies, and there is no communication delay. Fire events are indicated by black circles, and the corresponding interaction indicated by a red arrow. Note how the size of the phase jump depends on the current phase of the receiving node.

Since Mirollo and Strogatz' proof, several efforts have been made towards reaching a synchronised state for real systems. Specifically, the requirements of infinitely short impulses and no communication delay is problematic in real systems. An increased effort has been seen in recent years, as decentralised computing systems have become more popular.

Mathar and Mattfeldt showed that the problem of transmission delays and transmission pulses of finite length can be tackled by implementing a *refractory period* directly after firing [259]. Similarly to how neurons enter a refractory state directly after firing, in which it is unable to receive new input, a pulse-coupled oscillator does not respond to fire events received within the refractory period.

Babaoglu et al. addressed the challenging scenario of synchronising clock cycles in certain types of sensor networks which lack a central reference clock to synchronise with [21]. A number of efforts have been made to tackle the challenge of "deafness" that occurs when a node is sending and receiving a fire event at the same time. Werner-Allen et al. [411] and Leidenfrost and Elmenreich [233] suggested the *Reachback firefly algorithm* for this purpose, where instead of making immediate phase jumps, the received impulses are accumulated during each period, and the corresponding phase jump is made at the beginning of the next cycle.

Klinglmayr et al. target the problem of robustness against faulty nodes in a system. That is, nodes that become defective, or malicious intruding nodes that may potentially disturb the operation of the network [220, 221]. Their approach is rather than to push the phase of receiving nodes forward (excitatory coupling), to decrease the phase of the receiving node (inhibitory coupling).

7.3.1.2 Decentralised Phase Synchronisation in Active Music Systems

We use as example a collaborative active music system where each node is implemented on a mobile phone. The nodes communicate using audible sound. In other words, when reaching maximum phase a node outputs a short sound through its loudspeaker that other nodes pick up through their microphone. Node frequencies correspond to the duration of musical notes (eight notes, quarter notes, full notes, etc.). To exemplify, in a tempo of 120 beats per minute, a quarter note has a duration of 500 ms, a sixteenth note 125 ms, and a full note 2000 ms, corresponding to 2 Hz, 8 Hz and 0.5 Hz, respectively.

Given the low frequencies of the nodes in the application, the problem of transmission delay and finite duration of the sounds may easily be solved by implementing pulse-coupled oscillators with a refractory period. As descried by Klinglmayr et al. [220], the refractory period t_{ref} has a corresponding *refractory phase* interval $[0, \phi_{ref}]$ within which no phase jumps occur. Thus, when a node i not in a refractory state perceives a fire event from a node j, it immediately increases its own phase according to the *phase update function*, $P(\phi_i(t))$. More precisely:

$$\phi_j(t) = 1 \Rightarrow \begin{cases} \phi_j(t^+) = 0 \\ \phi_i(t^+) = \phi_i(t^+) & \text{if } \phi_j(t) \in [0, \phi_{ref}] \quad \forall i \neq j \\ \phi_i(t^+) = P(\phi_i(t)) & \text{if } \phi_j(t) \notin [0, \phi_{ref}] \quad \forall i \neq j \end{cases} \tag{7.5}$$

where t^+ is the time step immediately after t. The phase update function is given by:

$$P(\phi) = (1 + \alpha)\phi, \tag{7.6}$$

where the *pulse coupling constant*, α, denotes the coupling strength between nodes.

One example of a musical system with the above implementation has been implemented on mobile phones using Mobmuplat [191] and PD [326] at the University of Oslo.[2] In the example, each node has a fixed frequency of 1 Hertz and phase coupling constant $\alpha = 0.1$, and the refractory period is set to 40 ms. The source code is available online.[3]

7.3.1.3 Obtaining Frequency Synchronisation Through Self-awareness

The section above discussed a well-known solution to the problem of synchronising decentralised musical nodes with *equal* frequencies. Synchronisation of nodes with different starting frequencies, however, is a much more challenging task. In this section we will show how nodes are able to synchronise when higher levels of self-awareness are implemented.

---

[2] http://vimeo.com/67205605
[3] http://fourms.uio.no/downloads/software/musicalfireflies

To reflect that periodic patterns in a musical system may occur at multiple simultaneous levels (quarternotes, sixteenths etc.) we define a new target state for the system: *harmonic synchrony*. Harmonic is borrowed from the concept of *harmonics* in various waveforms, where the frequency of every harmonic is an integer multiple of the lowest (fundamental) frequency. Correspondingly, harmonic synchrony is a state where the frequency of each node is an element of $\omega_{low} \cdot 2^{\mathbb{N}_0^+}$, where ω_{low} is the lowest frequency of all nodes in the group. As an example, please refer to the illustration in Figure 7.5, where the most suitable adaptation would be if Node 1 adjusts to twice the frequency of Node 2, and Node 3 to four times the frequency of Node 2.

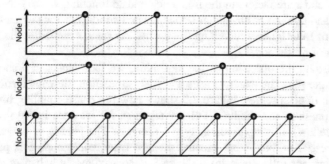

Fig. 7.5 Illustration of how *harmonic synchrony* may be beneficial in a collaborative active music system. The nodes may converge to integer-ratio frequencies, rather than equal frequencies.

In order to obtain synchrony, each node must be able to reason about its own level of synchrony with the rest of the group. A node calculates an error measure, $\varepsilon \in [0, 1]$, whenever a fire event is received. ε is at its highest value when $\phi = 0.5$, and lowest value when ϕ is equal to 0 or 1.

$$\varepsilon = sin(\pi\phi(t))^2 \tag{7.7}$$

with the special case that $\varepsilon = 0$ if a fire event is received within the refractory period. The sequence of error measures calculated by a node make up a discrete function $\varepsilon(n)$ describing the error measures at the nth received fire event. *Goal-awareness* is realised by applying a running median filter to $\varepsilon(n)$:

$$s = \text{median}\{\varepsilon(n), \varepsilon(n-1), ..., \varepsilon(n-m)\}, \tag{7.8}$$

where s is the self-assessed degree of synchrony within the node, and $m - 1$ is the length of the median filter. Thus, s takes a high value when the node is out of phase with the past received fire events, and a low value when the node is in phase with the past perceived fire events. A node uses this self-assessment of synchrony to scale the impact of received fire events from other nodes upon its own frequency adjustments, and as such the goal awareness is the base of the node's *meta-self-awareness*.

This self-awareness mechanism aids in giving less influence to occasional erroneous firings from other nodes or environmental noise.

Frequency adjustment is done as follows: Upon receiving a fire event from an external node, the receiving node calculates ρ, indicating whether frequency should be increased or decreased.

$$\rho = \sin(2\pi\phi(t)) \tag{7.9}$$

ρ is positive when $\phi < 0.5$ and negative when $\phi > 0.5$, meaning that if a node is less than half-way through its cycle when a fire event is received, it increases its frequency to "catch up" with the firing node.

Both ρ and s are factors in the frequency update function, where ρ indicates the direction and amount of change in frequency and s indicates the degree to which a node listens to or ignores incoming fire events. As such, s embodies the meta-self-awareness of firefly nodes.

To illustrate the operation of the algorithm, we look closer at a short excerpt from a single run. The example uses a Matlab simulation, showing a short excerpt from a run of six fireflies. Figure 7.6 shows how the phase coupling immediately pulls the phase of the oscillators closer together, for instance in the fire event of node 6 (yellow) after approximately 6.4 seconds. The phases of nodes 1, 2 and 4 are increased to maximum, causing them to reset, while node 5 is pulled back towards (but not fully down to) 0. Node 3 is approximately half-way through its cycle, and only a very small phase adjustment is made. The frequency coupling effect is shown between 7 and 8 seconds, where upon the firing of node 4 at 7 seconds, the phase of node 6 is reset back to 0. In node 6, this is interpreted as being too early, and a negative frequency adjustment does take place the next time node 6 reaches its peak value (after ~7.6 seconds). After approximately 12 seconds, the system reaches a close-to-synchronised state. That is, the oscillator frequencies are not quite in harmonic relation to each other, but due to the phase coupling, the fire events occur as they should. A video showing the evolution of phases and frequencies of the fireflies is available online.[4]

The chapter on active music systems presents several examples of how nature-inspired learning may be used to obtain self-awareness in active music systems. Please refer to Chapter 14 for examples of how the principle of pheromone trails of ants may be exploited to develop efficient gesture recognition algorithms and paths between musical sections.

7.3.2 Benefits and Challenges at Levels of Self-awareness

The firefly algorithm could operate with different levels of computational self-awareness. In this section we analyse the behaviour of the algorithm given a certain level of self-awareness.

---

[4] http://vimeo.com/72493268

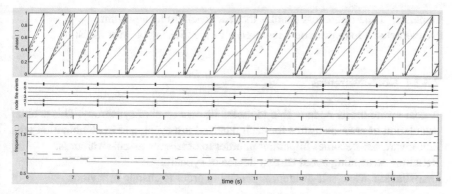

Fig. 7.6 A short excerpt from a Matlab simulation of six fireflies. The top plot shows the phases (ϕ) of all the nodes and the bottom plot shows the internal oscillator frequencies (ω). The fire events are shown in the middle plot, which also acts as colour legend for the phase and frequency plots.

7.3.2.1 Stimulus-awareness

A stimulus-aware firefly node is only aware of simple stimuli. The sender of fire events or the size of the group is unknown to each node. Each node in the example given above will react immediately to incoming stimuli (phase adjustment), and phase synchronisation may be realised at this level for nodes with equal frequencies.
Benefits: Simple and efficient implementation. No handshaking between nodes in a potentially large group.
Drawbacks: Nodes are unaware of their goal state and thus no self-assessment of synchrony is possible.

7.3.2.2 Interaction-awareness

An interaction-aware firefly node is aware of the neighbouring nodes from which it receives fire events. For this to be possible, nodes must be identified using, for instance, a networking protocol.
Benefits: The sender of incoming fire events could be identified.
Drawbacks: Networking protocol required, which makes entering and leaving a musical interaction more difficult, especially if acoustic musical instruments are involved in the group.

7.3.2.3 Time-awareness

Time-aware firefly nodes possess knowledge of previous events. Nodes are allowed to monitor a sequence of incoming fire events rather than only react immediately to a received fire event.

Benefits: Time-awareness enables the reachback firefly algorithm.
Drawbacks: Goal-awareness required for time filtering of error measure.

7.3.2.4 Goal-awareness

A goal-aware firefly node is able to assess its own level of synchrony. This enables calculation of error measures for each received fire event, and in combination with time awareness becomes important in order to obtain meta-self-awareness.
Benefits: Reasoning about current degree of synchrony.
Drawbacks: Meta-self-awareness needed in order to change behaviour.

7.3.2.5 Meta-self-awareness

Meta-self-aware firefly nodes change their behaviour based on the knowledge from the other levels of self-awareness. Firefly nodes adjust their own sensitivity to fire events from other nodes based on their self-assessed degree of synchrony. Furthermore, nodes may use other frequency update mechanisms if they get "stuck" at a frequency where they never reach maximum phase [295].
Benefits: "Insecure" nodes make larger changes to oscillator frequency, keeping nodes in sync.
Drawbacks: Nodes are still only self-aware and not aware of internal parameters of other nodes.

7.3.3 Other Related Techniques

Nature-inspired systems take inspiration from a natural process in order to obtain some goal. In our example above, the goal was to reach a certain *system-wide* state, which is often the case in such systems. Nature-inspired systems may take inspiration from how collective behaviour in certain species of animals reach a global goal. In the example above, the global goal of fireflies would be flashing in synchrony in order to enhance the light beyond that which each individual firefly is able to produce on its own. Variants of the firefly algorithm have been applied in a number of reports in order to synchronise devices in various types of self-aware and self-expressive system [411, 21, 74, 233]. A related algorithm has been used for a similar purpose, inspired by Japanese tree frogs [163]. The mating call of male Japanese tree frogs are not simultaneous, but rather spread out over time such that female frogs can locate the individual males.

Ant Colony Optimisation (ACO) is another type of nature-inspired learning useful for optimisation [108]. ACO takes inspiration from the way in which ants forage for food. In their search for food, ants lay behind a trail of pheromone for other ants to follow. When more ants pick the same trail, more pheromone is deposited, attracting even more ants. As such, trails to food sources become "ant highways" between the food source and the ant hill. When the food source runs out, the pheromone eventually evaporates, and new food sources are located. Dorigo et al. [109] showed how the use of a group of agents (ants) often would result in optimal solutions to problems without getting stuck at local optima. Examples of exploiting the pheromone mechanism from ACO in self-aware and self-expressive systems are given in Chapters 13 (multi-camera networks) and 14 (gait recognition in active music systems).

In optimisation, the field of Evolutionary Computing should also be mentioned in the context of nature-inspired methods. While several types of Evolutionary Algorithms (EA) exist, the common denominator is to search for a solution to a problem using mechanisms inspired by natural evolution [113]. An optimisation process is typically started by initialising a *population* of solutions to a problem. The individuals in the population are selected for reproduction through *recombination* and *mutation*, resulting in offspring of which some survive and other perish based on some criterion. The process is repeated until some termination condition is met (e.g., if a satisfactory candidate has been evolved). Certain Evolutionary Computing techniques can also be used to deal with dynamic optimisation, i.e., when the environment, including the objective function, the decision variables, the problem instance, the constraints, and so on, may vary over time [424]. Techniques from Evolutionary Computing have been used for dynamic and multi-objective optimisation in SA/SE systems. Specifically, Chen et al. [65] applied a dynamic multi-objective evolutionary algorithm in a self-adaptive system for temperature management in multi-core FPGAs.

7.4 Socially-Inspired Learning in Collective Systems

In multi-agent systems, the research on simulating societies in interaction with their environment has been taking shape, particularly the modelling of collective decisions. Collective (or distributed) learning is referred to as learning that is carried out by a group of agents instead of a single agent on its own, e.g., by exchanging knowledge or by observing other agents. It is an important concept in decentralising the system and reducing system complexity. It can be viewed as a social behaviour of the individual nodes in the collective. Having social behaviour allows us to create a social network. This furthermore enables each individual to focus its efforts on making a collective decision towards this smaller set of entities in the collective. This reduces the complexity in a collective decision making process when resources are limited. This section will give a thorough explanation of socially-inspired learning techniques in collective systems, through our SA/SE system example—the distributed smart camera network.

7.4.1 Example Application

In recent years 'dumb' cameras have evolved into embedded smart cameras [415, 341], combining a processing unit with an image sensor on a single platform. These processing capabilities, even though limited, allow the smart cameras to pre-process video data on-site and transmit only aggregated information, or a complete analysis of a scene, instead of plain images. Modern smart cameras are even capable of accomplishing processing-intensive tasks, such as object tracking. In object tracking, a description of the object of interest is initially provided to the camera. The camera thereafter attempts to re-identify this object in consecutive frames of its own field of view (FOV). There are various tracking algorithms to locate objects in each frame matching the given description with the highest probability. While we employ tracking algorithms to identify and locate moving objects, we do not elaborate on these fundamental tracking techniques here.

Soon enough, single smart cameras have been connected to distributed smart camera systems [342, 336]. Tracking objects in multi-camera systems can be approached in two different ways. The first approach uses all cameras to track various objects and the gathered information is fused at a central control. When tracking objects within multiple cameras concurrently, cameras need to align their FOVs to ensure the gathered data is coherent. To do so, a calibration process is employed to remove geometric distortions caused by the camera lens. This calibration process requires additional work before the system can go online. This extra effort might be feasible with small numbers of cameras but could be highly problematic in larger systems with tens, hundreds or even thousands of cameras. Furthermore, in case one of the camera's parameters is changed, new cameras are added or cameras are removed from the network, cameras might need to be re-calibrated to ensure proper functionality. While this might result in a very high network-wide utility, visual object tracking is a resource-intensive task and our facilitated cameras are rather resource constrained.

We refer to the second approach as *distributed tracking*, where each object is only tracked by a single camera at any given time. This means, the collective of cameras has to know where each object currently is located—given it is visible within the field of view of any of the cameras. Therefore the network needs to decide which camera is responsible for keeping track of a specific object at any time. Furthermore, the tracking camera has to decide when and to which camera it should transfer the tracking responsibility. This process of transferring a tracking responsibility is know as *handover*. Each camera is self-aware, acts autonomously and tries to maximise its own utility. Nevertheless, every single camera is influenced and affected by the other cameras in the network and as a collective system all cameras share a common goal: keeping track of an object of interest with a certain utility. This utility is generated locally by each camera by tracking objects of interest. We consider the size as well as the confidence of the employed tracking algorithm as a common measurement of utility among all cameras. Inspired by market mechanisms, tracking responsibilities are treated as goods and our cameras act as independent merchants able to buy and sell tracking responsibilities autonomously.

This distributed handover process described above is depicted in Figure 7.7. This handover process has first been introduced by Esterle et. al [125]. Whenever a camera decides to sell an object, it initiates an auction by transmitting an object description to the other cameras in the network (cp. Figure 7(b)). Thereafter, the receiving cameras, generally willing to participate in such an auction, try to value this possible tracking responsibility (cp. Figure 7(c)). This valuation is based on an instantaneous utility of the corresponding object. To allow for a equivalent evaluation among all cameras, all cameras have to apply the same approach to calculate this instantaneous utility for the observed objects. The auctioneering cameras transfer the tracking responsibility at the end of the auction to the highest bidder (cp. Figure 7(d)). Instead of requesting the highest bidder's valuation, only the second highest bid has to be paid. This Vickrey auction mechanism makes truthful bidding the dominant strategy for all participants and hence less vulnerable to malicious cameras. Additionally, this distributed approach allows continuous tracking of objects without any central component, analysing all data and coordinating tracking responsibilities.

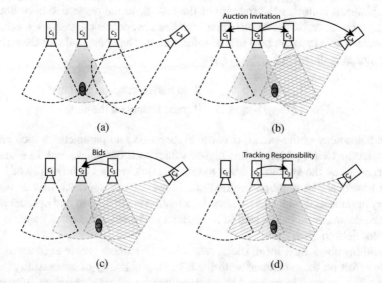

(a)   (b)

(c)   (d)

Fig. 7.7 Illustration of the market-based handover approach. An object of interest is tracked by camera c_2 in Illustration (a) as indicated by the dashed green line, with the shaded area representing the FOV. c_2 initiates an auction for an object as it is about to leave the FOV in Illustration (b). Bids for taking over tracking responsibility are sent in Illustration (c) by cameras c_3 and c_4, which have the object within their FOV as indicated by the orange dashed lines and the hatched area representing their FOV. c_3 wins the auction and the tracking responsibility is transferred from c_2 to c_3 in Illustration (d). A link in the vision graph is created (indicated by the red line between c_2 and c_3).

This distributed handover approach gives rise to two important questions:

1. Which cameras should be invited for an auction in order to maximise the success of the auction while minimising the marketing effort?

2. At what time should the auctioneering camera try to sell the object of interest?

7.4.1.1 Building and Exploiting the Neighbourhood Relations

While cameras could offer their objects to all cameras in the network, only those cameras having a neighbouring field of view would actually have a chance to "see" and hence valuate the object. To optimise their marketing efforts, cameras build up knowledge about their neighbourhood by reasoning about the received bids after initiating an auction. Those neighbours having the object in their field of view will most likely have a field of view quite close to their own. As cameras might misidentify objects and hence mistakenly submit a bid with an valuation for the wrong object, but also due to possible changes in the network topology, we require a technique capable of forgetting previously or incorrectly learnt neighbourhood relations. Inspired by the ant foraging process, we use artificial pheromones to build up a local neighbourhood graph on each camera. With every trade a selling camera i completes, it increases the strength of the link τ_{ix} to the respective neighbouring camera x by the value Δ. At the same time links evaporate continuously, ensuring cameras are able to *forget* now invalid information. The update rule for the artificial pheromones is shown in Equation 7.10.

$$\tau_{ix} = \begin{cases} (1-\rho) \cdot \tau_{ix} & \text{if no trade occurs on the edge} \\ (1-\rho) \cdot \tau_{ix} + \Delta & \text{if trade occurs on the edge} \end{cases} \tag{7.10}$$

As in ant colony optimisation, ρ is the evaporation rate parameter, which can be understood as a forgetting factor; higher values lead the pheromone to evaporate faster, enabling the system to adapt to changes quicker, but at a penalty of losing more historical vision graph information. However, our approach here is not ant colony optimisation, since pheromone information is not used to find optimal routes through the network, but instead to represent a social environment of cameras with adjacent fields of view.

Learning about their local, social network, each camera is able to focus its marketing effort on those cameras with the highest probability of successfully participating in auctions. To ensure new cameras can be included; cameras with a low or no link in the neighbourhood graph receive an auction invitation with a smaller probability.

As the vision graph is built up, the initial communication behaviour can be adapted. Specifically, when advertising an object that other cameras may wish to buy, a camera i sends a message to camera x with probability $P(i,x)$; otherwise it does not communicate with camera i at that time. We consider three different policies for determining these communication probabilities:

1. BROADCAST, which communicates with all available cameras in the network. This approach does not miss any camera but also generates a high overhead since it includes cameras which are not likely to respond.

2. STEP, in which an advertisement is sent to a camera if the strength of the link
   to that camera is above a certain threshold; otherwise the camera communicates
   with the other camera with a very low probability.
3. SMOOTH, in which the probability of communicating with another camera is
   based on the ratio of its link strength to that of the strongest link in its graph.

More formally, when employing a policy, the probability of camera i communicating with another camera x is given by

$$P_{\text{BROADCAST}}(i,x) = 1 \qquad (7.11)$$

when using the BROADCAST policy. For STEP, the probability is given by

$$P_{\text{STEP}}(i,x) = \begin{cases} 1 & \text{if } (\tau_{ix} > \varepsilon) \vee (\tau_{im} = 0) \\ \eta & \text{otherwise} \end{cases} \qquad (7.12)$$

where $\varepsilon = 0.1$ and m is the camera with the highest strength value, i.e.,

$$m = \underset{y}{\text{argmax}}\ \tau_{iy}, \forall y.$$

For the SMOOTH policy, the communication probability is given by

$$P_{\text{SMOOTH}}(i,x) = \frac{1+\tau_{ix}}{1+\tau_{im}} \qquad (7.13)$$

where m is again the camera with the highest strength value.

7.4.1.2 When to Hand Over

The timing of initiating auctions in a system observing a highly dynamic environment can be quite crucial. In a very straightforward approach, we enable our cameras to initiate an auction in very short intervals, trying only to maximise the utility for each object without considering the possible processing overhead for other cameras in the network. We call this an ACTIVE auctioning schedule.

In contrast, cameras can only initiate auctions when the object is about to leave the field of view of a camera. In this approach, the camera would only try to sell the object when its own utility is very low. On one hand, the camera might have been better off selling the object earlier. On the other hand, this results in a possibly lower processing overhead for the other cameras in the network. This approach is called PASSIVE auctioning schedule.

Combining these two auction initiation schedules with the three previously discussed communication policies results in six different strategies. An operator trying to balance the trade-off between communication overhead and tracking performance in the system can select one of these six strategies. To start our distributed tracking application, objects of interest have to be defined. In our system, this basic mechanism is accomplished by a human operator who has to connect to a remote camera and select the object or person to be tracked in a user interface. This user interface is only required to initiate the tracking process and does not act as a central component or is not needed in any way, besides for initialisation, to support our approach.

In addition to the two proposed approaches, the cameras could also learn their individual timings for handover. Starting with an ACTIVE approach, each camera tries to sell an object in very short intervals. Whenever an object has been sold to a neighbouring camera, the selling camera can keep track of this information. As objects are unlikely to always use the exact same trajectory, the camera has to approximate the time of sending out an action invitation for each object based on its current trajectory and speed. This introduces an exploration-exploitation dilemma in finding the optimal timings. We propose using a probability for exploration and reduce this value over time if the sale for the respective object is successful. With non-successful trades, the camera can assume a change in the network and increase the probability. This increase in exploration allows the camera to adapt faster to changed topology conditions.

7.4.1.3 Dynamic Strategy Selection

As discussed in the previous section, there are six different behavioural strategies available for our smart camera system. Each of this strategies gives rise to one out of two conflicting objectives: minimising network-wide communication or maximising system-wide tracking performance. In Lewis et al. [239, 238] we compare the performance of a system where all cameras employ the same strategy and one where the cameras may use strategies differing from the other cameras in the network. A system is said to have a *homogeneous* configuration when all cameras apply the same strategy. Alternatively, a camera network has a *heterogeneous* configuration when at least two cameras use differing strategies.

Permitting this heterogeneity in the network enables nodes to specialise to their local situation and hence allows for a wider range of global outcomes when compared to the homogeneous case. We speculate that optimal heterogeneous assignment of strategies can lead to the global performance of the network being strictly better in terms of both of the considered objectives. This would extend the previous Pareto efficient frontier with respect to the network-wide observed trade-off between communication overhead and achieved tracking performance. However, heterogeneity itself does not necessarily lead to better outcomes. It is also possible that nodes specialise wrongly, leading to a strictly worse global outcome when compared to any homogeneous case. Indeed, when considering all possible heterogeneous configurations for a given network, the number of possible configurations increases greatly compared to the homogeneous-only case. Having γ different strategies and n cameras in the network, an operator has to pick one configuration out of γ^n possible configurations.

The selection of an appropriate configuration, using a variety of strategies in the smart camera network, turns out to require knowledge of the camera setup, the environment as well as the movement patterns of the objects of interest. To avoid a priori knowledge of these parameters, we implemented multi-armed bandit problem solvers in every camera of the network [20]. This problem is analogous to being faced with n slot machine arms, where each pull of an arm returns a random reward drawn from an unknown distribution associated with that arm. Given m total arm pulls, the task is to select which arms to pull such that the total reward obtained is maximised. If the player were to know the distributions behind each arm, then the player could simply select the best arm for every pull. However, since the distributions are unknown, the player must sample from each arm in order to gain some knowledge of each arm's reward distribution. The multi-armed bandit problem therefore encapsulates another classic *exploration vs. exploitation* dilemma.

We consider three well-known bandit solvers from the literature: the simple EPSILON-GREEDY [399], UCB1, which is known to perform well in static problems [20], and SOFTMAX [380]. EPSILON-GREEDY requires an ε value to determine the amount of exploration. A low ε value ($\varepsilon \to 0$) results in random selection of algorithms while a high value ($\varepsilon \to 1$) selects greedily the best algorithm, based on the previous rewards. UCB1 requires no parameters. SOFTMAX uses a temperature value τ to steer the probability of selecting an arm based on the expected reward. This means that high temperatures ($\tau \to 1$) result in a random selection where each arm has nearly the same probability while lower temperatures ($\tau \to 0$) tend to select the arm based purely on the expected reward.

While we aim to achieve global, network-wide objectives in terms of communication overhead and tracking utility, we rely only on corresponding local metrics in order to avoid exchange of information among the cameras. On one hand we use the number of messages sent by a camera at a given time step and denote this by *communication*. On the other hand, we have *utility*, which is the sum of the obtained tracking performance over all objects tracked by this camera in the current time step, plus its balance of payments from all trading activity during this time step. We use these two local measurements in a linear combination as a reward function for the bandit solvers on each camera:

$$reward = \alpha \times utility - (1 - \alpha) \times communication. \qquad (7.14)$$

Here the value α allows a network operator to balance the trade-off between required communication and achieved tracking utility.

The use of bandit solvers on the camera level allows the individual cameras to learn on their own which strategy fits them best, given a priority on either minimising communication or maximising tracking performance.

7.4.2 Benefits and Challenges at Levels of Self-awareness

In this subsection we analyse our socially-inspired learning techniques based on the different levels of computational self-awareness, targeting the application explained in Section 7.4.1. A detailed discussion of the different levels of computational self-awareness is given in Chapter 4.

7.4.2.1 Stimulus-awareness

In a stimulus-aware system, a node has knowledge of only a single stimulus. The node is not able to identify the source of a stimulus. Furthermore, the node does not have any knowledge about previous stimuli and hence cannot infer possible future stimuli. For the discussed smart camera system, in a stimulus-aware system only a single camera would be responsible for tracking objects as handover interactions would not be possible.

Benefits: No Benefits.

Drawbacks: A stimulus-aware smart camera system corresponds to a camera system with only a single camera. A tracking coordination is not possible.

7.4.2.2 Interaction-awareness

An interaction-aware node knows that its actions trigger specific reactions from the local environment. In the smart camera system, ACTIVE and PASSIVE BROADCAST represent a simple interaction-aware system. Each camera triggers auction invitations based on the applied strategy expecting responses from neighbouring cameras. Conversely, upon receiving such an invitation, a neighbouring camera will try to identify the object and submit a bid, again expecting the initiating camera to assign the tracking responsibility. Building up knowledge about its local communities, each camera could also exploit this knowledge in order to focus marketing efforts. Nevertheless, as an interaction-aware system does not possess knowledge about historical or future phenomena, the system is not able to use an ant-inspired approach to build up the local neighbourhood graph or to facilitate the STEP or SMOOTH communication policy.

Benefits: Possible coordination of tracking responsibilities.

Drawbacks: No focussed marketing efforts. No robustness towards changes in the topology. Only fixed strategy assignment.

7.4.2.3 Time-awareness

With the introduction of time-awareness, a system gains knowledge about past and future events. In the smart camera system this relates to the ant-inspired neighbourhood graph and the evaporation of artificial pheromones, allowing the network to "forget" changed environmental or social conditions. Only with time-awareness can a system take full advantage of the proposed communication policies STEP and SMOOTH.

Benefits: Focussed marketing efforts possible when coordinating tracking responsibilities. Robustness towards changes in the network topology.
Drawbacks: Only fixed strategy assignments. No reasoning about current performance.

7.4.2.4 Goal-awareness

Goal-awareness requires a node to possess knowledge about current goals and objectives and also about preferences and constraints. Having a local performance function combining the required communication effort and achieved tracking utility enables the nodes of our smart camera system to reason about their current performance.
Benefits: Reasoning about current performance, allowing us to optimise local behaviour.
Drawbacks: Reasoning possible, but not changing of behaviour.

7.4.2.5 Meta-self-awareness

Having a node change its own level of computational self-awareness is considered meta-self-awareness. While our example of a smart camera system does not change these levels explicitly, we can reason about the performance of the different strategies representing different levels of self-awareness. The exploration of the different strategies and the dynamic selection of a specific strategy during runtime by means of a multi-armed bandit problem solver is an implementation of meta-self-awareness.
Benefits: Cameras can reason about the currently applied strategy and explore others possibly more beneficial ones.
Drawbacks: Cameras observe only their very own performance measurements and do not consider possible impacts on cameras in their neighbourhood.

7.4.3 Other Related Techniques

Socially-inspired learning in collective systems tries to enable individuals to learn about their social background and exploit this knowledge for their future actions and decisions. We focussed on building a social community on each individual node and the interactions within this community. While this allows exchange of information to a focussed group within the collective, it does not necessarily require consensus about certain states among all nodes of this group. As with dynamic environments, these individual social networks can change over time and strategies are needed to keep track of these changes. In our example, we induce our cameras with social behaviour by using market mechanisms. Furthermore, we learn the social network of each individual camera online using an ant-inspired approach, which also allows us to forget these physical neighbourhoods again in case the environment changes. Nevertheless, the individual nodes learn from each other only passively, rather than actively exchanging knowledge. A probably obvious example for active knowledge exchange is consensus mechanisms where nodes exchange information to achieve a common knowledge base. Olfati-Saber [299] proposes a distributed Kalman-filter [207] in order to exchange information and achieve a common knowledge base among nodes with different observation matrices. The approach allows us to reduce disagreement of the estimates among nodes in a target tracking task.

An alternative has been introduced by Michalski [265], known as *inductive learning* technique, which is also known as *transfer learning*. Here, an explicit teacher or even the environment provides examples, observations or facts about some phenomenon to nodes in the collective. The nodes make an inductive inference to achieve an accurate generalisation from multiple scattered facts and observations. Shaw and Sikora [358] introduce a distributed take on inductive learning. They break up the samples and use inductive learning on each single one of them in a distributed fashion. Using a genetic algorithm, they synthesise the results from each learning program into a final concept. Ontañón and Plaza [302] present an augmentation-based approach on multi-agent inductive learning in order to improve the individual learning capabilities of each node in the collective.

Stone [373] uses a layered learning approach in a multi-agent system. His reinforcement learning approach, called Team-Partitioned, Opaque-Transition Reinforcement Learning (TPOT-RL), deals with collaborating agents not necessarily able to observe the actions of another agent in the collective. By observing the long-term effects of their actions, agents can simultaneously learn collaborative policies.

In order to learn how multiple agents should coordinate their actions, Claus and Boutilier [78] extend Q-learning to a multi-agent setting. Here the Q-value function valuates the agents taking certain *joint actions* at a given state. Therefore, these learners are also known as *joint action learners*. In our smart camera example this would allow each social group to perform certain actions in a given situation in order to optimise the performance of the entire group. Transitively this would benefit the entire network. Nevertheless, joint action learners are incapable of applying mixed policies—whereby they select an action out of a set of available actions based on a certain distribution for a given situation.

Part III
Nodes and Networks

Part III presents the design of nodes and networks which provide self-aware and self-expressive capabilities. Many modern compute nodes are heterogeneous multi-cores that integrate several CPU cores with fixed function or reconfigurable hardware cores. In Chapter 8 we present a node architecture, programming model and execution environment for heterogeneous multi-cores, and show how the components of the reference architecture can be implemented on top of the operating system ReconOS. Chapter 9 describes how to build a heterogeneous cluster that can adapt to application requirements. Chapter 10 presents flexible protocol stacks as a promising alternative to today's static Internet architecture. Self-aware and self-expressive network nodes cooperate to select the protocol stacks that fulfil all communication requirements at the minimum cost at run-time. Finally, Chapter 11 compares different middleware paradigms and their suitability to support self-awareness in distributed applications and briefly describes a dedicated middleware implementation.

Chapter 8
Self-aware Compute Nodes

Andreas Agne, Markus Happe, Achim Lösch, Christian Plessl, and Marco Platzner

Abstract Many modern compute nodes are heterogeneous multi-cores that integrate several CPU cores with fixed function or reconfigurable hardware cores. Such systems need to adapt task scheduling and mapping to optimise for performance and energy under varying workloads and, increasingly important, for thermal and fault management and are thus relevant targets for self-aware computing. In this chapter, we take up the generic reference architecture for designing self-aware and self-expressive computing systems and refine it for heterogeneous multi-cores. We present ReconOS, an architecture, programming model and execution environment for heterogeneous multi-cores, and show how the components of the reference architecture can be implemented on top of ReconOS. In particular, the unique feature of dynamic partial reconfiguration supports self-expression through starting and terminating reconfigurable hardware cores. We detail a case study that runs two applications on an architecture with one CPU and 12 reconfigurable hardware cores and present self-expression strategies for adapting under performance, temperature and even conflicting constraints. The case study demonstrates that the reference architecture as a model for self-aware computing is highly useful as it allows us to structure and simplify the design process, which will be essential for designing complex future compute nodes. Furthermore, ReconOS is used as a base technology

Andreas Agne
Paderborn University, Germany, e-mail: agne@upb.de

Markus Happe
ETH Zurich, Switzerland, e-mail: markus.happe@tik.ee.ethz.ch

Achim Lösch
Paderborn University, Germany, e-mail: achim.loesch@upb.de

Christian Plessl
Paderborn University, Germany, e-mail: christian.plessl@upb.de

Marco Platzner
Paderborn University, Germany, e-mail: platzner@upb.de

for flexible protocol stacks in Chapter 10, an approach for self-aware computing at the networking level.

8.1 Heterogeneous Multi-cores

For about a decade single core performance has not been scalable for three reasons. First, pushing the clock frequency further leads to enormous power consumption, which causes thermal problems for chips and confronts data centres with cooling problems, horrendous energy bills and, increasingly, environmental concerns. Second, increasing core speeds would also require us to scale up the memory bandwidth. Although most of the silicon real estate of modern chips goes into caches, already current architectures have difficulty sustaining the bandwidth requirements. Third, instruction-level parallelism (ILP) is limited. The immense efforts invested in the past in microarchitectures and compilers to extract more ILP from applications has shown diminishing returns.

In reaction to these problems of scaling single core performance, two main trends have emerged in computer architecture: the transition from single to multi-core and even many-core architectures and the adoption of customised computing cores. Multi-core and many-core architectures increase processor performance by exploiting data-level and thread-level parallelism on a single chip. Originating in high-performance computing, multi-core architectures have rapidly found use in general purpose and finally also embedded CPUs. The abundance of available silicon real estate provided by Moore's law for future semiconductor generations will not only allow for integrating an increasing number of homogeneous CPU cores in a single chip but will also allow for including customised, i.e., specialised, cores to form so-called heterogeneous multi-cores. It is believed by many experts that such heterogeneous multi-cores will provide advantages in performance, energy efficiency and, to a certain extent, memory requirements over homogeneous multi-cores since the different cores can be tailored to a specific application domain [17].

Among customised cores, reconfigurable hardware cores play an important role as they can be reprogrammed to implement different hardware functions. Reconfigurable hardware cores leverage programmable hardware, typically field programmable gate arrays (FPGA). This technology allows for customising hardware structures to implement highly specialised and efficient coprocessors. The reconfiguration process is controlled by software and can occur either once at the startup of the system or even during runtime. Augmenting multi-core processors with reconfigurable cores is attractive, since the reconfigurability allows for tailoring the multi-core after fabrication to particular applications and system states. Thus, such heterogeneous multi-cores can be considered an enabling technology for building future self-aware adaptive computer systems.

Models of self-awareness and self-expression are highly relevant for the design and operation of future heterogeneous multi-cores, which form complex systems that are required to adapt to changes in their environment and system state. Changes in the environment are caused by changing workloads or user requests that subsequently cause scheduling and migration of tasks to different cores and core types at runtime to optimise for performance and energy. Furthermore, task scheduling and migration as well as the activation and deactivation of cores will be increasingly driven by thermal and fault management. Already today's multi-cores rely on thermal management techniques to avoid thermal hot spots and excessive temperature cycles for increasing reliability. In future, due to larger process variations in nano-electronic technology we will have to deal with an increasing number of faults. The corresponding fault management system will have to detect faults and react properly to maintain system operation, possibly with degraded performance.

This chapter presents work on self-aware compute nodes, particularly on heterogeneous multi-cores. Section 8.2 discusses related work on self-aware compute nodes. Section 8.3 presents the refinement of the generic reference architecture presented in Chapter 3 for self-aware compute nodes. In Section 8.4 we provide an overview of ReconOS, our architecture, programming model and execution environment for heterogeneous multi-cores. Section 8.5 shows how we use ReconOS to implement different case studies that demonstrate self-awareness and self-expression in compute nodes. Finally, Section 8.6 concludes this chapter.

8.2 Related Work on Self-aware Compute Nodes

Agarwal et al. [4] defined a self-aware computer to be introspective, adaptive, self-healing, goal-oriented, and approximate. The authors implement self-awareness through an observe-decide-act control loop similar to the MAPE-k model. Such a control loop found use in various prototypes [349], including a heterogeneous system comprising a workstation and an FPGA accelerator [363], a multi-processor inside a workstation [364, 27] and novel computer architectures for exascale computing [172]. The observation phase employs monitors to measure metrics such as performance and power consumption. Performance monitoring mostly relies on the Heartbeats framework [171], where a running application announces the completion of an application-specific amount of processing by issuing a heart beat. Simple performance goals can then be specified in terms of required heart beat rates. The decision phase allows for using performance models and learning components to determine a suitable adaptation whenever the goals are not met. The action phase actually performs the adaptation by turning "switches" or "knobs". For example, the switches might represent different implementation variants of an application's tasks and the knobs might represent parameters such as the clock rate or supply voltage of the cores.

Closely related to our work, Sironi et al. [365] presented an FPGA-based self-aware adaptive computing system based on heterogeneous multi-cores. Their system supports performance monitoring through the Heartbeats framework, decision making and self-adaption. Performance adaptation is enabled by mapping an application to a CPU, a reconfigurable hardware core, or both. The authors used an encryption algorithm as application and implemented their system on a Xilinx Virtex-II Pro FPGA. The experiments covered only static measurements of the application's performance without swapping between implementations at run-time and provided self-adaptation only at the conceptual level. More recently, Sironi et al. [364] discussed a heterogeneous system that consists of a multi-core processor and a reconfigurable device. Experiments were done on a platform with an Intel Core i7 and a Xilinx Virtex-5 FPGA. An application that hashes data blocks was instantiated four times with certain performance goals. Using a hot-swap mechanism that switches between a software and a hardware implementation, all performance goals were met.

Further related work focuses on general methodologies to allow for autonomous self-adaptation on embedded multicore systems. For instance, Zeppenfeld et al. [429] have developed an autonomic homogeneous multi-processor system-on-chip architecture, where each processor is connected to several monitors, actuators and a single learning classifier table evaluator. The table stores condition-action rules where the fitness is learned at run-time using reinforcement learning. Multiple optimisation goals were combined in a single objective function. Whenever the system performs a self-adaptation the used strategy receives a reward or a penalty, which depends on how much the system state has changed according to the objective function. The authors applied their general approach on a networking scenario with two levels of adaptation. On the one hand, the frequencies of each processor could be altered and, on the other hand, tasks could be migrated between the processors. The paper shows that it is beneficial when the autonomic network processors share workload information, defined as frequency times core utilisation. Diguet et al. [102] proposed a generic self-adaptation methodology for heterogeneous multicore architectures, which distinguishes between *algorithmic* and *architectural* adaptations. A global configuration manager controls the architectural adaptations, which optimise the hardware/software partitioning of the task set in order to deal with trade-offs at the system level, e.g., minimising the overall power consumption versus maximising the performance of the applications. Furthermore, each application is continuously optimised by a local manager, which can choose between different application-specific algorithmic adaptations. The parameters for the algorithmic adaptations have to be defined by the application developer at design time, for instance by using simulations.

Several related works focussed on balancing performance with power/energy consumption and thermal constraints in compute nodes. Although these works have not been using or stressing the term self-aware computing, they share the same scenarios, objectives and often also the methods. For example, Niu et al. [293] combined processors with reconfigurable hardware cores by equipping nodes of a compute cluster with FPGA accelerator cards. The computational intensive part of an N-Body simulation was mapped to the FPGA accelerator cards using multiple instances of a hardware core on a single FPGA accelerator. A central controller receives status information of all compute nodes over a wireless network, including temperature and power consumption readings for the FPGAs. Based on this information, the controller can enable/disable hardware cores on the FPGA accelerator cards and re-distribute workload inside the cluster to maximise the performance at given thermal and power budgets.

Jones et al. [205] proposed an adaptive FPGA-based architecture that switches between a low and a high clock frequency in order to decrease the latency of an application while maintaining a given thermal budget. The system uses a high clock frequency when the measured temperature is within the thermal budget and the application generates load; otherwise, the clock frequency is lowered. The authors demonstrated their approach on a Field Programmable Extender (FPX) platform, where they compared their adaptive strategy switching between 25 MHz and 100 MHz with a thermally-save static solution at 50 MHz. For longer workload bursts they reduced power consumption by 30% and doubled the performance while maintaining a given temperature threshold of 70°C.

Chen and John [64] designed a scheduling technique for heterogeneous multicore processors where the cores differ in instruction-level parallelism, branch predictor size, and data cache size. The scheduling technique profiles an application to find the best mapping with respect to high throughput and reduced energy consumption. Compared to a naïve scheduling technique, they could reduce the energy delay product by 24.5% on a 64-core system.

Table 8.1 Comparison of related work on self-aware compute nodes

Related work	Self-expression strategies	Goals and constraints
Agarwal et al. [4]	application parameters	performance
Sironi et al. [365] [364]	swapping HW and SW implementations	performance
Zeppenfeld et al. [429]	task migration	core utilisation
Diguet et al. [102]	algorithms and HW/SW partitioning	performance and power
Niu et al. [293]	enable/disable hardware cores and workload distribution	performance, power, and temperature
Jones et al. [205]	frequency scaling	performance, power and temperature
Chen and John [64]	heterogeneous scheduling	throughput and power
Our approach	partial reconfiguration of hardware cores	performance and temperature

Table 8.1 compares major related work in self-aware compute nodes with our approach. The table presents the strategies used for self-expression, even if the authors did not use this term in their work, and the goals and constraints. The main difference between related work and our approach is our use of more elaborate models of self-awareness and self-expression inspired from psychology (see Chapter 2). Further, in contrast to [364] we provide actual measurements of a system that dynamically adapts the number of used hardware cores in order to meet performance constraints. Unlike [364], we target an embedded architecture where the entire system is implemented on a single chip and, in addition to respecting performance constraints, our system can also respect thermal constraints. In contrast to [429] our self-expression strategies do not yet include on-line learning. Moreover, we focus on heterogeneous multicore systems with a CPU and multiple reconfigurable hardware slots. Finally, in contrast to [102] our approach does not differentiate between

application-specific and system-level adaptations, although both forms of adaptation could be easily modelled and integrated into our system.

8.3 Reference Architecture for Self-aware Compute Nodes

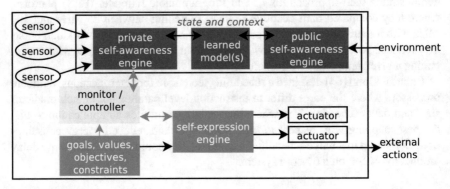

Fig. 8.1 Refined reference architecture for a self-aware compute node [32]

We base our work on self-aware compute nodes on the concepts of computational self-awareness and self-expression discussed in Chapter 2 and the generic reference architecture for designing self-aware and self-expressive computing systems presented in Chapter 4. To that end, we instantiate and refine the generic reference architecture shown in Figure 4.2 to achieve the reference architecture of Figure 8.1. The architectural primitives explained in Section 4.3.1.1 correspond to the following building blocks: Internal phenomena of the heterogeneous multi-core are captured by internal sensors, e.g., utilisation counters or thermal sensors, and handled by a private self-awareness engine. External phenomena, e.g., workload changes and varying quality of service requirements, are recognised through the environment by external sensors and handled by a public self-awareness engine. Both engines interact with models of the system and the environment which are optionally learned through this interaction, and altogether they provide the system with state and context (dark grey boxes in Figure 8.1). Self-expression (medium grey boxes in Figure 8.1) is the ability of a node to adapt to changes in the system or the environment. To that end, the framework foresees a self-expression engine containing either a single adaption strategy or multiple ones. The self-expression engine takes the system state and context as input in order to decide on an adaptation action. The adaptation itself is done by internal actuators, e.g., power management or thermal management by migrating threads in a heterogeneous multicore, or by taking external actions via external actuators, e.g., communicating with other compute nodes or a user. For goal-aware implementations of self-aware compute nodes we let the system goals, values, objectives and constraints drive the self-expression engine. The

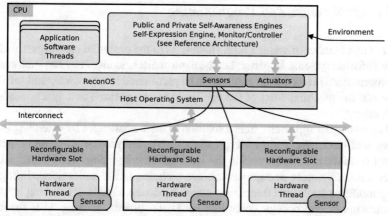

Fig. 8.2 A ReconOS heterogeneous multi-core architecture

goals, values, objectives and constraints might be given at design time or, in rather involved systems, dynamically updated at run-time.

8.4 ReconOS

ReconOS is an architecture, programming model, and execution environment for FPGA-based heterogeneous multicores [6, 250, 7]. ReconOS integrates reconfigurable logic cores with CPU cores and runs hardware/software multithreaded applications on top of a common host operating system. In this section, we demonstrate the suitability of ReconOS as architectural basis for implementing self-aware compute nodes according to the reference architecture shown in Figure 8.1. To that end, we provide an overview over the ReconOS v3 architecture and programming, discuss the feature of partial reconfiguration, present sensors and actuators, and comment on the availability of ReconOS.

Under ReconOS we use a thread-level implementation of the reference architecture. Figure 8.2 displays a ReconOS architecture with one CPU core and three reconfigurable hardware cores that are also denoted as reconfigurable hardware slots. Each such reconfigurable hardware slot can accommodate one hardware thread at a time. The self-awareness and self-expression engines as well as the monitor/controller are wrapped into regular software threads. ReconOS itself enables the creation of self-aware compute nodes mainly through sensors and actuators integrated into the ReconOS runtime system and into the hardware threads.

8.4.1 Architecture and Programming

There are a number of architectural challenges in the design of reconfigurable hardware/software systems. In particular, the main challenges are: How can such systems be programmed? How can an application be partitioned into hardware and software? How can the resulting parts of the application communicate and synchronise with each other?

The ReconOS approach offers a solution to these challenges by extending the existing, well-understood and widespread multithreaded programming and execution model from the software world to the world of heterogeneous multi-cores. Classic hardware accelerators or coprocessors are included in the multithreaded programming model as additional threads, so-called hardware threads. By turning hardware accelerators or coprocessor cores into hardware threads, existing APIs for multithreading, such as the POSIX API, can be offered to the programmer, and existing methods for inter-thread synchronisation and communication can be used for both, the hardware as well as the software parts of the application. The hardware/software partitioning follows the natural thread borders, where a thread is implemented either completely in hardware or in software.

In a ReconOS architecture, hardware threads can either be mapped to reconfigurable hardware slots statically at design time, or dynamically at runtime through the feature of partial reconfiguration. In ReconOS, all threads share the same address space and may synchronise and communicate with other threads through the use of POSIX-compliant operating system services, such as mutexes, semaphores, and message queues. Whether a specific thread resides in hardware or software is fully transparent to other threads. Operating system services such as the multithreading API, scheduling, and memory management are provided through a host operating system. For instance, a Linux system can be used to host the ReconOS runtime environment, providing the desired services and functionality. On such a system, support for multithreaded software is already available out of the box.

While software threads have sequential execution semantics, typical hardware accelerators extensively use fine- and medium-grained parallelism offered by FP-GAs. As shown in Figure 8.3, we split hardware threads into three main blocks. The user logic comprises the accelerator datapaths of the thread and implements the main computations of the thread. The operating system synchronisation state machine (OS-FSM) controls the datapaths via a set of handshake signals and interacts with the host operating system via the OS interface in a sequential manner. Access to the system's memory hierarchy via the memory interface is also controlled by the OS-FSM. Finally, hardware threads may contain local RAM to buffer data blocks for processing.

ReconOS hardware threads are coded in VHLD. We provide a VHDL function library with OS calls that is used to program the OS-FSM. Table 8.2 gives an overview of the functions currently available with ReconOS v3.

In order to include hardware threads with the desired level of transparency and functionality, ReconOS uses a delegate mechanism where each hardware thread is represented to the host operating system by a light-weight software thread, the so-

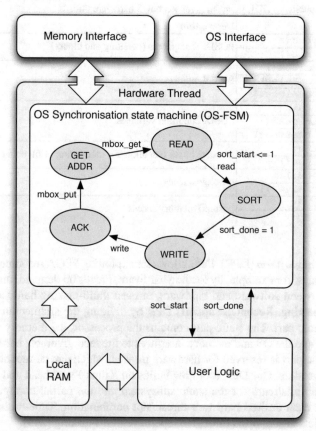

Fig. 8.3 An exemplary ReconOS hardware thread with OS-FSM, user logic, an internal local RAM as well as interfaces to memory and the OS

called delegate thread. Whenever a hardware thread's OS-FSM issues an OS call, the delegate thread becomes active and performs the OS call to the host operating system on behalf of the hardware thread. Through this concept neither the host OS kernel nor the user threads have to know whether other threads run in software or hardware. This transparency with respect to thread implementation greatly eases design space exploration and is key to adapting the system by changing thread mappings across the hardware/software boundary at runtime.

8.4.2 Partial Reconfiguration

ReconOS defines a standardised interface for hardware threads which simplifies exchanging them, not only at design time but also during runtime using *dynamic*

Table 8.2 Supported VHDL procedures for ReconOS hardware threads

Function	Description
osif_sem_post() osif_sem_wait()	POSIX semaphores (counting and binary)
osif_mutex_lock() osif_mutex_trylock() osif_mutex_unlock()	POSIX mutexes
osif_cond_wait() osif_cond_signal() osif_cond_broadcast()	POSIX condition variables
osif_mbox_get() osif_mbox_put()	message boxes that allow for monitoring of fill count and fill rate
osif_rq_send() osif_rq_receive()	message queues
memif_read() memif_write()	(virtual) memory access

partial reconfiguration (DPR). DPR allows for exploiting FPGA resources in unconventional ways, for example, by loading hardware threads on demand, moving functionality between software and hardware, or even multi-tasking hardware slots by time-multiplexing. ReconOS supports DPR by dividing the architecture in a static and a dynamic part. The static part contains the processor, the memory subsystem, peripherals and the OS and memory interfaces to the reconfigurable hardware slots. The dynamic part is reserved for hardware threads, which can be reconfigured into the hardware slots. Our DPR tool flow builds on Xilinx PlanAhead and creates the configuration bitstream for the static subsystem and the partial configuration bitstreams for each desired hardware thread/slot combination. Time-multiplexing of hardware slots is supported through cooperative multi-tasking [249].

The transparency with respect to the execution mode of a thread (hardware or software) is maintained by ReconOS at all times, even when hardware slots are reconfigured with different threads at runtime. This is achieved by handling reconfiguration in the context of the delegate threads. In a cooperative multi-tasking scheme, a hardware thread signals its delegate thread whenever rescheduling is possible. The delegate thread then calls the ReconOS scheduler to find out whether there are other threads that want to run in the respective slot. If so, the hardware slot is reconfigured and the delegate thread continues serving the new hardware thread.

8.4.3 Sensors and Actuators

In our current ReconOS version we provide internal sensors for capturing performance and utilisation as well as temperature data. Sensing utilisation is straightforward and supported by corresponding OS services, monitoring software threads and extensions to hardware threads, e.g., activity counters that measure times where a hardware thread is idly waiting for data or synchronisation. Performance is typically defined and captured in an application-specific way. We augment user-level threads with calls to message boxes to send events or performance data to the thread implementing the private self-awareness engine. While resembling Heart-beats [171], our approach is more flexible. Many services, for example FIFOs (mes-sage queues) in ReconOS, are provided by operating system objects implemented in software. Data such as current fill levels can be retrieved via the ReconOS API at any time by the self-awareness engines. Periodic measurements of fill levels can be used to derive related utilisation and performance data, such as fill rates.

Sensing temperature data is far more involved. In case only one temperature reading per time unit is required without any need for capturing the spatial tem-perature distribution, the FPGA's built-in thermal diode can be used. The diode is pre-calibrated and can be read out using the system monitor core provided by Xilinx. In addition, ReconOS allows the designer to instantiate a thermal sensor array for finer-grained temperature measurements, in particular to measure the core temper-atures in a multi-core design. The sensor array spans over the total reconfigurable area of the chip and–after a calibration phase–delivers highly accurate, high fre-quency temperature readings [158]. Figure 8.4 displays a 10×15 sensor array with an exemplary measured heat map. ReconOS includes device drivers for both ways of sensing the temperature.

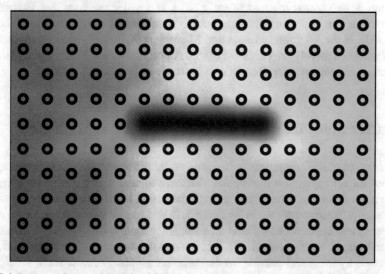

Fig. 8.4 Instance of a 10×15 ring oscillator based temperature sensor grid for fine-grained mea-surement of on-chip temperature distributions. The exemplary heat map overlay shows a measured temperature gradient of 7K across the FPGA.

In general, the OS layer of a heterogeneous multi-core can react to changes in the system state and environment by thread and thermal management, includ-ing, for example, starting, stopping and migrating threads. ReconOS supports in-ternal actuators that drive thread scheduling operations such as creating and killing threads. Besides the POSIX-specified functions for software threads, there are currently two hardware thread-related API functions for that purpose: The function `reconos_hwt_create()` is called to create a new hardware thread in a specific reconfigurable hardware slot, while `osif_thread_exit()` exits the calling hardware thread, freeing the operating system resources as well as the reconfigurable area associated with that thread. Technology-wise, ReconOS relies on partial reconfiguration (see Section 8.4.2) to implement the thread management functions for hardware threads.

8.4.4 Availability of ReconOS

ReconOS has been actively developed since its inception in 2006. Since then it has gone through three major revisions and has been ported to several operating systems and hardware platforms. The first version of ReconOS used the eCos operating system running on PowerPC CPUs embedded in Xilinx Virtex-2 Pro and Virtex-4 FPGAs. Version 2 improved on the original version by providing FIFO interconnects between hardware threads, adding support for the Linux operating system, and offering a common virtual address space between hardware and software threads. Version 3, which was released in early 2013, is a major overhaul that streamlines the hardware architecture towards a more lightweight and modular design. It brings ReconOS to the Microblaze/Linux and Microblaze/Xilkernel architectures and has been used extensively on Virtex-6 FPGAs. A port to the new Xilinx Zynq platform has been released recently. ReconOS is available as open source [335].

8.5 Case Study for a Self-aware Heterogeneous Multi-core

The concept of self-awareness in compute nodes provides a way to structure applications on heterogeneous multi-cores that have to deal with unpredictable system dynamics at runtime. The resulting implementation is then an instantiation of the reference architecture shown in Section 8.3. In this section we report on a case study involving two applications running on a single compute node at the same time, sorting and matrix multiplication. We first present the applications and the workload generated by them and provide implementation details. Then, in Section 8.5.1 we discuss self-expression strategies for dealing with performance constraints and in Section 8.5.2 we examine self-expression strategies combining temperature constraints with performance constraints. Finally, Section 8.5.3 compares the presented strategies.

Figure 8.5 outlines our case study setup. For the sorting application we generate 8 kilobyte blocks of 32 bit integers at a varying rate and insert them into the application's input FIFO. We vary the rate to mimic a fractal workload W_s exhibiting a degree of self-similarity that is commonly observed in a number of application domains, such as networking [234]. The matrix multiplication operates on matrices of size $2^7 \times 2^7$. Using Strassen's algorithm [374] for matrix multiplication, larger matrices of size $2^n \times 2^n$ with $n >= 7$ can be handled by performing 7^{n-7} multiplications of matrices of size $2^7 \times 2^7$. This reduces the total number of scalar multiplications at the cost of additional memory. We insert generated matrices in the application's input FIFO. We assume the workload W_m for the matrix multiplication to be infinite, i.e., there will always be matrices for the system to multiply.

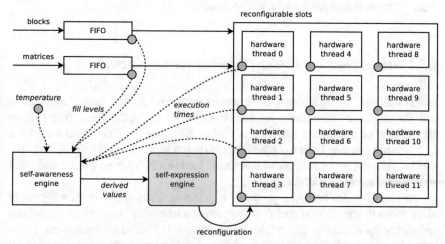

Fig. 8.5 Case study involving two applications, sorting and matrix multiplication, running on a ReconOS architecture with one CPU and 12 partially reconfigurable hardware slots

We have implemented both applications on a ReconOS system running on a Virtex-6 FPGA (XC6VLX240T). The FPGA area is divided into 13 regions. One large region contains the static system, including a Microblaze CPU, the memory controller and a UART device used to transfer data to a workstation. Additionally, 12 partially reconfigurable regions, i.e., reconfigurable hardware slots, are allocated for the hardware threads. Any slot may contain either one sorting thread or one matrix multiplication thread at a time. We use the internal configuration access port (ICAP) of the Xilinx FPGA to partially reconfigure these regions, which enables us to time-share FPGA resources between the two applications.

The system is highly heterogeneous because it contains two fundamentally different kinds of computational cores: The instruction set-based main CPU runs the operating system and scheduling components, as well as the matrix subdivision step of the Strassen algorithm, while dedicated hardware cores perform integer sorting and fixed-size matrix multiplication.

The resource requirement (post-synthesis data) of a sorting hardware thread amounts to 1452 6-LUTS, 424 flip-flops, and two block RAMs for local memory. A matrix multiplication hardware thread uses 1406 6-LUTS, 754 flip-flops, three multiply-accumulate blocks (DSP blocks), and 17 block RAMs for local storage.

The complete system, including CPU, hardware threads, and memory bus, runs at 100 MHz clock frequency.

The typical execution time for sorting a block of integers is around 82 ms, and around 108 ms for a matrix multiplication. However, due to limited shared memory bandwidth these numbers may vary during runtime depending on the system's load. The best achievable reconfiguration delay D_{min} for a hardware thread is also determined by the available bandwidth to system memory. While in an otherwise idle system we have measured a D_{min} of 42 ms, the reconfiguration delay increases by a factor of more than 9 for heavily loaded systems.

8.5.1 Self-expression Under Performance Constraints

In this scenario, we combine a performance constraint for the sorting application with the objective to maximise the number of matrix multiplications. With L_{max} as the capacity of the sorting application's input FIFO, we want the FIFO's fill level $L_s(k)$ at any time step k not to exceed the maximum level $\forall k : L_s(k) \leq L_{max}$, i.e., the FIFO should not overflow. Blocks that would lead to a FIFO overflow are discarded and counted as constraint violations.

The challenge for finding a good self-expression strategy, that is, an assignment of hardware threads to slots and a corresponding schedule meeting the constraint and optimising the objective, arises from i) the workload imposed on the system, which is not known in advance, and ii) the internal dynamics of the system caused by mutual interference between memory accesses for reconfiguration and processing. Since the reconfiguration interface and the hardware threads share a single bus to main memory, bandwidth used for transferring reconfiguration data is not available for transferring processing data and vice versa. Moreover, when several hardware slots undergo reconfiguration, some of the hardware threads will be delayed because there is only one ICAP reconfiguration interface. The resulting system dynamics are hard to model analytically, which motivates the self-aware computing approach.

We generate the sorting workload W_s deliberately to temporarily exceed the system's maximum sorting rate R_{max}, i.e., the maximum continuous rate at which blocks can be inserted without overflowing the input FIFO. Our implementation would achieve R_{max} if all 12 hardware slots were configured with sorting threads. Workloads temporarily exceeding R_{max} stress the system's self-expression strategy, because spikes in the workload must be compensated for in order to meet the performance constraint.

According to our reference architecture, we implement a software thread containing the private self-awareness engine. This engine collects and maintains information about the system state, such as the FIFO fill level $L_s(k)$, the current FIFO in-rate, and the last measured reconfiguration delay D_i for each reconfigurable hardware slot i. The collected information is then used by the self-expression engine implementing the self-expression strategy. We invoke the self-expression engine at discrete time steps with the interval Δt. The engine decides on the number of threads required for sorting and for matrix multiplication and runs the scheduler that stops threads and triggers reconfiguration if necessary. In terms of the reference architecture, the scheduler drives the internal actuators. The choice of Δt has a significant impact on the behaviour of the self-awareness engine. A lower Δt enables the system to respond faster to workload changes at an increased computational overhead. In our experiments we use a Δt of one second, which reduces the load caused by the self-awareness engine to a negligible level while it still allows the system to react to workload changes sufficiently quickly.

In the following, we discuss two fundamental and complementary self-expression strategies and one meta strategy that decides which fundamental strategy will be used. The meta-strategy tries to incorporate the strengths of both sub-strategies while avoiding their weaknesses.

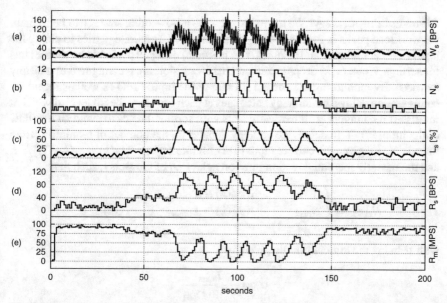

Fig. 8.6 Experimental evaluation of the *proportional* strategy. The figure is structured as follows: (a) workload, in blocks/s; (b) number of sorting threads; (c) input FIFO fill level, in percent; (d) the rate at which blocks are sorted, in blocks/s; (e) the rate at which matrices are multiplied, in multiplications/s.

Proportional strategy: This strategy sets the number of required sorting threads $N_s(k)$ in time step k proportional to the current fill level of the FIFO $L_s(k)$,

$$N_s(k) := k_p \cdot L_s(k)$$

with the proportional factor

$$k_p := N_{\max}/L_{\max}$$

where N_{max} is the number of available hardware slots and L_{max} is the capacity of the input FIFO. The proportional strategy handles most workloads well that do not surpass the maximum continuous sorting rate R_{max}. It will, however, quickly violate the performance constraint in case the workload peaks greater than R_{max}. Figure 8.6 summarises the results for the proportional strategy. Figure 8.6(a) displays the modulated fractal sorting workload W_s in blocks per second (BPS). Figure 8.6(b) depicts the number of currently configured sorting threads, ranging from 0 to 12, and Figures 8.6(c) to 8.6(e) present the fill level of the input FIFO, $L_s(k)$, the rate $R_s(k)$ at which blocks are sorted in blocks per second, as well as the rate $R_m(k)$ at which matrix multiplications are performed in multiplications per second (MPS). All measurements have been taken on a single run of the application over a period of 200 seconds. After 200 seconds, the proportional strategy has dropped 126 blocks in violation of the performance constraint.

All-or-nothing strategy: We have designed the all-or-noting strategy to better handle workload peaks in excess of R_{max}. Once engaged, the strategy tries to empty the input FIFO completely and as quickly as possible, using all available hardware slots for sorting. When $L_s(k)$ drops to zero, all hardware slots are reconfigured to multiply matrices. By monitoring the current fill rate of the input FIFO as well as recent reconfiguration delays, the strategy calculates at each time step the fill level at which it must start reconfiguring all hardware slots with sorting threads to avoid an overflow of the input FIFO. In detail, the strategy decides on the number of required sorting threads $N_s(k)$, given the current input FIFO fill rate $L'_s(k) = (L_s(k) - L_s(k-1))/\Delta t$, and the recently measured reconfiguration delays D_i of the slots i, such that

$$N_s(k) := \begin{cases} 0 & \text{if } L_s(k) = 0 \\ N_{max} & \text{if } L_s(k) > L_{trigger} \quad \text{or} \quad L'_s(k) > \alpha \\ N_s(k-1) & \text{else} \end{cases}$$

with α being the critical fill rate given by

$$\alpha := \frac{L_{max}}{\sum_{i=1}^{N_{max}} D_{i-1}}$$

The purpose of the condition $L_s(k) > L_{trigger}$ is to avoid the FIFO slowly becoming full without $L'_s(k)$ ever surpassing α. In our experiments we have found it sufficient to set $L_{trigger}$ to $L_{max}/4$. Figure 8.7 presents the experimental results for the all-or-nothing strategy. While this strategy handles the sorting workload without constraint violations, it also performs fewer multiplications than the proportional strategy, as the comparison in Table 8.3 shows.

Meta strategy: We have developed a meta strategy that tries to leverage the advantages of both fundamental strategies while avoiding their weaknesses. Both, the proportional and the all-or-nothing strategy prefer certain workloads over others. The first strategy leads to high multiplication performance over a wide range of workloads, but handles sorting workload spikes rather poorly. The second strategy copes with workload spikes without constraint violations at the price of an overall

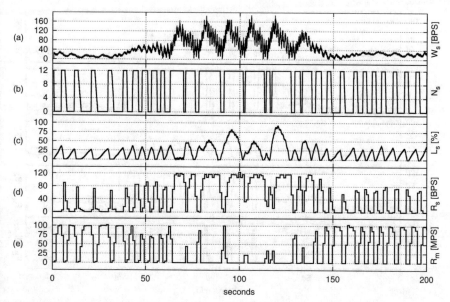

Fig. 8.7 Experimental evaluation of the *all-or-nothing* strategy. The figure is structured as follows: (a) workload, in blocks/s; (b) number of sorting threads; (c) input FIFO fill level, in percent; (d) the rate at which blocks are sorted in blocks/s; (e) the rate at which matrices are multiplied, in multiplications/s.

decreased number of matrix multiplications. The definition of the meta strategy follows straight from an examination of the experimental results. The key insight is that the proportional strategy will meet the performance constraint as long as the input FIFO's fill rate is at most R_{max}. For higher fill rates, the all-or-nothing strategy which is more conservative regarding FIFO space will minimise constraint violations.

Self-expression strategy *meta*:

(rule #1) proportional: $(\forall j \in [(k-p)...k]: \quad L'_s(j) \leq R_s(j))$
(rule #2) all_or_nothing: (else)

If the FIFO fill rate $L'_s(j)$ is below or equal to the sorting rate $R_s(j)$ for the last p iterations, the meta strategy will use the proportional strategy. The switch to all-or-nothing happens when this condition is violated. The purpose of the parameter p in the meta strategy is to reduce oscillations between the two single strategies in the face of a noisy workload. We found a value of 20 to work well in our experiments. Figure 8.8 shows the experimental evaluation of the meta strategy. The switch from the proportional to the all-or-nothing strategy happens at the beginning of the first workload spike surpassing R_{max}, which is also indicated by the abrupt changes in the number of sorting threads. After the workload smoothes, the meta strategy switches back to the proportional mode.

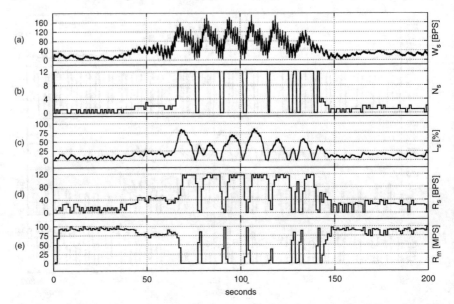

Fig. 8.8 Experimental evaluation of the *meta* strategy. The figure is structured as follows: (a) work-load, in blocks/s; (b) number of sorting threads; (c) input FIFO fill level, in percent; (d) the rate at which blocks are sorted, in blocks/s; (e) the rate at which matrices are multiplied, in multiplications/s.

8.5.2 Self-expression Under Conflicting Constraints

Often, computing systems have to deal with conflicting objectives or constraints. In the case of our heterogeneous multi-core, a typical conflict exists between performance and temperature management. To exemplify a situation with conflicting constraints, we introduce thermal constraints to our case study. We extend the private self-awareness engine to continuously monitor the chip temperature using the Xilinx system monitor. When the temperature exceeds a threshold θ_{low}, the self-expression strategy stops hardware threads of applications that do not have any performance constraints, i.e., the matrix multiplication in our case study. When the temperature rises above the threshold θ_{high}, all remaining hardware threads are stopped for a quick temperature reduction. If no thermal constraints are violated, the system applies the meta strategy for self-adaptation.

In order to decrease the chip temperature the application might have to disable hardware threads, which, potentially, leads to a violation of the performance constraint. As a consequence, our strategy favours thermal constraints over performance constraints. While thermal management of FPGA-based systems will become increasingly important in the foreseeable future due to shrinking device structures and increasing densities [42], our sorting and matrix multiplication applications on today's FPGA technology, however, do not generate significant heat. In order to emulate the thermal situation of a future FPGA multi-core, we have integrated dedicated logic into the hardware threads that creates heat whenever the thread is active.

This dedicated heat-generating logic comprises a number of ring oscillators, each implemented in a single LUT [5]. For the sorting hardware thread, we have integrated 150 ring oscillators and for the matrix multiplication hardware threads 50 ring oscillators.

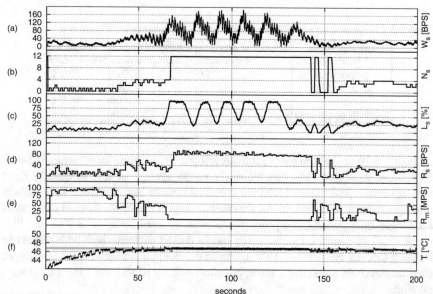

Fig. 8.9 Measurement results for the thermal-aware meta strategy: The figure is structured as follows: (a) workload, in blocks/s; (b) number of sorting threads; (c) input FIFO fill level, in percent; (d) the rate at which blocks are sorted, in blocks/s; (e) the rate at which matrices are multiplied, in multiplications/s; and (f) the chip temperature over time, in °C. The thermal constraints are highlighted by a grey horizontal bar.

Figure 8.9 displays the experimental results for the thermally-aware strategy. The thermal thresholds were set to $[\theta_{low}, \theta_{high}] = [46.5°C, 47°C]$, indicated by the grey band in Figure 8.9(f). Since the thermal thresholds are quite tight, the system often violates the performance constraint of the sorting application and drops 309 data blocks in 200 seconds. Furthermore, compared to the results achieved by the meta strategy without thermal constraints, the matrix multiplication performance decreases from 75 to about 25 multiplied matrices per second once the chip temperature exceeds θ_{low}; compare Figure 8.8(e) and Figure 8.9(e).

8.5.3 Comparison of Self-expression Strategies

Table 8.3 compares for all self-expression strategies the performance constraint violations in number of dropped sort blocks, the matrix multiplication performance, and the maximum measured temperature. Among the thermally-unaware strategies, the meta strategy clearly excels by respecting the performance constraint while processing almost as many matrix multiplications as the proportional strategy. The thermally-aware meta strategy can respect additional thermal constraints at the price of more performance constraint violations and a lower number of matrix multiplications.

Table 8.3 Comparison between different self-expression strategies over 200 seconds

self-expression strategy	number of dropped sort blocks	number of matrix multiplications	maximum temperature
proportional	126	12747	50.2°C
all-or-nothing	0	7467	48.9°C
meta	0	11606	49.7°C
thermally-aware meta	309	5646	47.5°C

Figure 8.10 depicts the temperature development over time for all presented self-expression strategies. Only the thermally-aware strategy is able to quickly react to temperature peaks that exceed the specified bounds and, therefore, successfully manages the chip temperature. All the other strategies violate the thermal constraints for longer time periods; the proportional strategy even by up to 3.2°C, see Figure 8.10(a). In contrast, the *thermally-aware* design performs best with respect to meeting thermal constraints. Although even the *thermal* strategy exceeds the upper thermal threshold sometimes, the application quickly adapts and lowers chip temperature.

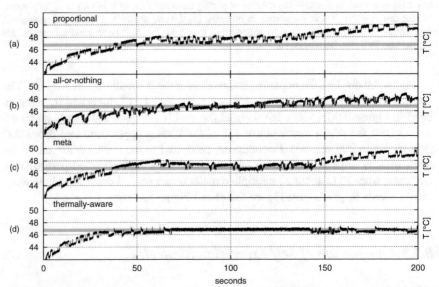

Fig. 8.10 Temperature development over time for all experiments. The thermal constraints are highlighted by a grey horizontal bar.

8.6 Discussion and Conclusion

In this paper we have shown how we refine the reference architecture for self-aware and self-expressive computing systems presented in Figure 4.2 to a model suitable for operating FPGA-based heterogeneous multi-cores. Our multi-core is based on the ReconOS architecture, programming model, and execution environment, which provide multithreaded programming as a unified abstraction for hardware and software threads. In addition, ReconOS supports dynamic partial reconfiguration which allows us to implement a unique self-expression technique, starting and terminating hardware threads at runtime.

We have presented a case study in this chapter that maps a sorting and a matrix multiplication application to a heterogeneous multi-core with a CPU and 12 reconfigurable hardware slots. We have implemented the basic components of the refined reference architecture of Figure 8.1 and examined the system under constraints in performance and temperature using a varying, noisy workload. We have demonstrated several self-expression strategies that start and terminate hardware threads at runtime, relying on partial reconfiguration as actuator technology. The self-expression strategies perform rule-based adaptations and are able to operate the system with one or several, possibly conflicting, constraints. Finally, we have shown that none of the presented strategies is dominant in the sense that it delivers the best performance for all points of operation, but the system's awareness of its current operational state and its goals can be exploited in a meta strategy that selects a good self-expression strategy for the current operation point. In this sense, we can clearly state that the meta strategy provides our system with goal-awareness.

The reference architecture for self-aware and self-expressive computing systems has proven very useful in separating concerns and structuring both the engineering process as well as the component architecture of the runtime system. While adaptive computing systems with limited complexity and basic adaptation strategies could arguably also be implemented using an ad hoc approach, presumably relying on a simple monitor/control loop, such ad hoc designs become certainly infeasible for complex systems that comprise many sensors, actuators and more sophisticated algorithmic techniques in their self-awareness and self-expression components.

Chapter 9
Self-adaptive Hardware Acceleration on a Heterogeneous Cluster

Xinyu Niu, Tim Todman, and Wayne Luk

Abstract Building a cluster of computers is a common technique to significantly improve the throughput of computationally intensive applications. Communication networks connect hundreds to thousands of compute nodes to form a cluster system, where a parallelisable application workload is distributed into the compute nodes. Theoretically, heterogeneous clusters with various types of processing units are more efficient than homogeneous clusters, since some types of processing units perform better than others on certain applications. A heterogeneous cluster can achieve better cluster performance by adapting cluster configurations to assign applications to processing elements that fit well with the applications. In this chapter we describe how to build a heterogeneous cluster that can adapt to application requirements. Section 9.1 provides an overview of heterogeneous computing. Section 9.2 presents the commonly used hardware and software architectures of heterogeneous clusters. Section 9.3 discusses the use of self-awareness and self-adaptivity in two runtime scenarios of a heterogeneous cluster, and Section 9.4 presents the experimental results. Finally, Section 9.5 discusses approaches to formally verify the developed applications.

Xinyu Niu
Imperial College London, UK, e-mail: niu.xinyu10@imperial.ac.uk

Tim Todman
Imperial College London, UK, e-mail: timothy.todman@imperial.ac.uk

Wayne Luk
Imperial College London, UK, e-mail: w.luk@imperial.ac.uk

9.1 Overview of Heterogeneous Computing

9.1.1 Heterogeneous Clusters: Performance

Modern society relies more and more on high-performance computing services to improve the quality of life, such as weather forecast, climate modelling and disease diagnosis based on analysing large-scale data. In contrast to the increase in the requirements for high-performance services, the performance increase rate of existing computer systems has slowed down significantly, mainly due to power density and dissipation issues. While Moore's Law continues to hold, the clock frequency and the power of Intel Central Processing Units (CPUs) have been almost stagnant since 2005 [379].

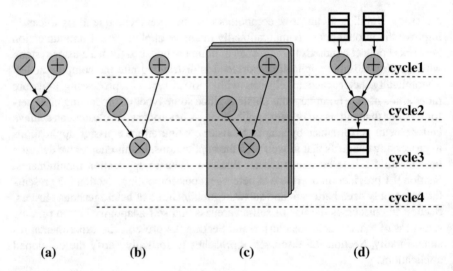

Fig. 9.1 Execution of (a) a simple data-flow graph, in (b) control-flow architecture, (c) vector machine, and (d) data-flow architecture

To further improve computing performance, researchers have proposed various computer architectures that support vector computation [420], out-of-order execution [99] and customised streaming applications [164]. Figure 9.1 shows the execution model of different architectures.

- When following control-flow programming models, the implementation in Figure 9.1(b) processes one datum in four clock cycles (we assume arithmetic operations take one clock cycle, and instruction overhead is neglected). When iterating over a large data set, this implementation takes four clock cycles per datum on average.
- Vector computer architectures concurrently apply the same instructions on multiple data in a vector. As shown in Figure 9.1(c), the architecture replicates the

processing elements three times to process three datum in parallel. On average, this leads to a throughput of 3/4 datum per clock cycle.

- Out-of-order execution models execute (or fire) arithmetic operations as soon as all dependent data are ready. For operations that can be executed in parallel, this can reduce the number of clock cycles to process a program. For the example in Figure 9.1(d), the processing of one datum takes three cycles instead of four cycles. In addition, when data are streamed into the customised data-path, the processing of the second input datum starts as soon as the addition and subtraction operators are free, i.e., one datum is processed per clock cycle.

The overall speedup S_{ove} of a program can be expressed as the ratio of original execution time T_{ori} to accelerated execution time T_{acc}. A program often contains parallelisable portions (P_{por}) and non-parallelisable portions ($1 - P_{por}$). Suggested by Amdahl's law, the upper bound of program performance is determined by the less parallelisable portion. With P replicated data-paths processing the parallelisable portions, the overall speedup can be expressed as:

$$S_{ove} = \frac{T_{ori}}{T_{acc}} = \frac{T_{ori}}{\frac{T_{ori} \cdot P_{por}}{P} + T_{ori} \cdot (1 - P_{por})} = \frac{1}{\frac{P_{por}}{P} + (1 - P_{por})} \quad (9.1)$$

where $\frac{T_{ori} \cdot P_{por}}{P}$ indicates the execution time of parallelisable portions after acceleration, and $T_{ori} \cdot (1 - P_{por})$ indicates the execution time of the remaining portions. In order to achieve maximum performance for various applications, a computer system requires heterogeneity and adaptability. As an example, for the execution model in Figure 9.1, the architecture performance highly depends on application characteristics, available resources, and architecture properties. Table 9.1.1 lists the parameters to estimate the performance of various computer architectures.

Table 9.1 Overview of model parameters

Parameters	Descriptions
$f_{req,arc}$	Architecture clock frequency
U_{res}	Resource usage of a single PE
A_{res}	Available resources in a computer system
N	Number of replicated PEs
P_{por}	Percentage of parallelisable portions
$C_{por,arc}$	Number of clock cycles to process one datum in parallelisable portions, in architecture arc
$C_{non,arc}$	Number of clock cycles to process one datum in non-parallelisable portions, in architecture arc

Based on these parameters, the execution time T of a program can be expressed as follows.

$$T = \frac{1}{f_{req,arc}} \cdot \frac{P_{por,arc} \cdot ds \cdot C_{por}}{N_{arc}} + (1 - P_{por}) \cdot ds \cdot C_{non,arc} \qquad (9.2)$$

where ds is the size of problem data set, and $ds \cdot C$ indicates the number of clock cycles to process ds items of data. Given an architecture arc, (1) $\frac{1}{f_{req,arc}}$ is the cycle time, (2) $C_{por,arc}$ and $C_{por,arc}$ indicate the number of clock cycles to process one datum in parallelisable and non-parallelisable portions, respectively, and (3) N_{arc} indicates the number of replicated data-paths, and is calculated as the ratio between available resources A_{res} and resource usage of a PE (Processing Element) U_{res}. In the example problem, $P_{por} = 1$, $C_{por} = 4$ and $N_{arc} = 3$ for Figure 9.1(c), and $C_{por} = 1$ and $N_{arc} = 1$ for Figure 9.1(d).

In order to demonstrate the benefits of heterogeneous computing, we estimate the performance of a program running on CPU, Graphics Processing Units (GPUs) and Field-Programmable Gate Arrays (FPGAs). Table 9.1.1 summarises the properties of CPU, GPU and FPGA developed in the same technology (65-nm). The $f_{req,arc}$ and P of an FPGA design depends on application characteristics. Typically, for applications running on a Virtex-6 SX475T FPGA, the clock frequency of an FPGA design is between 0.1 and 0.3 GHz [290], and the number of replicated PEs is between 24 and 256 [72]. The number of clock cycles to process one datum in Table 9.1.1 is specific to the example program, with the simplified execution model shown in Figure 9.2. If we set the P_{por} of the program to be 0.9, an adder/subtracter takes one resource unit, and a multiplier takes 1.5 resource units (resource usage is based on the measured results from Xilinx 32-bit floating-point operators); the P is determined by the number of available resource units. A CUDA core implements a fused multiply-add operator. In this example, we assume the CPU, GPU and FPGA have the same available resources. A GPU core uses 2.5 resource units to implement the multiply-add operator. Given the available resources to accommodate the 448 GPU cores, and the resource usage of an FPGA PE, the P for FPGA is estimated to be 320. Figure 9.2 shows the speedup of the GPU and FPGA designs for the example algorithm. In general, the more parallelisable portions the example program has, the more speedup GPU and FPGA designs can achieve due their higher number of replicated PEs (cores). Similarly, as the depth of the program pipeline increases, the performance of FPGA designs increases.

Table 9.2 Overview of model parameters

Architecture	Model	$f_{req,arc}(GHz)$	N	$C_{por,arc}$[1]
CPU	Intel Xeon X5650	2.66	6	4
GPU	Nvidia Tesla C2070	1.15	448	4
FPGA	Xilinx Virtex SX475T	0.3	application-specific	1

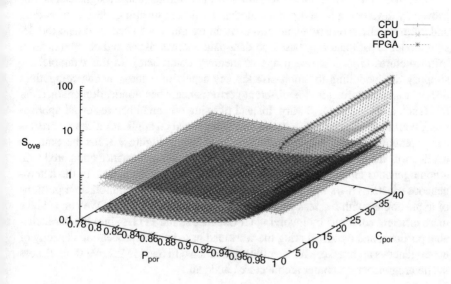

Fig. 9.2 Speedup of CPU and GPU designs compared with reference CPU design, with P_{por} and C_{por} increased

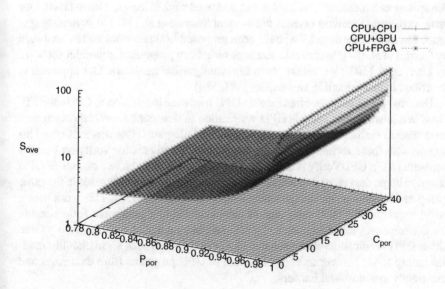

Fig. 9.3 Speedup of heterogeneous designs compared with reference CPU design, with P_{por} and C_{por} increased

Obviously, the performance estimation of the heterogeneous designs can be improved by developing more rigorous models for design mapping, design execution, and program characteristic exploration. As an example, we need to extend the design model to estimate U_{res} based on data-path characteristics and communication infrastructures (e.g., resource usage of memory controllers). In this example, we simplify the modelling to emphasise the key aspect of heterogeneous computing: different architectures provide different performance when application characteristics (such as P_{por} and C_{por}) vary. Instead of using one architecture for all applications, coordinating heterogeneous architectures based on application characteristics can potentially provide a better solution. As shown in Figure 9.3, for the example application, the heterogeneous clusters provide a better performance compared with a homogeneous cluster with the same number of computing nodes. In the heterogeneous cluster, GPUs and FPGAs are used to process the parallelisable portions of applications, with the less parallel portion processed by CPUs. We can achieve more efficient solutions by involving CPUs, GPUs, and FPGAs into the parallelisable portions, and by distributing the workload based on the processing capacity of heterogeneous architecture for specific applications. In Section 9.2, we shall discuss the heterogeneous computer architectures in detail.

9.1.2 Heterogeneous Clusters: Verification

Assertions as a means of verification can be traced back at least to Hoare [168]. Our static verification method extends the work of Susanto et al. [378] to streaming systems. In-circuit assertions [398] have been proposed in simulations and in hardware by Curreri et al. [87]. Statistical assertions have been proposed for parallel software by Dinh et al. [103]; we extend them to reconfigurable hardware. Our approach is described in more detail in two papers [387, 386].

Heterogeneous clusters often include GPU hardware in addition to CPUs and FP-GAs. While we do not address GPU verification in this section, several researchers have studied verification of some aspects of GPU programs. Our methods could integrate with these techniques as well as existing techniques for verifying parallel software [103]. GPUVerify is a tool for verifying freedom from two classes of GPU program bugs: data races and barrier divergence, where different threads in the same group reach different barriers, causing undefined behaviour [34]. This work uses a novel semantics for GPU programs to compile them to a predicated form suitable for decomposing into formulae to be checked automatically by a theorem prover. Other GPU verification work includes PUG [243], which uses a satisfiability modulo theory (SMT) solver to verify freedom of GPU programs from data races and incorrectly synchronised barriers.

---

[1] The $C_{por,arc}$ is highly application-specific. In this example, we simplify the execution process, and assume a CPU / GPU core takes n clock cycles to execute n arithmetic operations, and an FPGA data-path pipelines the n operations to generate one result per clock cycle.

Our techniques can aid the verification of self-adaptive systems, and could be combined with existing work, such as that by Calinescu et. al [56], which proposes runtime quantitative verification of systems with stochastic behaviour, our in-circuit statistical assertions could feed this approach. More details can be found in Section 9.5.

9.2 Architectures of Heterogeneous Clusters

9.2.1 Overview of Existing Heterogeneous Clusters

A heterogeneous cluster refers to a high performance cluster with heterogeneous computing devices. CPUs, GPUs and FPGAs are popular candidates for heterogeneous computing. Compared with homogeneous clusters, heterogeneous clusters provide better flexibility when various applications are mapped onto the cluster, and achieve better performance by distributing applications to suitable devices. Heterogeneity can also benefit other forms of computing, such as cloud computing.

The heterogeneity in CPUs comes from instruction set extensions and multi-core architectures. Instruction set extensions — such as MMX [314] and Streaming SIMD Extensions (SSE) from Intel — provide multi-way Single instruction, multiple data (SIMD) parallelism to execute the same instruction on multiple data concurrently. Two to six times speed improvements haven been achieved for various applications from neural network simulation [377] to matrix operations [1]. Moreover, as Moore's law continues, CPUs contain more and more parallel cores, from commonly used server-end Intel Xeon X5650 with six cores to recent Intel Xeon Phi coprocessor with 61 cores [85]. Distributing workload among the parallel cores often can improve application performance. However, the scalability of the approach is limited by the memory coherency and communication between cores in these architectures.

GPU-based heterogeneous clusters rely on CPUs for hosting operations and distribute parallelisable portions of applications among CPUs and GPUs. Tokyo-Tech Supercomputer and Ubiquitously Accessible Mass-storage Environment (TSUB-AME) contains 170 NVIDIA Tesla C1070 cards, and provides 77.48 TFlops throughput [120] for stencil applications. A Cray XK6 node contains an NVIDIA Tesla X2090 GPU and an AMD Opteron 6272 CPU. A Cray XK6 system [339] with 176 Cray nodes shows linear scalability for seismic imaging applications by overlapping the communication and computation operations. FPGA-based heterogeneous clusters adapt FPGAs as accelerators, and map customised hardware designs into FPGAs to accelerate computationally intensive portions. The Cray XD1 system integrates 12 Opteron CPUs and six Xilinx Virtex-II FPGA devices into a single motherboard [376]. Similarly, the SRC-7 MAPstation [368] integrates CPUs and up to six Stratix-II FPGAs into a single compute node. Recently, Maxeler Technologies released the MPCX systems where CPUs and FPGAs are kept separately in different compute nodes, with up to eight Stratix-V FPGAs in a compute node. The large-scale reconfigurable fabric on-chip enables FPGAs to accommodate replicated customised PEs, which provide orders-of-magnitude improvements in performance for applications in finance [385] and image processing [290].

In order to combine the advantages of different architectures, heterogeneous clusters with CPUs, GPUs and FPGAs are proposed. A Quadro Plex (QP) Cluster [360] contains 16 heterogeneous nodes, with each node integrating one AMD dual-core CPU, four NVIDIA G80GLs and one Xilinx Virtex-4 LX100. The QP cluster uses FPGAs for communication operations. The cluster achieved from 1.23 to 6.3 times speedup for applications such as molecular dynamics, weather prediction and cosmology. The Axel cluster [395] contains one CPU, one GPU and one FPGA in a compute node, with all heterogeneous devices used for communication operations. Up to 22.7 times speedup is achieved for N-Body simulation.

The interconnect network between the heterogeneous devices plays an important role in heterogeneous computing, since performance bottlenecks often come from communication operations. In a heterogeneous cluster, the heterogeneous devices within a compute node can be connected via on-board communication interfaces, such as PCI-Express and dedicated interfaces [376]. Start topologies [63] and Torus networks have been applied to improve the efficiency of intra-node communication operations. Shared memory between heterogeneous devices has been proposed to enable heterogeneous devices to read and write into a shared memory region [361]. Among compute nodes, the Message Passing Interface (MPI) library is commonly used to exchange data between concurrent processing tasks over a communication network such as Ethernet or Infiniband [395].

9.2.2 Software Layers in Heterogeneous Clusters

In order to coordinate the heterogeneous computing resources within a heterogeneous cluster, multiple software layers are required to handle the development, mapping, and scheduling of heterogeneous applications. Figure 9.4 demonstrates the development and execution process of a heterogeneous application.

- The compiler layer translates application descriptions into hardware and software designs optimised for underlying architectures.
- The resource layer virtualises the available resources.
- The scheduling layer coordinates communication and computation operations during runtime.

Compiler layer. At the compiler layer, the developers of heterogeneous clusters describe applications within the compiler infrastructure. The application descriptions are compiled as executable for the heterogeneous devices. Compiler techniques at this layer can be categorised into vendor-specific compilers, template-based compilers, and unified compilers.

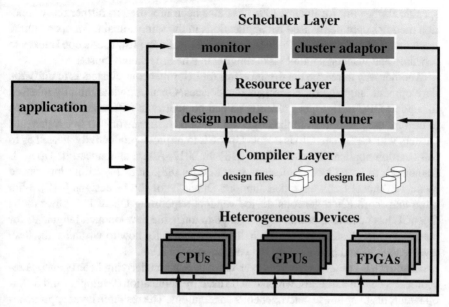

Fig. 9.4 Software layers of a heterogeneous cluster

Vendor-specific compilers refer to the programming models and tools provided by the device vendors. For CPUs with vector instructions, Intel provides specific libraries [192]. NVIDIA and AMD adapt CUDA [294] and OpenCL [150] to distribute the same instructions over parallel cores, and to synchronise between the parallel threads. For FPGA-based accelerators, conventional hardware description languages such as VHDL and Verilog are used to develop customised hardware designs [395]. Compared with the development tools for CPUs and GPUs, the low-level hardware language significantly limits the productivity for heterogeneous application development. In order to reduce the design effort for hardware designs, High-Level Synthesis (HLS) techniques are proposed to compile high-level descriptions into low-level circuit designs. The HLS tools can be traced back to more than 20 years ago, from early effort such as Handel-C [307] to recent commercial products such as Vivado. In order to develop applications for heterogeneous clusters, developers use the vendor tools to build CPU, GPU and FPGA designs separately, and link them in host programs. The linking and instantiation of heterogeneous programs are often under the OpenMP framework. Even with the high-level compiler tools, the amount of development effort required for a heterogeneous application still limits the use of heterogeneous clusters.

In order to improve the productivity of heterogeneous application development, template-based approaches reuse commonly used modules, and provide a framework to define the communication interfaces between heterogeneous devices. The Map-Reduce framework [93] groups the parallelisable portions of applications as a map phase, and distributes the workload into heterogeneous devices. A domain-specific framework for finance [394] has also been proposed to further reuse function modules commonly used for applications in the same domain. An open-source implementation of the Map-Reduce framework, Hadoop [15], uses a JobTracker to schedule and manage parallel tasks running in a heterogeneous cluster.

Another way to reduce the design effort for heterogeneous clusters is to unify the development language for heterogeneous devices. OpenCL, while initially released to support GPU designs, has recently gained popularity in CPU and FPGA designs to capture the parallelism in the target application. A framework for heterogeneous clusters with CPUs and GPUs adapts OpenCL to improve productivity as well as to fully exploit application parallelism [218]. In 2012, Altera announced its OpenCL framework to map OpenCL designs into FPGAs [88]. In 2014, Xilinx introduced its new SDAccel framework that supports OpenCL for FPGA designs [418]. With more and more heterogeneous device vendors supporting OpenCL in their tools, OpenCL has become a promising language to unify the development languages for heterogeneous clusters. However, challenges remains for how to virtualise the low-level resources and how to unify the vendor-specific APIs.

Resource layer. The resource layer virtualises the underlying heterogeneous resources as design variables, which can be used by application developers and development tools to optimise heterogeneous applications. The resource layer approaches can be divided as design models and auto-tuners.

Design models build the relationship between design parameters, resource usage, and performance of applications. The models update their design variables based on the specification of available heterogeneous resources. A hardware specification file is used in Axel [395] to specify the available hardware resources and communication bandwidth in a heterogeneous cluster. Integer Linear Programming (ILP) models [73], convex programming models, and analytical models [292] are used to exploit of the design space of hardware designs.

Auto-tuners enable non-expert developers to utilise the various heterogeneous devices. Parallel GPU codes are generated in [337] to optimise applications based on properties of GPU architectures [90]. CUDA design kernels are mapped into FPGAs [308], with design parameters automatically updated based on synthesised results. For specific application domains such as stencil applications, spatial blocking is optimised to balance workload among parallel threads [226] and auto-tuners are built to search for the optimal blocking strategies for various resources [89] and data structures [208]. Temporal blocking is supported with a blocking algorithm [286], and the design space is searched with various searching algorithms to minimise execution time for CPU and GPU designs. Due to its large search space, an auto-tuner often takes a long time to find an optimal solution.

Scheduler layer. Compared with the resource layer, which is used to exploit compile-time information, the scheduler layer in a heterogeneous cluster handles runtime issues, such as workload distribution, communication scheduling, and resource monitoring.

Workload distribution balances the distributed workload in each heterogeneous device with its processing capacity. In heterogeneous clusters, workload distribution is configured during compile time. In order to improve the adaptivity of the cluster, dynamic workload distribution is proposed, based on the measured runtime performance of involved heterogeneous devices [394].

The scalability of heterogeneous application is mainly determined by the communication operations. A common technique to improve the efficiency of communication operations is to build heterogeneous communication models that overlap computation operations with communication operations. This technique is widely used in CPUs [316], GPUs [120] and FPGAs [291].

Resource monitoring improves the adaptivity of heterogeneous clusters by collecting the status of heterogeneous devices. The variations in the resource status can be processed to adapt cluster configurations. In 2004, a distributed monitoring network called Ganglia was built [258]. Wired network resources are used to monitor the system status of CPU-centric clusters with small overhead. Recent developments in wireless sensor networks (WSNs) enable system operators to investigate server room environment. The RACNet [246] is a wireless sensor network for CPU-based data centres. Temperature variations in server rooms are detected, and operations of cooling systems are adapted to reduce their power consumption. A CPU status scheduler [183] was proposed for a CPU-based Beowulf cluster to reduce the cluster peak temperature. Moreover, adaptivity is introduced into a communication network to increase communication stability. A dynamic network load distribution strategy is presented in [201]. Probing loads are used to estimate network status.

In the next section, the workload distribution, communication scheduling and resource monitoring will be discussed in more detail to introduce self-awareness and self-adaptivity into a heterogeneous cluster.

9.3 Self-aware and Self-adaptive Applications for Heterogeneous Clusters

9.3.1 Self-awareness in Heterogeneous Clusters

The self-awareness in a heterogeneous cluster refers to (a) the awareness of resource status variations during application execution, and (b) the ability to predict the impacts of different cluster configurations on cluster performance. The self-adaptivity is introduced in heterogeneous clusters to tackle issues unpredictable during design time and compile time. The resource status discussed in this chapter includes power, temperature, and availability of heterogeneous devices, which can only be known during the execution time of applications. In order to adapt a reconfigurable cluster, first, a monitoring network reports the resource status variations into the scheduler layer. Second, in the scheduler layer, cluster models update design variables based on the measured status, and predict the impacts of various reactions. Third, once the scheduler decides to apply certain reactions, the heterogeneous cluster is reconfigured to adapt to the status variations. In this chapter, we focus on two runtime scenarios: device overheat and runtime design scaling.

9.3.2 Runtime Scenarios

Nowadays, stability is a serious issue for HPC systems. Given a large collection of heterogeneous computing nodes and intensive computation workload in a heterogeneous cluster, hardware failure has become more and more common. The Mean Time Between Failures (MTBF) for the ASCI Q constellation at Los Alamos National Laboratory (LANL) is less than 6.5 hours [182]. In 2003, a one-hour breakdown of HPC clusters can cost up to six million dollars [130]. As indicated by the Arrhenius equation, a 10°C increase in device temperature doubles its failure rate. The device temperature depends on initial device temperature, room temperature, cooling infrastructure, and distributed workload. These parameters cannot be known during compile time, and therefore need to be adapted during runtime. One another runtime issue for heterogeneous clusters is the elasticity. In a heterogeneous cluster, applications start and finish from time to time. The heterogeneous devices unavailable to an application can become available during the application execution. Sharing resources in a cluster adds complexity to the development process: applications must not only efficiently exploit a given set of compute resources, but also adapt dynamically to available resources at run time [291].

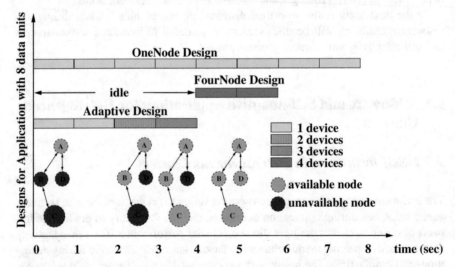

Fig. 9.5 An example of different designs executed in a heterogeneous cluster with the availability of devices being unpredictable, and each node being able to accommodate one application at each time (often the case for FPGAs). Node B, D and C become available at time 2, 3, 4, respectively.

The basic idea of cluster elasticity is illustrated with the following motivating example (Figure 9.5). In a reconfigurable cluster, four heterogeneous devices, A, B, C and D are released by other applications at time 0, 2, 3 and 4, respectively; node A, B and D possess one resource unit and process one data unit per second, while node C can process two data units per second; an application with eight data units to process is launched into the cluster. Linear scalability is assumed for executed tasks, i.e., execution time is halved if the number of utilised resource units doubles. In this scenario, two static designs are illustrated in Figure 9.5. The OneDevice Design will make use of only one node, so would take eight seconds to complete. The FourDevice Design will take all four nodes when all of them become available at time 4, and would take two seconds to complete. Only half of the computational capacity in node C is utilised, as the FourDevice Design pre-defines that one resource unit is used in each run-time node. The Dynamic Design, in contrast, can start at time 0 when node A becomes available; then at time 2, after node A processes two data units, node B becomes available too, so both nodes process another two data units in the next second. At time 3, nodes A, B and C are available, completing the processing of the four remaining data units.

9.3.3 Monitoring

A monitoring network provides awareness of resource status variations. Various Ethernet-based monitoring networks [258, 246] have been proposed to efficiently collect cluster status with minimal impact on cluster performance. Compared with the wired network, a wireless monitoring network [293] provides better flexibility and less performance overhead. Figure 9.6 shows the integration of a wireless network into a heterogeneous computing node. The wireless driver is implemented in FPGAs, with a front-end module collecting the temperature, power, and performance information of each heterogeneous device. This brings three advantages.

First, as the wireless communication channels do not use the bandwidth resources of wired connections in the cluster, the additional wireless channels circulate cluster status information without affecting application performance. Although the wired monitoring network is optimised to minimise the bandwidth requirements, a small increase in bandwidth usage will impact cluster performance when network bandwidth is saturated with application traffic. Moreover, the monitoring latency will be increased when there is not enough communication bandwidth to support both application traffic and monitoring traffic. By separating the application traffic and monitoring traffic into respectively wired and wireless communication channels, the wireless monitoring network provides better monitoring network scalability.

Second, implemented in FPGAs, the network driver is customised to use the left-over hardware resources in FPGAs. From the application perspective, this consumes no computation capacity of the monitored heterogeneous devices. If the network driver is implemented as software in host CPUs, it will inevitably introduce additional computational workload.

Third, while the network topology of a wired network is limited by the available network connections, a wireless network can be easily reconfigured into any topology. Moreover, separated monitoring can be built by using different carrier frequencies. When multiple applications are running in the cluster, involved heterogeneous nodes are dynamically grouped according to distributed applications. For

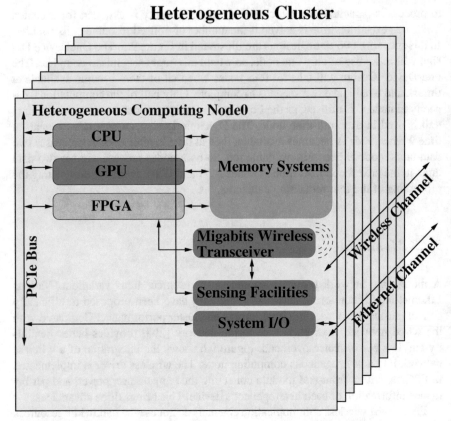

Fig. 9.6 The integration of a wireless sensor network into a heterogeneous cluster

the application-based monitoring operations, the network topology can be adapted when involved heterogeneous nodes are distributed to other applications or new nodes are assigned to the target applications.

Modelling. The monitored information includes the temperature, application performance, and power consumption of all involved heterogeneous devices. To properly react to device overheat and available devices during runtime, cluster models need to predict device temperature, as well as performance improvements and overhead for scaling over additional resources in a cluster.

Device temperature is affected by three parameters: the initial device temperature T_{ini}, temperature increase due to power dissipation T_{inc}, and temperature decrease due to cooling systems T_{dec}. The relationship between these parameters and the device parameters is non-linear. Modelling this non-linear relationship will consume a large amount of computational resources. To simplify the modelling, we use a linear model to predict the device temperature, and adaptively adjust the model based on the monitored information during runtime, expressed as follows.

$$T_{i,t} = T_{i,ini,t} + T_{i,inc,t} + T_{i,dec,t} \tag{9.3}$$

$$T_{i,ini,t} = T_{i,t-1} \tag{9.4}$$

$$T_{i,inc,t} = \alpha_{i,t} \cdot P_{i,t-1} \cdot ET_{i,t} \tag{9.5}$$

$$\alpha_{i,t} = 0.5 \cdot \frac{T_{i,t-1} - T_{i,t-2}}{P_{i,t-2} \cdot E_{i,t-1}} + 0.5 \cdot \alpha_{i,t-1} \tag{9.6}$$

where $T_{i,t}$ indicates the temperature of heterogeneous device i at time step t, the initial temperature $T_{i,ini}$ is updated based on measured temperature, and the temperature decrease due to cooling system $T_{i,dec}$ is considered as constant during runtime. The measured temperature at time step $t-1$ is used as the initial temperature for time step t, as shown in Equation 9.4. The temperature increase due to power dissipation is estimated based on power usage. $P_{i,t-1}$ indicates the measured power consumption in the last time step, and $ET_{i,t}$ indicates the measured execution time in the last time step. $\alpha_{i,t}$ indicates the non-linear runtime variable, and is updated based on measured temperature and power variations. During runtime, once the predicted device temperature exceeds the upper limit, the cluster adaptation process reduces the workload of overheated devices to prevent hardware failures.

Adaptively scaling a design over additional heterogeneous devices will increase the overall computational power of the designs. However, such adaptation comes with overhead. First, application context data need to be preserved and redistributed. Second, for heterogeneous designs that involve FPGAs, scaling current designs into new FPGAs requires chip reconfiguration. Third, when the number of devices involved in a design varies during runtime, the communication operations need to be rescheduled to ensure correct results are generated. With performance information monitored during runtime, the reduction in execution time after runtime scaling ET_{dec} can be expressed as:

$$ET_{dec} = \frac{wl_{cur,i} - wl_{nex,i}}{TH_i} + O_{ref} + O_{tra} \tag{9.7}$$

$$TH_i = \frac{wl_{fin,i}}{\sum_{t=0}^{n} ET_t} \tag{9.8}$$

$$O_{ref} = max(\frac{R \cdot \gamma}{\theta}, \frac{wl_{cur,i}}{\phi}) \tag{9.9}$$

$$O_{tra} = \frac{wl_{nex,i}}{\phi} \tag{9.10}$$

where $wl_{cur,i}$ and $wl_{nex,i}$ respectively indicate the distributed workload for heterogeneous device i before and after scaling, O_{ref} indicates the additional time taken by chip reconfiguration operations, and O_{tra} indicates the time to redistribute workload. We calculate design throughput TH_i as the ratio between finished workload $wl_{fin,i}$ and execution time $\sum_{t=0}^{n} ET_t$.

The computational capacity of each heterogeneous device is calculated as the ratio between processed workload and the current execution time. Chip reconfiguration time O_{ref} depends on two parameters, configuration file size and configuration throughput θ. Configuration file size is estimated as $R \cdot \gamma$, where R accounts for the size of configuration file for one resource unit in an FPGA, and γ accounts for the number of resource units consumed in the FPGA design. Workload redistribution can be divided into two phases: reading the current context data into a host device, and writing the distributed context data into assigned devices. For the reading time, since the operations to read context data can run in parallel with chip configuration, only the upper bound of configuration time and data loading time contribute to scaling overhead O_{ref}. The writing time is the time to redistribute workload O_{tra}, and is estimated as the ratio between redistributed workload $wl_{nex,i}$ and data transfer throughput ϕ.

During runtime, once the performance model determines the scaling the current design into available heterogeneous devices can reduce the overall execution time for the remaining tasks, the scaling process will start.

9.3.4 Adaptive Strategies in Heterogeneous Clusters

The self-awareness in a heterogeneous cluster enables the cluster to be aware of variations in cluster status and the impact of various reactions. After the temperature and performance models decide to react to the status variations, cluster configurations need to be adapted to meet the following constraints.

- Device temperature will not exceed its upper bound.
- Scaled designs still generate correct results.
- Expected performance improvements can be achieved after design scaling. Under the current runtime scenarios (overheat and adaptive scaling), the adapted cluster aspects include computational capacity, workload distribution, and communication scheduling.

9.3.5 Computational Capacity

Normally, the involved heterogeneous devices in a design run at full speed, i.e., all parallel cores in CPUs and GPUs are used, and PEs in FPGA designs are replicated to fully exploit available resources. However, for overheated devices (or devices to be overheated), their computational capacity is reduced to limit the power dissipation of these devices. The relationship between the power usage and the computational capacity of a device can be expressed as:

$$P_{i,n} = P_{ini} + \sum_{n=0}^{N} \tau_n \tag{9.11}$$

$$\tau_{n-1} = P_{i,n} - P_{i,n-1} \tag{9.12}$$

where $P_{i,n}$ indicates the power usage for a heterogeneous device i with n PEs, P_{ini} indicates the power usage of communication infrastructures such as memory modules, and τ_n indicates the increase in power usage when increasing the number of active PEs from $n-1$ to n. τ_n is updated based on measured power information. A PE in CPUs, GPUs and FPGAs respectively indicate a core, a streaming processor and a data-path. As discussed in Equation 9.3, reducing power consumption of a device (i.e., reducing the number of activated PEs in this device) directly reduce device temperature. With Eqs. 9.3 and 9.11, the number of PEs in heterogeneous devices is maximised under the constraints that the maximum device temperature is not reached.

9.3.6 Workload Distribution

Workload distribution balances the execution time of involved heterogeneous devices, so that all devices involved in a design finish their task at the same time. Application workload is distributed each time the computational capacity of involved heterogeneous devices changes. In the current runtime scenarios, variations in device computational capacity come from two aspects: (1) the number of activated PEs in a device is reduced due to temperature limit, and (2) additional devices are found available during runtime. Given N heterogeneous devices with computational capacity TH, the distributed workload for device i can be expressed as:

$$wl_i = \frac{wl}{\sum_{n=1}^{N} TH_n} \cdot TH_i \tag{9.13}$$

where wl indicates the overall remaining workload to process, and the computational capacity TH is updated based on measured performance and current design configurations (the number of activated PEs in devices).

9.3.7 Communication Scheduling

In order to eliminate performance bottlenecks due to communication operations, communication scheduling is introduced to minimise the impact of communication operations on design performance. Compared with synchronous communication models, asynchronous communication models [120, 291] are commonly used to overlap the communication operations with communication operations. As shown in Figure 9.7, with workload distribution balancing the execution time of heterogeneous devices, the dependent data, once updated, are collected into host CPUs via intra-node communication interfaces (often PCI-E channels). The collected dependent data go through inter-node communication interfaces (such as Ethernet and Infiniband) under MPI framework. Figure 9.7 demonstrates the communication operations under heterogeneous communication models: the communication operations run in parallel with computation operations. In this example, dependent data are collected and transferred before the data are required, eliminating additional communication time in the overall execution time. Given a distributed workload $wl_{i,j}$ of device j in heterogeneous node i, and the region of dependent data dep_i, the arrival time of the dependent time in heterogeneous node $i + 1$ can be estimated as:

$$Del_{i+1} + \frac{it * wl_{i,j} + dat_{i,j}}{TH_{i,j}} < Arr_{i,i+1} < Del_{i+1} + \frac{(it + 1) * wl_{i,j} + dat_{i,j}}{TH_{i,j}} \quad (9.14)$$

where $dat_{i,j}$ indicates the position of the dependent data in the distributed workload, $\frac{it*wl_{i,j}+dat_{i,j}}{TH_{i,j}}$ indicates the execution time to calculate the dependent time (as workload is distributed within a heterogeneous computing node based on the computational capacity, this time is the same for all devices in a node), and Del_{i+1} is the scheduled computation delay in the target node $i + 1$. As shown in Figure 9.7, when the communication bandwidth between nodes i and $i + 1$ cannot guarantee the dependent data arrive on time, an initial computation delay Del_{i+1} can be added in the target node $i + 1$ to meet the latest arrival time.

9.4 Evaluation Results

The self-awareness and self-adaptivity of a heterogeneous cluster enable the cluster to properly react to variations in cluster status. We evaluate the cluster adaptivity in two runtime scenarios: device overheat and device availability variations. We use five benchmark applications from different fields to evaluate the benefits of introducing self-adaptivity in clusters: N-body simulation, K-means clustering, Asian-option pricing, Reverse-Time Migration (RTM) and Bond-option pricing. For the overheat scenario, we run the experiments in a heterogeneous cluster with up to eight heterogeneous computing nodes, with each node containing an AMD Phenom Quad-Core CPU, an NVIDIA Tesla C1060 card and a Xilinx Virtex-5 LX330 FPGA. For the availability scenario, we adapt the heterogeneous designs in a commercial system: a MPC-C500 compute node from Maxeler Technologies with two Intel Xeon X5650 CPUs and four Xilinx Virtex-6 SX475T FPGAs. This system is shared by multiple users, with various applications launched from time to time. Therefore the resource availability becomes unpredictable.

9.4.1 Benchmark Applications

N-Body simulation is commonly used to study interactions between objects under gravitational force. The simulation is an iterative process where the 3D position and

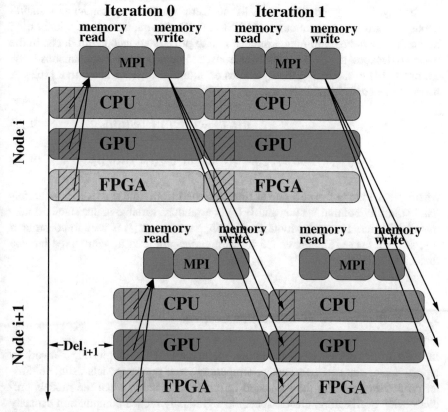

Fig. 9.7 Asynchronous communication operations and scheduled initial delay in heterogeneous computing nodes

velocity vectors are updated after calculating the total force acting on each individual particle.

The K-means clustering algorithm iteratively groups N data points into K clusters. The data point grouping process is based on the similarity of their attributes [293].

RTM is an advanced seismic imaging technique to detect terrain images of geological structures, based on the earth's response to injected acoustic waves. The propagation of injected waves in terrains is modelled with the isotropic acoustic wave equation [16], which is solved with a fifth-order stencil in space with three dimensions.

An option is a financial instrument which provides the owner of the option with the right to buy or sell a bond at a fixed price in the future. A call option allows owners to buy an asset, while a put option allows owners to sell an asset. Asian-option pricing calculates the value of the asset by taking the arithmetic value over the option's lifetime. Hence, it will be more difficult to manipulate asset price at expiry date.

The payoff of a bond option uses Monte Carlo simulation to model an instantaneous forward interest rate curve [162]. Monte Carlo simulations are widely used in the finance industry to model interest rate to price fixed-income products. In the past two decades, the field has evolved from modelling a single instantaneous interest rate [167] to modelling the dynamics of an entire forward rate curve [162]. A forward rate curve is modelled as:

$$\mu(t,T) = \sigma(t,T) \int_t^T \sigma(t,u)du \qquad (9.15)$$

$$df(t,T) = \sigma(t,T) \int_t^T \sigma(t,u)dudt + \sigma(t,T)^T dW(t) \qquad (9.16)$$

where $f(t,T)$ is the forward rate at time T started from time t; $\sigma(t,T)^T$ is the forward volatility column vector; and $W(t)$ is a random variable under standard normal distribution. For each Monte Carlo path, a random $W(t)$ is used to construct a forward rate curve. The generated forward curves are used to value fixed income financial products

9.4.2 Self-adaptive Temperature Control

In order to prevent device overheating, we use a wireless monitoring network to report cluster status. Performance, temperature and power models estimate cluster properties based on the measured information. Finally, when the models predict device overheating, cluster schedulers gradually reduce computation capacity of overheated devices, and relocate workload based on the adapted computational capacity. Figure 9.8 shows the device temperature of CPUs for three benchmark applications. With the adaptivity in computational capacity and distributed workload, the peak temperature of a CPU reduces by 12°C. Based on the Arrhenius equation [83], this halves the device hardware failure rate. Moreover, by assigning parallel workload into more suitable hardware architectures (GPUs and FPGAs for the three benchmark applications), the overall power efficiency of heterogeneous clusters is improved. Experiments show that the overall cluster throughput is respectively decreased by 4%, 17% and 2% for N-body simulation, K-means clustering and Asian-option pricing. Up to 2.17 times improvement is achieved in terms of cluster power efficiency, while energy efficiency is increased by up to 2.26 times.

9.4.3 Self-adaptivity for Resource Availability Variations

Unlike instruction-based architectures such as CPUs and GPUs, FPGAs can be shared by multiple applications at the same time due to complex design and synthesis procedures. Therefore it is often the case that FPGAs in commercial systems

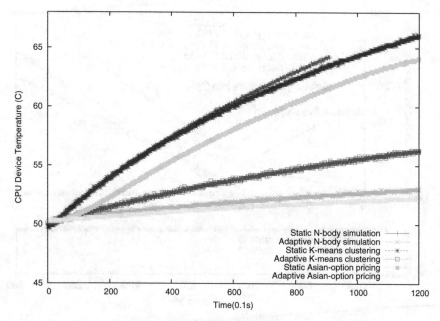

Fig. 9.8 CPU temperature variations, for both static and adaptive designs of N-body simulation, K-means clustering, and Asian-option pricing

can only take one application at each time. As demonstrated in Figure 9.5, when FP-GAs in a system become available at a different time, we introduce self-adaptivity to adjust design configurations to scale a design over FPGAs that become available during runtime.

We show the adaptation process in Figure 9.9. The lower section of the figure shows the resource availability of FPGA nodes during runtime, which is measured from the target system. As shown in the figure, node A is available from beginning, and nodes B and C become available at time 102 seconds and 119 seconds, respectively. Node D is not available throughout the runtime. In this example, we show the execution process of RTM designs. The performance of static designs is limited by the availability of FPGAs. For the static design with four FPGAs, it is not executed throughout the 300 measured seconds as node D is busy. The self-adaptive design, on the other hand, monitors the availability of cluster nodes, evaluates the benefits and overhead once new nodes become available, and adjusts design configurations (workload, communication operations) if the benefits outweigh the overhead (i.e., $ET_{dec} > 0$ in Eq.9.7).

With self-adaptivity introduced, the cluster resource utilisation is increased. We define the cluster resource utilisation as the ratio between measured application performance and the application peak performance if all computing nodes are utilised. As shown in Figure 9.10, compared with static designs that use one, two, three, and four FPGAs regardless of runtime resource availability, the self-adaptive design

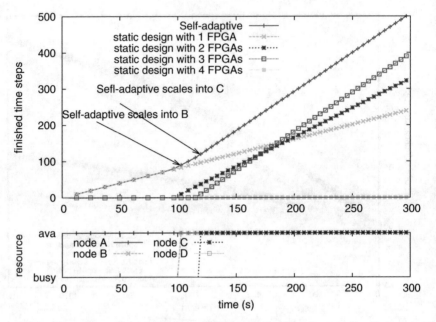

Fig. 9.9 Adaptation process of a self-adaptive RTM design, along with static designs that use one, two, three, and four FPGAs. 'ava' indicates an FPGA node is available, and 'busy' indicates the FPGA nodes is not available.

achieves 1.8 to 2.3 times higher throughput, i.e., after introducing self-adaptivity into designs, the cluster resource utilisation ratio is improved by 1.8 to 2.3 times.

9.5 Verification of Heterogeneous Clusters

This section shows our approach to the verification of heterogeneous clusters, possibly involving reconfigurable hardware accelerators. As the rest of this chapter has shown, heterogeneous clusters involving reconfigurable hardware can achieve greater performance, and lower power and energy consumption than homogeneous clusters, but reconfigurable hardware complicates the problem of verification. Software and hardware can be verified separately, using existing approaches, but many bugs may lie in the interaction between the two.

In our approach, designs are verified both statically (at compile-time) and dynamically (at run-time). At compile-time, we verify the functionality of a design, by comparing it to a golden reference design, such as a single core software design. Since not all bugs can be caught at compile-time, we verify designs at run-time, using in-circuit assertions to test design properties at full circuit speed.

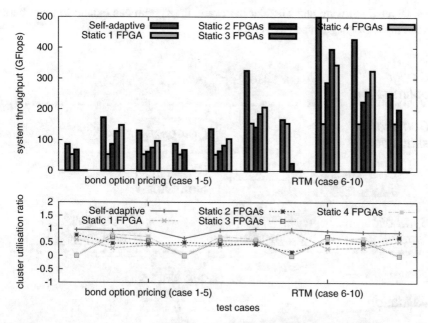

Fig. 9.10 Measured throughput of self-adaptive and static designs for bond option pricing and RTM, in ten different runtime cases. Each case records the measured resource availability during a certain period of a cluster. The self-adaptive and static designs are executed for each case to compare the performance. The cluster utilisation ratio is measured as the ratio between achieved design performance and the peak design performance if all available computing nodes can be efficiently used.

9.5.1 Verification of Hardware-Software Codesign

Figure 9.11 shows our general verification flow. We refer to the reference model as *source model* and to the optimised model which is being checked for equivalence as *target model*. Either design could be software, hardware, or a combination.

Given source and target designs, our approach has four modular phases: 1. *Design Translation*: source and target designs are translated to an intermediate representation, keeping application/input-specific parts as small as possible. 2. *Symbolic Simulation*: both designs are executed symbolically using the same inputs, using a symbolic simulator for our intermediate language. 3. *Output Combination*: to support weaker notions of equivalence than strict bisimulation-equivalence, such as stutter-equivalence, we pre-process the two output traces of the symbolic simulator in an output combiner before passing them to the equivalence checker. 4. *Validation*: an equivalence checker compares the symbolic outputs from source and target designs. The result is either successful verification, or a counterexample with symbolic inputs leading to different outputs.

Partitioning the workflow in these four steps, keeping the design translation small and straightforward, increases the modularity and generality of the approach. While we target C software and Maxeler hardware, the system can extend to other hardware input languages, such as Verilog.

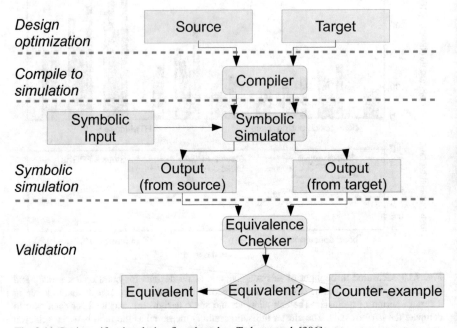

Fig. 9.11 Static verification design flow, based on Todman et al. [386]

To handle large systems, we model at word level. This limits the state-space explosion problem, but assumes correct operator implementations. We can still use bit-level correctness results for components in our system using a modular verification approach as operator implementations are combinatorial.

Our approach has some restrictions, which can be mitigated: 1. we assume synchronous hardware with one global clock; 2. some data such as array sizes must be numeric, not symbolic; 3. data-dependent control flow can still cause state-space explosion. Using numeric values for data that cannot be abstracted to symbolic values rarely poses a problem in practice: for example, there is only a limited set of useful image sizes which need to be verified. Data-dependent control flows can be addressed by verifying different modes separately.

Finally, our approach can symbolically compare hardware and software implementations of same algorithm—often the translation between the two is a source of bugs.

9.5.2 Runtime Verification by In-Circuit Statistical Assertions

We propose *in-circuit, statistical assertions*, compiled into the hardware part of a software-hardware design as a dedicated self-monitoring facility for self-adaptive systems, with a fast response time to adaptation. Compared to the proposed in-circuit assertions that can compute in parallel with the rest of the design, purely software-implemented assertions need to wait until the hardware has finished computing its results before they can process their own tasks. Moreover, efficient hardware designs are often deeply pipelined, operating on large batches of data, further prolonging the time until software assertions can start processing. Additionally, by preprocessing potentially large amounts of data, in-circuit data gathering can improve use of limited bandwidth between hardware and software of the self-adaptive system, triggering and controlling system adaptation. In summary: in-circuit assertions are the necessary precondition to realise fast response times to adaptation not realisable by pure software assertions.

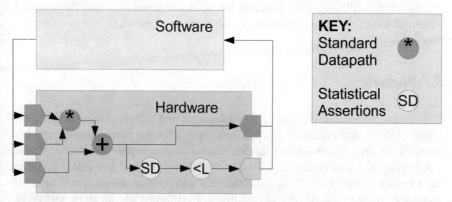

Fig. 9.12 Dynamic verification: in-hardware statistical assertions (based on Todman et al. [387])

Figure 9.12 provides a structural overview of our approach. A hardware datapath is instrumented by in-circuit statistical operators which compute relevant statistics about the design. These are then sent back to a software engine running a self-adaptive system. The software builds up the self-adaptive representation which is used to control reconfiguration of the system. It should be mentioned that whilst we target a software-hardware system setting, our approach is not limited to this setting at the outset. The software could likewise run on a soft processor within a Field Programmable Gate Array (FPGA) fabric.

9.5.3 Results

Static verification: we compare several implementations of a 1D convolver, with window size 3, comparing 1. streaming hardware, 2. software plus hardware, 3. software plus hardware reconfiguring the convolver weights.

(1): software versus hardware: the software produces one result per cycle, while the hardware also produces one result per cycle; this is handled by the output combiner and the designs are verified as equivalent; *(2)*: software versus software driving hardware: again the designs take differing numbers of cycles, but are verified as equivalent; *(3)*: software versus software plus reconfiguring hardware (changing convolver weights): again the hardware produces the same results as the software, but at a faster rate.

Dynamic verification: we choose a convolver circuit with assertions on the variance of the output at various stages. For 32-bit, single-precision floating-point data, our implementation uses about 2% to 3.5% of chip resources, and runs at 300 MHz (targeting Maxeler MaxCompiler 2012.1 and the Maxeler MAX3 board with Xilinx xc6vsx475t FPGA). This indicates low area and speed overhead for in-circuit statistical assertions. The calculated variances are as expected.

9.6 Summary

This chapter discussed the use of self-awareness and self-adaptivity in heterogeneous clusters, and presented verifications approaches for heterogeneous applications. We mainly discussed two runtime scenarios that require adaptation: device overheating and indeterministic resource availability. We use a wireless monitoring network to report cluster status variations, and develop design and scheduling models to react to the monitored variations. Experiment results show that adaptively adjusting device computational capacity and workload halves the hardware failure rate of a heterogeneous cluster. Moreover, by reconfiguring designs to adapt to computing devices that become available during runtime, the design throughput of a heterogeneous cluster can be improved by up to 2.3 times.

This chapter also showed our approaches to the verification of heterogeneous clusters including reconfigurable hardware. We verify applications both statically (at compile-time) and dynamically (at run-time). At compile-time, we verify the equivalence of two designs, such as a simple, single core software design and an optimised, possibly reconfiguring, hardware-software design. At run-time, we use in-circuit, statistical assertions to allow designers to express conditions which must be true if the design is working correctly as statistical conditions of the design signals. Experiments show that such in-circuit, statistical assertions can have a low overhead and run at high circuit clock speeds.

Chapter 10
Flexible Protocol Stacks

Markus Happe and Ariane Trammell-Keller

Abstract This chapter presents flexible protocol stacks as a promising alternative to today's static Internet architecture. Self-aware/expressive network nodes cooperate to select the protocol stacks that fulfil all communication requirements at minimal cost at run-time. Since the network conditions and the internal state of the communication peers change over time, the communication peers can dynamically adapt the flexible protocol stack to add, remove or replace protocols on-the-fly during communication. This stands in strong contrast to the traditional Internet architecture, which only supports a fixed set of protocol stacks that hardly adapt to the network dynamics. For our flexible network architecture, we adapt the concepts of self-awareness and self-expression for an autonomous observation of the node's internal system state and the network state and an autonomous run-time management of the protocol stacks. Our self-aware network node architecture also allows us to dynamically migrate the individual protocols between hardware and software at run-time to optimise the packet processing inside the network node to the current network traffic. In this chapter, we present the methodologies with which communication peers can negotiate the optimal protocol stacks amongst each other. Furthermore, we discuss the EmbedNet execution environment that supports an adaptive hardware/software mapping of flexible protocol stacks. Prototypes of the EmbedNet execution environment have been implemented on reconfigurable system-on-chip architectures. We demonstrate the efficiency of flexible protocol stacks in two case studies.

Markus Happe
ETH Zurich, Switzerland, e-mail: markus.happe@alumni.ethz.ch

Ariane Trammell-Keller
ETH Zurich, Switzerland, e-mail: ariane.trammell@alumni.ethz.ch

10.1 Introduction

The Internet architecture works well for a wide variety of communication scenarios. However, its flexibility is limited because it was initially designed to provide communication links between a few static nodes in a homogeneous network and did not attempt to solve the challenges of today's dynamic network environments. Although the Internet has evolved to a global system of interconnected computer networks, which links together billions of heterogeneous compute nodes, its static architecture has remained more or less the same. Nowadays the diversity in networked devices, communication requirements, and network conditions vary heavily, which makes it difficult for a static set of protocols to provide the required functionality. Therefore, there is a need for more flexible protocol stacks that can be optimised continuously during communication according to the current requirements.

We present a novel clean-slate network architecture that fundamentally differs from the current Internet architecture we have learned to cherish over the last few decades. In contrast to widely used static protocol stacks such as TCP/IP[1] and UDP/IP[2], we propose splitting the networking functionality into functional blocks which can be dynamically linked with each other to form flexible protocol stacks. At the beginning of a new communication between two peers, the nodes have to agree on a common protocol stack that respects the communication requirements dictated by the application, and takes into account the node-specific system constraints and the current network conditions. Changes in the node-specific system constraints or in the network conditions might result in an adaptation of the used protocol stack in order to optimise the communication behaviour of the nodes in a resource-efficient manner.

Due to the newly introduced flexibility in the protocol stacks, setting up and maintaining communication channels between communication peers introduces a novel class of challenges. Our novel network architecture employs the concepts of self-awareness and self-expression (as described in Part I of this book) to efficiently manage and optimise the protocol stack configurations.

10.2 Concepts and Methodologies

Figure 10.1 shows our two-step approach to transform the static Internet architecture into a network node architecture with flexible protocol stacks. The current Internet architecture is defined by a fixed set of protocol stacks that are partitioned into five layers: the application, transport, network, media access control (MAC) and physical (PHY) layers. Bypassing any of these layers is not permitted by design and the integration of new network functionalities e.g., a novel packet encryption algorithm, into the Internet architecture is a non-trivial task.

---

[1] TCP/IP: transmission control protocol (TCP) and Internet protocol (IP)

[2] UDP/IP: user datagram protocol (UDP) and Internet protocol (IP)

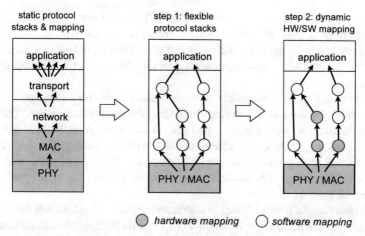

Fig. 10.1 Our two-step approach: First, we split up the network functionality of the network and transport layer into functional blocks that can be dynamically interlinked. Second, we dynamically map the functional blocks to hardware and software [159].

In the first step, our novel dynamic protocol stack architecture splits up the network functionality of today's Internet architecture into functional blocks that can be dynamically interlinked with each other. In this way, the dynamic protocol stack architecture might still support all protocol stacks of the current Internet protocol suite, but can also support arbitrary protocol stacks which might change their stack configurations during communication to adapt to a dynamic network environment. In our work, we focus on introducing dynamic protocol stacks (DPS) to the network and transport layer. New network functionalities can be implemented as modular functional blocks that have a generic interface, such that all functional blocks can be dynamically connected with each other. This approach allows for an arbitrary combination of functional blocks to form flexible protocol stacks.

In the second step, we map the functional blocks of the currently used protocol stacks dynamically to hardware and software to accelerate the protocol processing at the node-level. This step requires an implementation of the network node on a reconfigurable system-on-chip platform that allows us to map functional blocks dynamically to hardware and software. The reconfigurable hardware area is strongly limited for most reconfigurable systems. Therefore, only a subset of the used functional blocks can be mapped to hardware at the same time. Hence, it is the task of the network node architecture to find the optimal mapping of the used protocol stack. This requires that all functional blocks be implemented in an equivalent manner in hardware and software. In practice, this tight constraint might be relaxed, such that not every functional block in software has an equivalent hardware counterpart.

The functional blocks have unique identifiers and describe their properties using keywords. For instance, a functional block that encrypts packets using a certain encryption algorithm defines a `privacy` property and an idle-repeat-request protocol defines a `reliability` property. Application developers no longer use socket families with fixed protocol stacks such as TCP/IP or UDP/IP, but define the required communication properties using a set of keywords. We have developed a novel Berkeley Software Distribution (BSD) socket family in Linux, which enables C programmers to set up communication channels that are based on dynamic protocol stacks.

It is crucial that all communication peers speak a common language (i.e., use the common protocol stack) during the entire communication period. Therefore, the communication peers need to agree upon a common protocol stack at the start of a communication. In case a communication partner would prefer to change the protocol stack during communication, it needs to inform the other peers, which have to agree to this change. The configuration of protocol stacks has to be synchronised between all peers at all times.

The peers might have different preferences for protocol stacks and not all peers might support the same set of functional blocks. This heterogeneity complicates the task of finding an optimal protocol stack. The initial selection of a common protocol stack and all subsequent adaptation of the protocol stacks need to be synchronised until the communication terminates. Both, the selection and the subsequent adaptations are influenced by several constraints and goals, such as:

- the current network conditions (such as type of network, signal-to-noise-ratio, communication bandwidth),
- the communication requirements of the corresponding application (such as the reliability of the connection or quality-of-service constraints),
- user goals (such as the requirement of an encrypted channel), and
- system goals (such as increasing the packet processing performance and minimising the power consumption)

Handling all these constraints and concurrent user/application/system goals at the same time is a challenging task, since the goals have to be matched at network level (to optimise the protocol stack) and at node level (to map the used protocol stacks to the underlying hardware/software platform). Decisions at network level (i.e., protocol stack adaptations) can influence the decisions at node level (i.e., the mapping of the functional blocks).

The decisions about when and how to adapt a protocol stack are especially crucial tasks, since they generate a synchronisation overhead (protocol stack negotiation) and a configuration overhead at each peer. Adaptations should be avoided when the achievable benefits do not outweigh the costs. It is extremely hard or even impossible to foresee the run-time dynamics of the network communication of a network node at design time. Instead, a run-time approach is required that monitors the network conditions and the system state to dynamically decide at which times adaptations of the protocol stacks and/or hardware/software mapping are necessary.

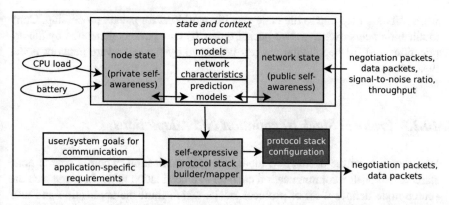

Fig. 10.2 The self-aware/expressive network node architecture (based on the reference architecture in Section 4.3): We distinguish between private self-awareness at the local node level and between public self-awareness at the network level. The node's and network's state and context drive the self-expressive protocol stack builder/mapper to define the optimal protocol stack configuration for given communication goals and requirements [212].

10.2.1 Self-aware/expressive Network Node Architecture

Figure 10.2 shows our self-aware/expressive network node architecture. This architecture is based on the generic reference architecture, which as been defined in Section 4.3, and uses a *private self-awareness* engine to capture the state and context of the local node and a *public self-awareness* engine to capture the network state and context. Both self-awareness engines can access various sensors. For instance, the private self-awareness engine can access system sensors that capture node-specific information such as the current load of the central processing unit (CPU) or the battery level. The public self-awareness engine monitors the network to gather information about the current network conditions such as signal-to-noise-ratio or packet throughput.

The self-awareness engines maintain several models of the local system and the global network behaviour. First, the protocol model contains a list of all available functional blocks and their properties. Second, the network characteristics contain information about the current network environment, and, third, prediction models evaluate past network traffic and networking conditions to speculate about the future.

User and system goals might be either defined explicitly or implicitly. An application developer who defines certain communication properties for an application with keywords sets explicit user goals. The self-expressive protocol stack builder/mapper might be programmed in a way that it implicitly tries to meet global system goals such as minimising the CPU overhead, maximising the overall packet throughput and/or minimising the power consumption.

The self-expressive protocol stack builder/mapper sets up the initial protocol stacks, adapts the protocol stacks to the network conditions and updates the hardware/software mapping to the node-specific workload for packet processing. Communication peers exchange two packet types: (i) data packets generated by the applications and (ii) negotiation packets generated by the self-aware network nodes themselves.

10.2.2 Protocol Stack Negotiation and Adaptations

We assume a source-destination model for communication, where a source node starts to establish a communication channel to a destination node. In a first step the source node defines a set of protocol stacks which fulfil the application and user requirements and sends this set to the destination node. Then, the destination node selects one protocol stack according to its own goals and sends this information back to the source node. Finally, both nodes configure the selected protocol stack. We assume that there is a default protocol that is supported by all nodes, which can be used for an initial protocol stack negotiation. After this initial negotiation phase, both nodes can propose adaptations of the used protocol stack. The nodes can use the currently selected protocol stack for such adaptations. The other node might accept or reject the proposal. If the proposal is accepted, both peers use the new protocol stack.

When a protocol stack is adapted on-the-fly, both nodes should support the old and the new protocol stack for a transitional period. The used protocol stack of an incoming packet can be identified by a unique hash value. For the computation of the hash value, we concatenate the unique identifiers of the individual functional blocks and apply a given hash function to the concatenation. Each functional block is identified by an *information dispatch point* (IDP), which is mapped to the function actually processing the packets. The complete protocol stack is identified by the sequence of IDPs of its functional blocks. We assume that each self-aware/expressive node uses the same identifiers for the functional blocks and the same hash functions. Hence, each self-aware/expressive node is able to recognise the used protocol stack using a single parameter.

For protocol stack negotiation and adaptation, we have investigated the *goal sharing pattern* and the *temporal goal-aware pattern* as introduced in Chapter 5. We define the self-awareness levels for protocol stack negotiation and adaptation as follows:

- **Stimulus-awareness:** Our protocol stack negotiation and adaptation is stimulus-aware. Nodes react to initial protocol stacks and adaptations that are proposed by other self-aware/expressive peers. Furthermore, protocol stack adaptations were usually directly triggered by some sensor input, such as the current link quality.
- **Time-awareness (optional):** Proactive self-expression techniques can be applied to learn regular traffic patterns. This implies a certain level of time-awareness. However, we believe that time-awareness is optional, since we can

also use reactive self-expression techniques, which possibly show worse results than their proactive counterparts.

- **Interaction-awareness:** Our self-adaptive networking infrastructure is inherently interaction-aware, since it interacts with other nodes in the environment and finds out which protocols these nodes prefer/support. Furthermore our networking infrastructure monitors the network conditions, which are a fundamental part of its environment. Our networking infrastructure can possibly integrate active sensors that insert probes into the network in order to study the current networking conditions.

- **Goal-awareness:** Our dynamic protocol stack architecture is inherently goal-aware, since the application's communication requirements are encoded using keywords, such as 'reliable', 'robust' or 'best effort'. These requirements are later used as incoming goals for our protocol stack builder. Our networking architecture tries to optimise towards or fulfil these application goals, while also trying to meet internal system goals, e.g., saving system resources. As part of the protocol stack negotiation, each node is also influenced by the goals of the communication peers.

10.2.3 Dynamic Hardware/Software Mapping

Once the peers have agreed on the protocol stack, a self-aware/expressive node can individually map the used functional blocks to hardware and software. The mapping might be non-trivial since many applications can use different protocol stacks that compete for the limited hardware resources. Hence, we also investigated self-aware/expressive mapping techniques in this scenario. The self-aware/expressive network node architecture profits from the vertical function migration capabilities of the ReconOS execution environment as described in Chapter 8, which allows us to dynamically start, stop and migrate hardware and software threads.

We implemented the *basic pattern* and also investigated the *temporal knowledge-aware pattern* which was introduced in Chapter 5. We define the self-awareness levels for our self-aware/expressive protocol stack mapper as follows:

- **Stimulus-awareness:** Most of our mapping algorithms are purely stimulus-driven. We usually map the functional blocks to hardware that provides the maximum benefit in packet processing time.

- **Time-awareness (optional):** If the system has learned regular traffic patterns, the self-aware/expressive node can pro-actively update the hardware/software mapping in order to provide the best possible performance in packet processing for the known traffic patterns. The network node needs time-awareness to learn such traffic patterns. It can provide benefits if there are significant regular traffic patterns.

10.3 EmbedNet Execution Environment

We use a system-on-chip (SoC) platform that is based on the self-aware compute node architecture presented in Chapter 8. Specifically, we implemented FPGA-based prototypes for our self-aware/expressive network node architecture using the ReconOS programming model and execution environment presented in Section 8.4. Our SoC platform combines a soft-core MicroBlaze processor with several hardware modules and peripherals on a single device. The Linux operating system runs on the MicroBlaze processor and the required file system is stored on external memory, i.e., a compact flash disk. The functional blocks can either be executed on the MicroBlaze processor or directly in the FPGA fabric. In order to implement our architecture on an FPGA board, we assume that it has external SDRAM, a physical Ethernet interface and a compact flash disk reader.

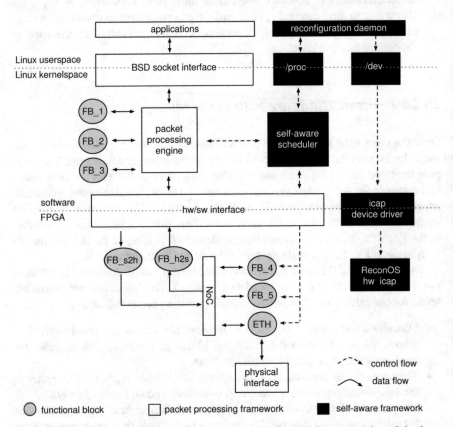

Fig. 10.3 EmbedNet system allowing for a self-aware hardware/software mapping of the functional blocks [212]

The complete EmbedNet execution environment is shown in Figure 10.3. The packet processing framework together with the functional blocks is responsible for packet processing, whereas the self-aware framework is responsible for the adaptation of the system. In the left part, user-space applications send and receive packets over a BSD socket interface. Since we don't use the TCP/IP protocol suite, we implemented a dedicated socket class that offers the well-known system calls like *send*() and *recv*(). Sending a packet through our socket inserts it in the packet processing engine, which forwards the packets to the correct functional blocks, no matter whether the block is currently mapped to software or hardware.

The self-aware framework in Figure 10.3 shows the infrastructure that is required for the self-aware mapping of functional blocks to either hardware or software. It is split between Linux user space, Linux kernel space and the FPGA itself. The central part is a `self-aware scheduler` which is responsible for selecting the hardware/software mapping of the protocol stack. While the framework does not specify a given algorithm to be executed, it provides an interface to access the statistics generated by the packet processing engine, such as the number of packets it processed for a given flow. After the self-aware scheduler decides on a new mapping, it performs the reconfiguration.

There are three modules that are always present in the FPGA fabric. One is the Ethernet functional block (ETH) that interfaces with the physical interface, and two are the modules responsible for transmitting the packets over the hardware/software boundary (called FB_H2S and FB_S2H, respectively). In addition to these statically configured modules, there are also *dynamic* modules. These modules can be reconfigured at run-time with the functionality of arbitrary functional blocks.

Functional blocks are connected by a network-on-chip (NoC) that forwards packets between them and also supports pipelined packet processing. The NoC consists of switches in a ring topology, where each switch connects to a configurable number of functional blocks. The total number of modules is a design-time parameter; this allows for the throughput of the NoC to scale appropriately by increasing the bandwidth between the switches and by allowing for more hardware modules to be connected to one switch.

For communication between hardware and software, we use the ReconOS extension to Linux. While the original ReconOS implementation provides transparent communication between Linux user space and the hardware, we have extended it to also support communication between Linux kernel space and the hardware. This extension is required because many software parts of our network architecture run in the kernel space for performance reasons.

To aid implementation of a functional block in hardware and software, we provide code templates: in hardware, this is a VHDL entity, and in software, this is a Linux kernel module. The templates consist of the code required for receiving and sending packets and configuration data as well as transferring internal state between a hardware and a software module.

Since there is no automatic translation from a functional block in software to one in hardware, it is the responsibility of the functional block's author to make sure that the respective implementations are equivalent, and also to provide the state that is required when resuming a hardware block in software or vice versa.

Run-time reconfiguration of the modules is done with the help of a ReconOS hardware ICAP[3] core [160], where the bit files for the partial reconfiguration are stored on an external flash card.

EmbedNet uses a dedicated addressing scheme for the functional blocks. The system updates these addresses on-the-fly whenever it needs to adapt the route a packet flow takes within a single node. Each functional block is identified by an information dispatch point (IDP). An IDP is mapped to the receive function of the corresponding functional block. If the functional block is currently executed in software, a pointer to its function is stored; if the functional block is executed in hardware, the hardware address of the functional block is stored. This address consists of the number of the switch to which the functional block is connected and the port number on this switch.

If a packet shall be forwarded, first the IDP of the next functional block is looked up. Second, it is checked whether the next block is implemented in hardware, and if so, the current hardware address is looked up. In software, this decision is made by the packet processing engine. In hardware, however, each functional block has to make this decision by itself. Therefore, each hardware functional block has a dedicated configuration interface. The hardware Ethernet functional block is slightly different, since in addition to making IDP-to-address lookups, it also has to make lookups from the hash identifying a given packet flow to IDPs (and vice versa for sending packets).

When adapting the packet flow due to either a new protocol stack or a new hardware/software mapping, conceptually, only the mapping of the IDP to the actual execution environment needs to be changed. Practically, the procedure is somewhat more complicated, since the FPGA also needs to be reconfigured in order to hold the new functional block.

10.4 Case Studies

In order to evaluate the benefits of a self-aware/expressive network node architecture, we show how our system autonomously adapts itself to changing network conditions. We will demonstrate the benefits of a flexible network node architecture as compared to a static network node architecture in two case studies: a sensor network and a smart camera network. For a sensor network, a sensor node adapts the protocol stack to the link quality to minimise the number of sent network packets. In the smart camera demonstrator, the nodes adapt the hardware/software mapping of the protocol stack according to the current communication scenario to improve the packet throughput.

The first case study was implemented as a software-only solution on commodity laptops running Linux that uses dynamic protocol stacks for communication [212].

---

[3] ICAP: internal configuration access port

The second case study is more elaborate and is implemented on FPGA boards that support the dynamic partial reconfiguration of the FPGA fabric [159].

10.4.1 Sensor Network

We developed a simple application that mimics a sensor that sends measurement data periodically to a server. We argue that transmitting a packet over a wireless interface costs energy, and therefore should only be performed when necessary. We implemented a protocol stack builder that includes an idle repeat request (IRR) reliability protocol in the protocol stack only when sensors report low link quality. The link quality is determined by a sensor that divides the current by the maximum possible wireless link quality.

We evaluated our architecture on commodity laptops in a mobile communication scenario. We recorded the link quality between two nodes while walking around in our office building; see Figure 10.4. Simultaneously, we measured that packets got lost when the link quality was below 35%. We have used this recording as realistic input for our emulation. In order to obtain reproducible results, we used a wired connection between the test machines and used the Linux traffic control tool `tc` with the `netem discipline` [285] to emulate packet loss.

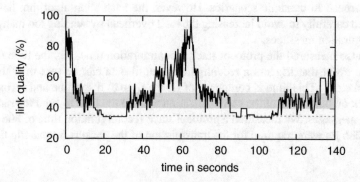

Fig. 10.4 Measured link quality over 140 seconds. Packets got lost when the link quality was below the dashed line. The DPS is updated when the graph crosses the grey bar [212].

Our protocol stack builder requests to be notified by the sensor daemon when the signal strength falls below a threshold of 40% or increases beyond 50%; see Figure 10.4. Upon such an event, it either inserts the reliability module or it removes the reliability module, and renegotiates the protocol stack with the neighbouring node. The lower threshold for renegotiation ensures that the reliability protocol is inserted into the protocol stack before the link quality reaches the critical value of 35%. The upper threshold is used to avoid frequent adaptations of the protocol stack.

For evaluation purposes, we compared the data loss rate and the total number of packets sent for (i) a protocol stack that dynamically adapts itself to the link quality, (ii) a protocol stack that never uses reliability, and (iii) a protocol stack that always uses reliability. We used these measured values to emulate the network conditions on a machine that connected the two test machines.

Table 10.1 Comparison between static and self-expressive configurations over 140 seconds [212]

| | | overhead | |
configuration	packet loss rate	reliability	reconfiguration
unreliable	31%	-	-
reliable	0%	128%	-
self-expressive	0%	60%	40%

Table 10.1 summarises our results. The configuration with no reliability lost on average 31% of the packets, whereas we did not observe packet loss in the other two configurations. However, this reliability comes at a price. The overhead (in terms of sent packets) for achieving reliability was 128% for the configuration that was statically configured to use the reliability protocol. The total overhead for the dynamic configuration was 100%, split into 60% for sending acknowledgement and retransmission packets and 40% for sending the protocol stack reconfiguration messages. This clearly shows that dynamic protocol stacks can reduce the total communication overhead in dynamic scenarios. However, the adaptation algorithm has to be designed carefully to avoid increasing the total overhead by sending too many stack reconfiguration messages.

We also measured the protocol stack reconfiguration time, i.e., the time it takes from an event that triggers a reconfiguration until data can be sent over the new protocol stack. This time is composed of (i) the time to determine and reconfigure the stack on both sides of the communication and (ii) the time to send the reconfiguration messages. We measured a protocol stack reconfiguration time of 806 μs, of which 286 μs were required for the transmission of the packets (round-trip time).

10.4.2 Smart Camera Network

Smart cameras combine several tasks such as video sensing, processing and communication on a single embedded device [342]. A typical task of a smart camera network is to track an object over the field of view of several camera nodes. In such a scenario they often need to process large amounts of data and communicate with each other under strict real-time constraints. Furthermore, smart camera nodes are usually implemented on embedded devices with limited resources, where the computationally expensive parts are implemented in hardware and the control tasks in software [428]. A flexible network node architecture that can adapt itself to the current network condition or communication scenario is therefore highly desirable in this context.

Video object tracking has been identified as one of prime case studies for the investigation of self-awareness and self-expression in computing platforms. Self-awareness and self-expression techniques can be used to improve the collaboration between the smart camera nodes and to improve the tracking performance of the network. Chapter 13 provides a detailed analysis on how self-awareness can be introduced into video object tracking in real smart camera networks. In this section, we show how smart camera networks can benefit from flexible protocol stacks on FPGA-based prototype platforms that are based on the EmbedNet execution environment. These prototypes have less functionality than the real smart camera network that will be presented in Chapter 13.

We use a basic video object tracking algorithm and only support tracking at most one colourful object per smart camera node. Only one of the smart camera nodes should be in charge of tracking a specific object at any given time. The smart camera nodes therefore need to collaborate in order to decide which camera node is in charge of tracking.

Initially a user selects the object to be tracked on a controller that runs on an external computer that is connected to the smart camera nodes over Ethernet. The controller computes the colour histogram of the object, which will be used to identify the object, and sends it to the smart camera node that currently sees it. This node is initially in charge of tracking. Whenever a smart camera node is tracking the object, it sends the current position and the approximate outline of the object in the form of a bounding box to the controller that stores all tracking information.

When the object approaches the edges of its field of view a Vickrey auction [400] is performed to determine which camera will continue to track the object. Our self-aware/expressive handover mechanism follows the adaptivity algorithm introduced in Section 6.2.1. The node originally tracking the object acts as an auctioneer and sends the object's histogram to the other nodes using a broadcast message. The other smart camera nodes try to find the object in their video input stream using a CamShift filter [71]. If a camera node sees the object, it bids for the object by sending a message back to the auctioneer. The height of the bid depends on how well the camera can see the object. The bid value corresponds to the detected size of the object, weighted by the position of the object with respect to the field of view of the camera. The smart camera node with the highest bid wins the auction and is informed by the auctioneer that it should continue tracking. The other nodes are informed that they have lost the auction. It is possible that the auctioneer decides not to hand over the tracking responsibility if the bids are low and it can still see the object.

Scenario

We assume that a smart camera node can only track a limited number of objects by themselves at the same time. In our case, the platform only allows for a single object to be tracked. Whenever a node is supposed to track multiple objects, it can transmit its video input stream to (idle) smart camera nodes, which take over the tracking responsibility. We assume that the number of smart camera nodes in the network is at least as large as the number of tracked objects. The selection of the camera that performs the remote tracking works as follows: (i) the camera that is currently responsible for tracking broadcasts a *remote tracking request* to all cameras in the network, (ii) all cameras that are currently not tracking any objects reply, and (iii) the initial camera randomly selects one of the idle cameras for remote tracking and transfers the colour histogram of the object to be tracked to this camera. The remote tracking camera then performs the tracking and sends the object's position and bounding box back to the camera that initially was tracking the object.

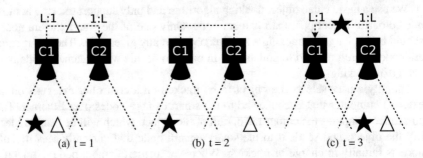

Fig. 10.5 Smart camera case study: Tracking two objects with two smart camera nodes. (a) Node C1 sees two objects and transmits its video stream to node C2, (b) both nodes see and track one object, and (c) node C2 sees two objects and streams its video to node C1. In (a) and (c) the ratio of sent vs. received packets per node is either L:1 or 1:L, L = #lines per frame [159].

Figure 10.5 shows an example with two smart camera nodes, C1 and C2, and two objects, a white triangle and a black star, at three different points in time. First, C1 sees both objects whereas C2 sees no object in Figure 5(a). In this case, C1 asks the currently idle C2 to track one of the objects, i.e., the white triangle. If C2 agrees, C1 transmits its input video to C2 over the network. C2 then tracks the white triangle and sends its current position back to C1. In Figure 5(b) the triangle has moved and each camera node can see a single object. Hence, no video stream has to be transferred. In Figure 5(c) the black star has also moved into the field of view of C2. Now C2 streams its video input to C1, which tracks the star for C2. A video frame is split into L packets, one packet per line. The object's position can be encapsulated into a single packet. Therefore, the ratio between sent and received packets per node is either L:1 or 1:L for a video transmission.

For this case study, we require a secure connection between all nodes since the video streams and the tracking results might contain sensible data. Therefore, we insert an advanced encryption standard (AES) block, called ENC, for encrypting outgoing packets, and another AES block, called DEC, for decrypting incoming packets into the protocol stacks. Since packet encryption/decryption is computationally expensive, it would be desirable if this could be accelerated in hardware. However, the

FPGA design only offers the space for one hardware accelerator, e.g., for either an encryption or a decryption block, but not for both.

Hence, a self-aware/expressive camera node needs to dynamically adapt its hardware/software mapping of functional blocks over time whenever it changes its role from transmitting a video stream to receiving a video stream (and vice versa).

In this case study, there is a high packet throughput between the nodes, and the ratio between sent and received packets can change drastically over time. Therefore, the multi-object tracking case study is a suitable testbed to evaluate the efficiency of the self-expressive hardware/software mapping of the EmbedNet architecture.

Smart Camera Node Architecture

Each smart camera node is implemented as a reconfigurable system-on-chip architecture that is configured on a single FPGA. The hardware/software architecture of a smart camera node is shown in Figure 10.6. The FPGA design consists of two parts, a video object tracking application (i, ii) and the EmbedNet network node architecture (iii, iv). The functionality of both parts is partially mapped to an embedded processor and partially to the FPGA fabric.

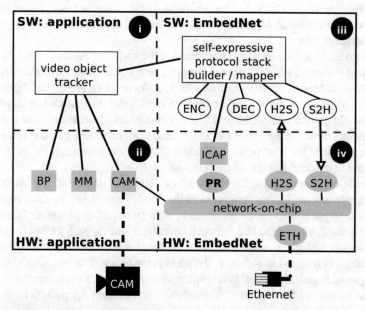

Fig. 10.6 Hardware/software architecture of a smart camera node: The left part represents the video object tracking application and the right part the EmbedNet network node architecture. The upper half of the FPGA design is mapped to software and the lower half is mapped to hardware [159].

We have implemented the smart camera node on a Xilinx Virtex-6 ML605 board. A full high definition Panasonic HC-V727 camcorder is connected to an AVNET HDMI Input/Output FPGA mezzanine card (FMC) module, which is plugged into the ML605 board. The camcorder provides a 1920×1080 video input stream at 50 frames per second (FPS). A Xilinx MicroBlaze processor has been instantiated on the FPGA fabric and runs the software parts of the smart camera node. The MicroBlaze processor and all hardware blocks are clocked at 100 MHz.

The architecture of the video object tracking application on a smart camera node is as follows: The video frames that are captured by the external camera are sent to a hardware module (CAM). The CAM block scales the input video down to a resolution of 320×180 at 25 FPS. This reduced frame size is then used for all further operations. The CAM module sends the captured frames to the software application, where the object tracking is started. We implement object tracking with the help of a CamShift filter [71], which requires two computationally expensive operations, namely histogram back projection (BP) and the computation of the moments (MM). Therefore, those two blocks are implemented as hardware modules.

To verify if the tracking works correctly, the software application sends the calculated location and size of the object back to the CAM module. The CAM module can send video frames together with the current location/s of the object/s over the EmbedNet architecture to the controller, which displays the video frames together with the objects position. The video transmission to the controller is optional and can be activated/deactivated by sending a specific control message to the corresponding smart camera node.

The EmbedNet architecture consists of a self-expressive protocol stack builder/-mapper in software, which forms the required protocol stack and dynamically maps the functional blocks either to hardware or to software. The prototype implementation of EmbedNet supports AES encryption (ENC) and decryption (DEC) of functional blocks in order to establish a secure communication channel. Our prototype contains one dynamic hardware slot (PR), which can be reconfigured at run-time to implement either the ENC or the DEC block. The partial bitstreams for both blocks are stored in external memory, i.e., SDRAM.

In addition to networking functional blocks, the CAM module of the video object tracking application is also connected to the NoC. This allows the CAM module to directly send packets to the protocol stack, without requiring processor interaction. For the transmission of packets between functional blocks, each packet is extended with a header that specifies the address of the next functional block. If the next functional block is mapped to software, the address corresponds to the H2S block. The self-expressive protocol stack builder/mapper can configure each functional block with the address of the functional block that should process the packets next.

Hardware/Software Mappings

We analysed the packet throughput supported by a single smart camera node for several hardware/software mappings of the encryption and decryption blocks in Table 10.2.

Table 10.2 Measurement for packet throughput in PPS, the packet loss rate in % and the video transmission rate in FPS [159]

send (encrypted) frames	PPS	Loss	FPS
HW:CAM → H2S → **SW:ENC** → S2H → HW:ETH	429	90%	2
HW:CAM → **HW:ENC** → HW:ETH	4 500	0%	25
receive and decrypt frames	**PPS**	**Loss**	**FPS**
HW:ETH → H2S → **SW:DEC** → tracking application	420	91%	2
HW:ETH → **HW:DEC** → H2S → tracking application	4 487	0.3%	25

If the node transmitting the video implements the encryption block in hardware, it is able to transfer at the requested rate of 25 FPS. However, if it implements the encryption block in software, it can only transmit two FPS. Similarly, for the node receiving the video: If it implements the decryption block in hardware, it can receive 25 FPS, but if it implements the decryption block in software it can only receive two FPS. In Table 10.2, we also see that the receiving node loses about 0.3% of the packets, even if the decryption module is implemented in hardware. This packet loss is introduced by the hardware/software interface, which has limited bandwidth. However, the video object tracking application is robust to a certain amount of packet loss. Whenever a video packet is lost, the application will use the pixel line of the previous frame instead.

In the multi-object tracking scenario not only is video data transmitted, but also control data (e.g., object selection, handover, and object position messages). These messages are encrypted as well and might flow in the opposite direction of the video data transmission. Therefore, a smart camera node might have to provide encryption and decryption at the same time. Control messages are expected to be way less frequent than video data messages, since for each received video frame (consisting of 180 packets), only one control message is sent back. Object selection and handover messages will occur even less frequently. From Table 10.2 we see that a node that encrypts in hardware can decrypt about 400 packets per second in software (and vice versa), which should be more than enough to process all control messages.

From this analysis we learn that for successfully tracking multiple objects in our smart camera network, the smart camera nodes need to adapt their hardware/software mapping of the encryption/decryption modules depending on whether they currently transmit or receive the video frames.

The total resource consumption of our self-aware/expressive network node architecture and different static architectures is given in Table 10.3. It can be seen that the self-expressive architecture with one reconfigurable hardware slot consumes a slightly higher amount of look-up tables (LUTs) and flip-flops (FFs) than the static architectures that contain either `HW:ENC` or `HW:DEC`. This comes mainly from the overhead introduced by the ICAP controller that is needed to perform the partial reconfiguration of the FPGA for the self-aware/expressive system. The ReconOS ICAP controller used in our implementation consumes 5% of the resources of an encryption/decryption block.

Table 10.3 Resource consumption [159]

mapping	LUTs	FFs	BRAMs
static: HW:ENC	34 297	25 318	314
static: HW:DEC	36 041	25 653	314
self-expressive: HW:ENC or HW:DEC	**37 534**	**25 949**	**314**
static: HW:ENC and HW:DEC	46 743	31 689	314
ReconOS ICAP module + FIFOs	690	255	-
partial module: HW:ENC	11 974	6 648	-
partial module: HW:DEC	14 165	6 983	-

We have also implemented a static architecture that contains both hardware blocks, which results in 25% more LUTs and 22% more FFs as compared to the adaptive architecture. It should be noted that an adaptive architecture can provide much larger savings in resources when there are more functional blocks that can be dynamically mapped to the FPGA fabric.

In addition to the slightly increased FPGA resources required for the adaptive system, also a time overhead for the reconfiguration as well as the need to store the partial bitstreams for the encryption and decryption blocks need to be considered. In our implementation the partial bitstreams have a size of 1.3 MB and the time overhead is 38 ms. This corresponds to the time that is required to process a single video frame. The time overhead consists of the following parts: First, the protocol stack builder/mapper maps all functional blocks to software. Second, the FPGA is partially reconfigured with the bitstream of the functional block that should be mapped to hardware. Third, the protocol stack builder/mapper configures the new hardware functional block (in our example with the encryption key) and adapts the protocol stacks to use the new hardware functional block.

To summarise, the static architecture that implements both `HW:ENC` and `HW:DEC` would provide the best packet processing performance because this architecture allows for a video transmission of 25 FPS without requiring any reconfiguration overhead. However, we believe that it is not always feasible to statically map all functional blocks to hardware since the FPGA resources are usually limited. Therefore, we need efficient techniques to dynamically switch between different hardware blocks at run-time (according to the current protocol stacks and network traffic mix). Our experimental results show that the HW/SW mapping of our adaptive architecture can be efficiently updated for a given tracking scenario such that the packet processing performance is only influenced marginally.

10.5 Comparison to Related Research Projects

There are several research projects that investigate novel, clean-slate networking architectures. In this section, we present related networking architectures that have a similar focus as our flexible protocol stack architecture. For each architecture we present the principles and show how it relates to the DPS architecture.

The Autonomic Network Architecture: ANA

The autonomic network architecture [43, 13] is a predecessor of the DPS architecture and has established several underlying concepts of flexible protocol stacks. In both architectures the communication protocols are negotiated when an application initiates a communication with another node. This allows for the coexistence of several network architectures in parallel. For each application and network condition, the best suitable architecture can be chosen, without the need that the application knows the available network architectures.

The mechanism for selecting the protocols differs between the ANA and the DPS architectures: While in ANA a publish/subscribe mechanism is used, in the DPS architecture a dedicated negotiation protocol is used. This has several benefits: (i) the number of control packets can be minimised, since all protocols can be negotiated in one step and the recursive negotiation used by ANA can be omitted, (ii) the initiating node can specify a set of accepted protocols from which the destination node can select, and (iii) the negotiation protocol can also be used to change the used protocols during communication. Especially the last point is important when the set of involved protocols should be adapted to time-varying network conditions, which is one of the goals stated in the introduction.

Recursive InterNetwork Architecture: RINA

RINA [91, 92] is based on the realisation that networking is nothing more than *Inter-Process Communication (IPC)*. The basic structure behind RINA is a Distributed IPC Facility (DIF). A DIF provides connectivity between all nodes that are a member of a given DIF, and provides functionality such as naming, access control, address resolution and data transfer. In order to divide the global network into smaller parts, DIFs can be layered on top of each other. This idea is similar to the idea of compartments in the ANA architecture. However, each DIF offers exactly the same functionality, and only policies have to be used to adapt a DIF to a certain scope. This also implies that only one data transfer protocol, called EFCP (Error and Flow Control Protocol), is available in the architecture. This protocol is based on Delta-T [410] which was developed in 1978. Delta-T is a timer-based protocol which does not require any handshake mechanisms. The EFCP protocol is split into two protocols: The Data Transfer Protocol (DTP) for addressing, fragmentation, etc. and the Data Transfer Control Protocol (DTCP) for flow control, retransmission control, etc.

This repeated use of the same DIF, with the same data transfer protocol on each layer, is fundamentally different from the current Internet architecture, where a layer is strongly coupled with the protocol that implements that layer. In order to evaluate how well RINA works in comparison to the Internet architecture, a dedicated project was funded: IRATI, Investigating RINA as an alternative to TCP/IP [193]. IRATI is a European Union research project which started in 2013. Within the IRATI project, RINA will be implemented from scratch on a Linux-based system. However, since the project only started recently, no concrete insights are available yet, but it will be interesting to see their conclusions at the end of the project.

Compared to our dynamic protocol stack architecture, RINA has a different focus. In the DPS architecture we focus on time-variant network conditions and how we can provide optimal service by adapting the protocol stack (the layers) to the current situation. In RINA, time-variant network conditions are not considered. With respect to the Internet architecture, in the DPS architecture we keep the diverse sets of protocols developed over the last few years and we simply suggest dynamically adapting the set of protocols used. RINA, in contrast, argues that only one protocol is required that can be configured for different layers. It would be desirable that the IRATI project shows that the RINA approach works well in real scenarios, since this would facilitate network operation considerably.

Recursive Network Architecture (RNA)

Similarly to RINA, RNA [343, 391] is a recursive network architecture that reuses a protocol on all layers of the protocol stack. However, in RNA, this protocol is not an actual protocol but a meta-protocol. The meta-protocol offers a static set of basic services, such as discovery, negotiation and template matching. Additionally, it offers hooks to configure the actual data processing functionality and a basic protocol to negotiate the required functionality between two nodes. The functionality is composed from individual modules such as buffering, reordering, and encryption. The combination of those modules can change over time.

RNA and the DPS architecture are similar with respect to the possibility of changing the protocol stack at run time. While in RNA a meta-protocol that offers hooks for arbitrary network processing is used, the DPS architecture has a more traditional approach and simply combines the individual network protocols.

Net SILOs

In contrast to recursive network architectures, the Net SILOS project [110] keeps a traditional understanding of layering. However, unlike in the Internet architecture, where the functionality on each layer is well defined, in Net SILOs, each application can define its own protocol stack. Each protocol stack is composed from a set of services. In addition to data processing, each service provides a set of *knobs* over which other services can interact with the given service. These knobs allow for easy cross-layer optimisation. With cross-layer optimisation it is possible that higher layer services can adapt their behaviour based on the observed characteristics of lower layer services.

Both Net SILO and the DPS architecture provide the possibility that each application can have its own protocol stack. While Net SILO focuses on cross-layer optimisation, the DPS architecture focuses on the run time optimisation of the protocol stack.

4WARD Node Architecture

In the 4WARD project Volker et al. [402, 403] built a node architecture to *let 1000 networks bloom* [402]. Therefore, a framework in which many network architectures can co-exist was developed. This allows for exploiting the possibilities of network virtualisation. The authors introduce the concept of a *Netlet*, which corresponds to a protocol stack. When an application wants to communicate, the node selects the Netlet (the protocol stack) that best fits the requirement of this application. Each Netlet is built from a set of functional blocks that are characterised by a set of properties. This allows the node architecture to evaluate the utility of a Netlet with respect to the application needs. The node can switch between different Netlets at run time.

Both the 4WARD node architecture and the DPS architecture allow for different protocol stacks to co-exist in one node. While the 4WARD node architecture focuses on the selection of entire protocol stacks, in the DPS architecture we focus on building protocol stacks from individual components and adapting those components at run time.

10.6 Conclusion

We tackled the difficult challenge of adding innovative changes to the Internet architecture by proposing a novel, clean-slate network architecture, called the DPS architecture. The DPS architecture allows for the inclusion of novel protocols at run time and even for the dynamic adaptation of the protocol stack during communication. On the one hand, this allows researchers to easily build a system where they can test their newly developed protocol (or integrate their protocols in an already running system). On the other hand, it allows for minimising the protocol overhead when a dedicated protocol stack is used, as opposed to a one-size-fits-all protocol stack.

We introduced a novel self-aware/expressive network node architecture that uses dynamic protocol stacks which are composed of functional blocks. In contrast to network communication in the current Internet with its fixed protocol stacks, we propose using keywords that define the communication requirements for an application. A self-expressive stack builder then identifies a set of suitable protocol stacks that fulfil all requirements for specific network conditions.

Two self-aware/expressive network nodes that want to communicate will select a common protocol stack in a negotiation phase that respects the user-defined application requirements for the current networking condition. This protocol stack can be updated on-the-fly by both nodes whenever the network conditions change. Our EmbedNet platform implements the proposed self-aware network node architecture on an FPGA, where the functional blocks can also be dynamically mapped to either a soft-core CPU or to reconfigurable hardware modules. A self-aware/expressive scheduler adapts the hardware/software mapping of the used functional blocks in order to minimise the overall packet loss and the CPU load caused by packet processing.

We demonstrated the efficiency of our self-aware/expressive network node architecture in two case studies. The first case study used a software-only implementation of our self-aware/expressive network node architecture and could reduce the communication overhead compared to static protocol stacks. The second case study showed the performance benefits of using a self-aware/expressive hardware/software mapping compared to a static hardware/software mapping.

Chapter 11
Middleware Support for Self-aware Computing Systems

Jennifer Simonjan, Bernhard Dieber, and Bernhard Rinner

Abstract The implementation of a distributed self-aware computing system (SACS) typically requires a substantial software infrastructure. A middleware system with dedicated services for self-awareness and self-expression can therefore support the development of SACS applications. In this chapter we show the advantages of using a middleware system as the basis for a self-aware computing system. We identify requirements for middleware systems to support the development of self-aware applications. By providing facilities for communication, decoupling and transparency, middleware systems can provide essential features needed in SACS. We compare different middleware paradigms and their suitability to support self-awareness in distributed applications. We argue that the publish/subscribe paradigm is very well suited for this application area since it supports modularisation and decoupling. Units can be added to and removed from existing applications and may well be reused in new applications. Thus, SACS can be constructed by recombining existing publish/subscribe modules. In addition, we present details of publish/subscribe and introduce our middleware implementation called Ella. We describe how different aspects of a SACS and patterns for self-aware applications can be represented using Ella. We present different communication paradigms in Ella (broadcasting, peer2peer) as well as decoupling mechanisms provided by the middleware. We argue that SACS applications can be developed (i) faster, (ii) more efficiently and (iii) more reliable with Ella. Finally, Chapter 13 presents a self-aware and self-expressive multi-camera application which has been implemented with Ella.

Jennifer Simonjan
Alpen-Adria-Universität Klagenfurt, Austria, e-mail: jennifer.simonjan@aau.at

Bernhard Dieber
Alpen-Adria-Universität Klagenfurt, Austria, e-mail: bernhard.dieber@county.at

Bernhard Rinner
Alpen-Adria-Universität Klagenfurt, Austria, e-mail: bernhard.rinner@aau.at

11.1 Introduction to Middleware Systems

Self-aware computing systems (SACS) [310, 212, 241] are often distributed systems i.e., networked computing systems running on multiple network nodes. Distributed applications are inherently more difficult to design, develop and maintain than applications running on single nodes. This is also true for distributed SACS. Middleware systems are therefore employed in networked application architectures to ease the development by abstracting parts of the networking from the application. These concepts can also be reused in the context of SACS. However, for self-aware computing we see additional requirements for a middleware system.

In this chapter we look at different middleware paradigms and discuss their suitability to support SACS. We show which middleware functions can support architectural primitives of self-aware systems and focus on the publish/subscribe paradigm. Further, we present a specific implementation—the Ella middleware [100]—and describe how its properties can support the development of distributed SACS.

11.1.1 Middleware Basics

A middleware system is a software layer which is located between the operating system and the application. Therefore, it serves as a bridging layer connecting distributed applications. Middleware thus provides similar services as an operating system but for distributed applications rather than for a single computer. The distinction between operating system and middleware functionality is, to some extent, arbitrary. While core kernel functionality can only be provided by the operating system itself, some functionality previously provided by dedicated middleware systems has been integrated into operating systems nowadays. A typical example is the TCP/IP stack for networking, nowadays included virtually in every operating system. As shown in Figure 11.1, every device in a network runs middleware which is located between the application and the transport layer. The logical communication is established between the corresponding layers of different devices and is depicted by the horizontal dashed lines. The physical communication takes place in a vertical manner through the communication stack of the devices and is depicted by the solid lines.

In a distributed system, the applications typically face heterogeneity in multiple dimensions. Such applications run on different physical locations, using different hardware platforms, networking technologies, operating systems or programming languages. A middleware system provides services for a distributed execution of applications and therefore eases the application development. Key aspects are hiding the complexity and heterogeneity of a distributed system and providing a messaging service that enables communication across all platforms within the system. A further important aspect of middleware systems is reusability.

An example of modern heterogeneous systems is a smart environment application. Smart environments are typically composed of different sensors, such as visual, acoustic or infrared sensors [278, 332]. In these networks, each node performs local

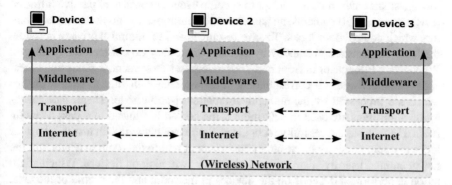

Fig. 11.1 Layered network architecture of multiple devices in a network

actions such as sensing and global actions including communication and cooperation with neighbouring nodes.

Summarised, a middleware system should provide the following functionalities:
(i) hiding distribution, (ii) hiding heterogeneity and low-level details, (iii) providing uniform language and platform-independent interfaces to application developers, (iv) providing common services such as messaging. Applying the design patterns presented in Chapter 5, middleware can support an application in accessing
external sensors and activating external actuators. Depending on the paradigm, middleware can also provide support in implementing self-awareness principles and
self-expressive behaviour. The reference architecture introduced in Chapter 4 can
be implemented by means of a middleware system if it is realised as a distributed
application.

11.1.2 Application Example of a Distributed Self-aware
Computing System

To explain SACS and middleware usage in such systems, we introduce a sensor
network example. A distributed sensor network consists of various sensors and actuators which have some communication capabilities. There is no central entity controlling the network nodes. The actions taken by nodes rely on environmental information and on interaction with local neighbours. Sensors capture certain events (i.e.,
light intensity, temperature) and push them as stimuli into the application. This can
mean either that the sensor uses this data itself to change its own behaviour or that
it informs other nodes in the network (i.e., by a broadcast).

We use a sensor network within a smart environment as an example. It is responsible for controlling the access of persons to rooms within a building. In addition,
it informs the infrastructure within a room of who has entered it, in order to al-

low the smart environment to adapt to specific users. It consists of passive infra-red (PIR) sensors, RGB cameras, an authentication/authorisation node as well as actuators which control door locks. To save resources and to minimise privacy invasion, the RGB cameras are in standby by default, where they do not record any images. Whenever a PIR sensor in front of a door recognises movement, it sends out a message which activates all corresponding cameras. They will start to stream images which are processed by the authentication/authorisation node. There, face recognition algorithms are used to determine if the person is allowed to access a certain room. If a face has successfully recognised, the sensor network will instruct the door locks to open. In addition, the smart environment within the room may adapt to the newly entered user by adapting the lighting or switching on devices. Which adaptation is performed depends on the devices in the room and the profile of the user. Figure 11.2 shows a visualisation of the example application using the reference architecture introduced in Chapter 4.

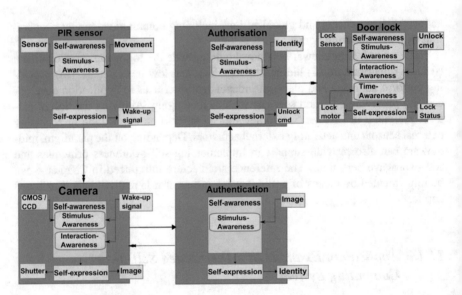

Fig. 11.2 Visualisation of the example application using the reference architecture introduced in Chapter 4

The PIR sensor detects movement in the environment with its external sensors. Upon a movement detection, the self-awareness module of the PIR sensor is informed by the sensors. This module is the prerequisite for self-expression capabilities, which allow a node to react properly to certain events. Thus, the self-expression module is responsible for deciding on how to act based on learnt models. In the case of the PIR sensor, this module issues a wake-up signal in order to inform the appropriate cameras of a movement in their area of interest. The wake-up signal received by a camera triggers its self-awareness module. Since cameras are aware not only of stimuli from the environment but also of stimuli from other agents, they are stimulus- as well as interaction-aware. The self-expression module of a camera reacts properly by starting to stream images to the authentication node. This node performs a face detection and forwards the identity information to the authorisation node, which decides if a door should be unlocked. If so, it issues an unlock command which is received by the door lock node. The door lock node is also time-aware, since it is able to lock an unlocked door after a certain time period. The self-expression module of the door lock node decides on the lock status of the door based on the learnt knowledge.

This application poses several challenges and can profit from SACS design principles. First, there are components with different purposes which change their states based on input from internal and external sources. Second, time dependencies exist in this system (e.g., biometric information must be current in order to open the door for the right person). Third, the various components should not constantly poll their data producers for recent data. Instead, new data should be provided in a push-based manner.

This networked application can be supported by a middleware system. First, multiple networked devices have to be connected to an integrated system. Second, the same application may run in different infrastructural circumstances, i.e., some rooms may not have PIR sensors (requiring the cameras to run permanently) or there may be additional authentication devices such as fingerprint or card readers. Further, different smart environment devices need to react differently to new persons. This requires the application to adapt and scale. Most of those circumstances can be hidden to the application using a middleware system. Finally, networked applications are typically rather communication intensive. Middleware reduces the amount of time required to implement such an application and can also be used to transparently incorporate different networking technologies like WiFi and ZigBee. In addition, middleware systems which provide space decoupling (i.e., hide the exact location of application modules) enable a restructuring of the application to a single node without programming effort.

We will use this application scenario consistently throughout this chapter to explain self-awareness requirements, middleware paradigms and implementation concepts.

11.2 Middleware Requirements

In order to be useful for distributed applications, a middleware system must satisfy several general requirements. First, it must provide meaningful services to the application (e.g., communication and access to data). Second, it must be flexible enough to allow for application restructuring and quick exchange of functionality. Thus, applications should be constructed from several modules or plug-ins. Further, the middleware system should provide mechanisms to decouple individual modules, i.e., make them independent from each other wherever possible. Also, a middleware system should ease application development by hiding complexity regarding hetero-geneous hardware or networks. As an example, the size of a network can be hidden by a middleware system, i.e., a developer should not be concerned if the application runs on two or twenty nodes. Thus, a middleware system must transparently adapt its operation according to the network size. This also includes changing the network size, i.e., when nodes join or leave the network during the application runtime.

Finally, it is required that the middleware system be simple, both in structure and usage. This reduces the effort for understanding and using its concepts in developing applications, makes the execution faster and eases maintenance.

Distributed self-aware computing systems pose additional requirements to a mid-dleware system. While it is not required for the middleware itself to be self-aware, it should support the application in being self-aware. It must provide facilities which make it easy for developers to asses the self-state of the system and to express it in the application behaviour.

We derive feature requirements for a SACS-supporting middleware from the ar-chitectural primitives (see Chapter 2) of a self-aware computing system as shown in Table 11.1.

Table 11.1 Middleware requirements derived from SACS architectural primitives

Architectural primitive	Supporting middleware feature
stimulus-awareness	Ability to push a new stimulus to the application
interaction-awareness	Ability to enable the application to participate in interactions with other application parts
time-awareness	Ability to relate stimuli or events to time instances or knowledge of his-torical/future phenomena
self-expression	Ability to re-route and modify data flows and ability to be split into com-ponents and exchange modules during runtime

For *stimulus-awareness*, the application needs a comfortable way to receive a stimulus. For a video-processing application, this might be a new image coming from a camera. Thus, a middleware system should provide a mechanism to push stimuli into the application or actively notify the application of a new stimulus.

To support *interaction-awareness*, the middleware system should make it easy for the application to participate in interactions like auctions, communication groups and multicast channels. This includes easy dissemination of messages and data and push-based message and data reception. Therefore, nodes need to know with whom they are communicating.

Supporting *time-awareness* means providing time information to the application. This can be done in various ways and on different levels: (i) time-stamping events, (ii) keeping the intervals between events proportional to the real time intervals, (iii) temporally ordering events and (iv) maintaining knowledge of historical data or pre-dicting future phenomena. A middleware system can support each of these levels. For time-stamping the middleware system could by default add a header including a time-stamp to each application message. Keeping the right time intervals and enabling temporal ordering would require the middleware system to buffer messages. To increase knowledge of historical data, the middleware system needs to store past events and enable nodes to access this data. Additionally, many distributed applications should run without any synchronisation. This means that the interacting parties do not need to be actively participating in the interaction at the same time. The middleware system should therefore support unsynchronised communication and interaction (known as time-decoupling).

Self-expression allows a node to take actions appropriate for learnt models or knowledge. A middleware system can support the application in re-routing data flows or by activating and deactivating system parts based on self-awareness. The latter can be realised in middleware systems which support a developer in modularising an application and thus make it easy to exchange modules at runtime.

Summarised, developing a distributed SACS application can be noticeably simplified by the support of a middleware system. At start-up of the application, node discovery is the main issue. A middleware system should take care of discovery of neighbouring nodes to enable later interaction between them and support interaction-awareness. Further, the middleware system should take care of the communication—forwarding of messages/events, communication channel, buffering of messages, etc. Taking care of delivery of messages and events means supporting stimulus-awareness of the application. In the sense of time-awareness, a middleware system could take care of temporal ordering of events (lowest form of time-awareness) or time-stamping of application messages (highest form of time-awareness). Also, self-expression can be supported by a middleware system.

11.3 Middleware Paradigms

Keeping the design issues and application goals in mind, middleware systems have to fulfill certain requirements in order to be useful in developing SACS. We classify middleware systems into the following two approaches: host-centric and content-centric systems.

A host-centric middleware system allows distributed applications to exchange general-purpose messages between specified hosts. This is used to remotely call functions, transport data and send notifications to known hosts. A key aspect of host-centric approaches is that for a single call typically there is a single previously known remote host. Those kinds of systems were originally intended to support only synchronous communication. However, language-specific extensions also allow asynchronous communication. Message delivery is enabled by knowing the host to communicate with and is ensured using reliable message queues.

In content-centric middleware, the nodes or applications are interested in specific data, rather than in the origin or sink of the data. A producer publishes data by making it accessible to the other nodes of the network. Consumers interested in this type of data can then consume the data. Content-oriented systems rely on asynchronous communication, which means that processes do not need to be blocked while waiting for a reply. Using this approach, senders have no guarantee that their messages are read and no indication in the transmission time. The main advantage is that communication can be realised in either a one-to-one or a one-to-many manner, whereby neither producers nor consumers care about that.

11.3.1 Host-Centric Middleware

Host-centric systems are interested in the host they are communicating with. Those systems typically support synchronous communication with a possible extension to asynchronous communication. In synchronous communication, the sender is blocked until a reply message is received. This means that the sender cannot execute any other functions while waiting for a reply. Further, connection overhead is introduced, since each call requires, for instance, marshalling (transformation of the stored object to data which is suitable for transmission). Another drawback is that senders need to know receivers and their locations or addresses in advance. Using asynchronous communication, sender and receiver are loosely coupled, meaning that a sender continues its work after sending a message. When the call returns (i.e., it has been processed on the remote host), the sender can collect the return value.

11.3.1.1 Remote Procedure Calls (RPCs)

A well-known example of host-centric middleware systems is the so-called remote procedure call (RPC) [300]. RPC enables the invocation of a procedure or function on a remote node in the same way local procedures are called (as if they belong to the same process). RPC uses synchronous communication, meaning that the calling process blocks until the remote procedure replies to the call. Changing interfaces of procedures on one side will therefore result in the need for changes of the calls to the procedures, making RPC inflexible. A remote procedure call is a one-to-one communication, i.e., to call a procedure on multiple remote hosts, multiple calls are necessary. RPC calls work similarly to local function calls: the calling arguments are passed to the remote procedure and the caller waits for a response. A remote procedure needs to be uniquely identifiable within the whole network. An example implementation of RPC is the so-called message passing interface (MPI). In MPI, the sender sends a message to a process rather than invoking a function directly by name. This standard defines syntax and semantics of core library routines to ease application development. It supports one-to-one as well as collective communication and is the most widely used model in high-performance computing.

11.3.1.2 Object-Oriented Middleware (OOM)

Object-oriented middleware systems extend the pure method-oriented concepts of RPC to object-oriented programming. Here, not only a functional interface but whole objects are provided to a remote interaction partner. Within an object-oriented language, remote and local objects can ideally be used in the same way.

Prominent implementations of OOM systems are CORBA, JavaRMI, SOAP and .NETRemoting. Common Object Request Broker, known as CORBA [401], allows method calls between application objects of different applications. CORBA hides the complexity of different operating systems and hardware platforms from the application developer, enabling an easier development of distributed applications. An interface definition language (IDL) is used to specify the interfaces represented by objects to the network. Mappings from IDL to programming languages such as Java or C++ are also specified by CORBA. CORBA data such as integers or arrays are passed by value, while CORBA objects are passed by reference. Although CORBA supports flexible data types, data-by-value passing enforces strong data typing. Object Request Broker is the essential concept used by CORBA. ORB allows clients to make requests for services without the knowledge of the server's location or interface. To enable correct delivery of messages and replies, the IIOP protocol is used. The Internet Inter-ORB Protocol (IIOP) enables applications of different nodes and different operating systems to communicate via the Internet.

Java Remote Method Invocation (JavaRMI) [357] performs the object-oriented equivalent of RPC, with support for direct transfer of serialised Java classes and distributed garbage collection. Like CORBA and JavaRMI, .NET Remoting [331] allows an application to make an object available across remote boundaries, which includes different application domains, processes or even computers connected by a network.

11.3.1.3 Service-Oriented Architectures (SOAs)

Service-oriented architectures are a general pattern of structuring applications. Single parts of applications are modelled as independent reusable services. Nowadays, SOA is typically implemented using web technologies, but also other realisations of this concept exist.

Simple Object Access Protocol (SOAP) [152] is a protocol specification for exchanging structured information in the implementation of web services in computer networks. It uses the XML information set for its message format, and relies on other application layer protocols, most notably hypertext transfer protocol (HTTP) or simple mail transfer protocol (SMTP), for message negotiation and transmission. A SOAP service typically provides a description of its methods and structures by means of a WSDL document (web service description language) [75]. This enables developer tools to generate client-side code for accessing a SOAP service automatically, which results in less error-prone and faster development.

However, the use of XML as transport envelope results in a large overhead in messages, thus making SOAP a technology rather unsuitable for embedded software. REST web-services (representational state transfer) [135] have proven to be a lightweight alternative to SOAP. REST uses standard HTTP methods such as GET, POST, PUT, DELETE to communicate to a remote server (i.e., GET is used to retrieve information, POST is used to send a certain payload, and so on). Methods to invoke are identified by URL paths. As an example, GETing *http://www.myservice.com/Items* would return a list of items available while POSTing to this URL may append one item. GETing *http://www.myservice.com/Items-/Item42* returns the item with index 42 while DELETEing would remove the corresponding data package. The payload in REST communication is typically serialised as XML or JSON data (Java Script Object Notation [86]). The advantage of REST is that nearly every connected device and platform has HTTP-accessing capabilities. Today, REST is the main communication paradigm used in smartphone apps.

11.3.1.4 Consequences of Host-Centric Middleware Systems for the Application Example

For a demonstrative explanation, we refer you to the example described in Section 11.1.2. Using a host-centric middleware system, the sensors need to be able to address the nodes they want to communicate with. If they want to share sensed data, they have to directly transmit them to a specific receiver. If synchronous communication is used, the sender is blocked until a response is received. This means that the sending node cannot continue sensing during message exchange. If for instance a PIR sensor wants to notify other parts of the system about detected motion, it needs to know the receivers and their addresses in advance. In this case, an application extension from cameras to cameras and audio sensors would explicitly affect the behaviour of PIR sensors.

The discovery of newly joining nodes is cost intensive, since those nodes have to distribute their address/location through the whole network and need to learn those of neighbouring nodes. Referring to our application example, supporting a changing smart environment within a room is cost-intensive (e.g., various devices are active at different times of the day, which is rather complicated to realise).

An RPC-based implementation of a sensor network first requires a definition of the functional interface of each node (i.e., the procedures available to be called by remote nodes). Further, the application needs mechanisms enabling nodes to find addresses of remote nodes. In a rather static network, this can be achieved by predefined host lists. Dynamic networks, in which nodes may join or leave, require a discovery mechanism or a centralised lookup registry (a previously known host where nodes can register on start-up and look up addresses of other nodes).

In an object-oriented implementation of our example sensor network, the middleware system first needs to define the object structures. Then the node addresses need to be resolved either by registry or pre-definition.

A sensor network which uses *REST* services in communication requires each sensor to run an HTTP server in order to receive incoming stimuli. A well-defined set of REST API calls has to be known already at design-time. To push a stimulus to multiple nodes, multiple HTTP requests have to be performed. This is hardly feasible for communication between all sensor nodes but may make sense in case multiple sensors report to the same sink node.

Summarised, host-centric systems were intended for synchronous operations which require sender and receiver to know each other and to block senders during data transfer. Further, synchronous operations support only one-to-one communication. Neighbour discovery is rather expensive as we can see on the example of REST, where services can only be accessed using fixed URLs. The benefit is the intuitive and transparent usage. Host-centric systems also support SACS in certain aspects. A detailed description of awareness requirements which are fulfilled by host-centric systems can be found in Section 11.3.3.

11.3.2 Content-Centric Middleware

In content-centric middleware systems, sending and receiving messages is not the central architectural viewpoint. Instead, the focus lies on accessing data independently of who is producing and who is consuming it. Typically, a data-oriented middleware system is used to transparently decouple content sources and content sinks (producers and consumers). While also host-centric paradigms are used to exchange data, they typically assume that the location of this data is known in advance (i.e., the URL of a SOAP service). In addition, they typically lack the capabilities to deal with multiple producers (i.e., multiple nodes would host a certain web service) compared to content-centric systems. Content-centric middleware systems can be realised in either a centralised or a distributed way.

11.3.2.1 Blackboard Systems

A blackboard system is a centralised content-centric middleware solution. Blackboard systems are distributed implementations of Linda spaces [8], where multiple processes exchange data via a centralised tuple matching system. In this approach, a producer provides its data by releasing it to a central point in the network (publishing the data on a blackboard). Consumers that are interested in that type of data can simply collect the data from the central point. A prominent example is the so-called tuple space. A tuple space can be interpreted as a distributed shared memory. It provides a repository of tuples that can be accessed concurrently. Producers post their data as tuples in the space, and the consumers then retrieve data from the space that match a certain pattern.

11.3.2.2 Message Queuing

Message queuing is another centralised host-centric middleware paradigm which uses the message queuing principle. Message queuing enables asynchronous communication and provides decoupling of senders and receivers, allowing them to be active at different times. Senders place their messages at a central point in a message queue, where they are stored until the receiver collects them. Since such a queue has restrictions on size, the number of messages in the queue and the size of messages are limited. Message queuing is a programming pattern allowing asynchronous applications in a distributed system to communicate with each other via message queues. A queue manager and the queues reside on a server node. Application nodes can send or receive messages only from message queues belonging to queue managers to which they are connected.

Message queue telemetry transport [26], known as MQTT, implements the message queuing pattern. MQTT is used for the transport of telemetry data and is well suited for wireless sensor networks or mobile-to-mobile communication since it is very lightweight [188]. In MQTT, the sender continuously transmits its sensor data to the message broker. Nodes which are interested in this kind of data collect it from the message broker. MQTT runs on connection-oriented transport (TCP).

11.3.2.3 Publish/Subscribe

Publish/subscribe [126] systems work in a similar way as blackboard systems do, especially if they are realised as a centralised solution. Publishers offer data and subscribers show interest in specific data by subscribing to that type of data. Thereby, publishers publish their data without any knowledge of the sink, and subscribers consume data without knowledge of the source. However, an important difference with blackboard systems is that in publish/subscribe the middleware system pushes data to subscribers whereas in a blackboard they have to pull data on demand. Publish/subscribe can be realised in a centralised as well as a distributed manner. In the centralised solution, a central unit is in charge of matching publishers with subscribers. The distributed solution does not need any central node to publish the data to or consume the data from. The middleware system is in charge of delivering the data to subscribers of matching data type. A detailed explanation of publish/subscribe can be found in Section 11.4.

11.3.2.4 Consequences of Content-Centric Middleware Systems for the Application Example

Referring to the example in Section 11.1.2, a sensor network based on the blackboard concept or on a message queue requires a central node. The blackboard system or the message queues and the queue manager would reside on this central node, which is responsible for delivering messages or data to the appropriate receivers. Thus, every node within the network has to communicate with the central node if it wants to interact with other nodes of the network. The central node may thus become a bottleneck of the system. In the case of message queuing, there has to be a queue for PIR notifications, a queue for camera images, a queue for authorisation information and a queue for user information. If the application is extended to more devices, they can use the already existing queues as input and define their own queues as output.

In publish/subscribe, there may or may not be a central entity, depending on the implementation. In any case, nodes only need to indicate their interest in specific data in order to start receiving it. However, without additional meta-data or middleware functionality, the node may not be able to distinguish between data of the same type received from different publishers. As an example, devices which want to process user information subscribe to this type of data and the middleware system will take care of finding suitable publishers and data transport.

Summarised, in content-centric middleware systems the focus lies on the data itself rather than on the data source or sink. This enables decoupling of data producers and consumers. Further, communication can be done in a one-to-one or one-to-many manner, since the data producers are not interested in who is consuming the data. As with host-centric systems, content-centric systems also support several SACS aspects, which can be found in Section 11.3.3.

11.3.3 Requirements Conformity of Middleware Paradigms

Table 11.2 shows which awareness requirements are fulfilled by the different types of middleware systems.

All middleware paradigms support stimulus-awareness whenever they push data into an application module. Using host-centric middleware, such as RPC, OOM and SOA, nodes have to directly address other nodes in order to push a new stimulus. In content-centric middleware concepts, nodes push their stimuli to the network by offering it to other nodes. Nodes which are interested in the information can then collect it. Since one of the main features of middleware systems is to provide communication and data exchange facilities to distributed applications, this makes them inherently interaction-aware.

Additionally to message exchange, a middleware system also enables nodes to express their selves. Host-centric paradigms explicitly model the data flow (they directly address each host) and can thus easily change this flow based on the self-awareness state. Data-centric approaches provide facilities to change the data produced and consumed by application modules and thus enable a change in the application behaviour. This means both types of middleware support applications in changing their behaviour based on the self-awareness state and are thus self-expressive.

Host-centric middleware solutions are not time-aware, since they typically do not care about time-stamping, temporal ordering of events or time-decoupling. However, content-centric solutions can support time-awareness in different ways. Publish/subscribe and message queuing systems can support time-awareness. In both paradigms, messages can be buffered in queues. This means they can be delivered in right intervals and/or in temporal order. A similar explanation holds for blackboard systems, since the data is stored at a central point where it can be buffered or time-stamped.

Interaction-awareness is supported by all host-centric middleware systems, since they need to know the host they are communicating with. In content-centric middleware systems, senders and receivers do not know each other, since they are only interested in the data which is exchanged. Without knowing other nodes, a sender is not able to interact with one specific node. Thus, content-centric solutions do not directly support interaction-awareness. To enable the application to participate in interactions with other application parts, a content-centric middleware system requires an additional communication channel. This channel could then be used by nodes to interact with specific neighbours.

Table 11.2 Support for awareness primitives in various middleware paradigms

SA/SE Primitive	RPC	OOM	SOA	Blackboard	Message queuing	Pub/Sub
stimulus-aware	✓	✓	✓	✓	✓	✓
interaction-aware	✓	✓	✓	✗	✗	✗
time-aware	✗	✗	✗	✓	✓	✓
self-expression	✓	✓	✓	✓	✓	✓

11.4 Publish/Subscribe

In this section we present the publish/subscribe paradigm in more detail to show how it can support SACS.

The publish/subscribe paradigm is a type of content-centric middleware, defining two different roles: a publisher and a subscriber. Publishers are modules which produce data (e.g., capturing images or fetching data from a specific source) to then pass it on to consumers, the so-called subscribers, which process the data. A module may be publisher and subscriber at the same time, i.e., it processes data from another publisher and publishes the results itself.

Publish/subscribe is a mechanism which allows for elegant decoupling of functional elements within applications. A component for publish/subscribe management takes care of the decoupling. Instead of directly connecting the publisher and subscriber modules, a publisher announces its events and a subscriber can announce interest in certain types of events. The publish/subscribe management takes care of matching published events and subscriptions and is also responsible for delivering the published data to all interested subscribers. This means that publishers provide their data to the network, rather than sending their events to specific receivers.

A key requirement of publish/subscribe is that neither publishers nor subscribers need to be aware of each other. A publisher does not need to keep track of where its data is sent and how many subscribers exist for its events, and a subscriber does not need to care about where publishers are located and where their data is coming from (i.e., the local node or a remote node). All this is transparently handled by the publish/subscribe middleware system.

In the case of a centralised publish/subscribe solution, the publish/subscribe management resides on the central node. We prefer a decentralised solution since centralised networks are hardly scalable. Figure 11.3 illustrates a distributed publish/-subscribe architecture, where a small publish/subscribe management component resides on each node in the network and is responsible for its local modules and for connecting them to remote nodes.

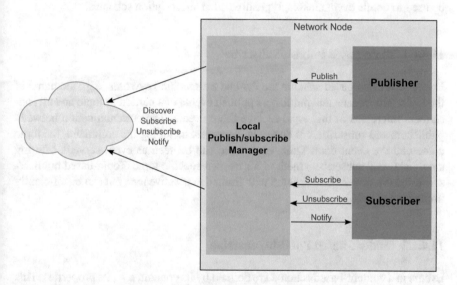

Fig. 11.3 A simple distributed publish/subscribe system. Each node keeps track of other nodes and subscriptions in the network.

The publish/subscribe management is responsible for notifying subscribers whenever data of their interest is published. To enable these notifications in this many-to-many interaction, publish/subscribe relies on asynchronous communication. The messaging model of publish/subscribe systems is very powerful, connecting senders and receivers of messages anonymously. It provides many-to-many and one-to-many interaction, enabling a sender to send its messages to either one or multiple receivers. There is no restriction on the role of a node within a publish/subscribe system. Each node can be a publisher, a subscriber or both.

There are three methods that are typically provided by a publish/subscribe middleware system: subscribe(), unsubscribe() and publish(). The subscriber calls the subscribe() method to show its interest in specific event data and the unsubscribe() method whenever it is no longer interested in this data. The publish() method is called by publishers to notify that there is new event data ready to be offered to subscribers. To enable the matching of published events and subscriptions without the need for the source to know the sink and vice versa, different subscription schemes had been defined. Those are explained in the following section.

11.4.1 Publish/Subscribe Flavours

Since subscribers are usually interested in some specific events, rather than in all events, a way of defining different events is required. Three different schemes can be used to define event classes, typically called subscription schemes.

11.4.1.1 Topic-Based Publish/Subscribe

Events in a topic-based scheme are defined by specific keywords, e.g., the name of the topic. This means that publishers publish events of a particular topic and all subscribers interested in this topic receive these events. The communication between publishers and subscribers is done via messages, including the command, the topic name and the event data. Thus, each topic can be seen as event service, offering a publish() and subscribe() method for its own specific topic. Topic-based publish/-subscribe is a rather static scheme with limited expressiveness, but can be efficiently implemented [126].

11.4.1.2 Content-Based Publish/Subscribe

Events in a content-based scheme are defined by its content, e.g., its properties. This means that subscribers subscribe to content (properties of an object) they would like to receive rather than to a topic name. Events are thus not predefined by keywords, which makes the scheme dynamic. Properties can be internal attributes or meta-data related to events. Subscribers subscribe to an event by typically using a key-value pair. Furthermore, some systems provide a mechanism for event correlation. In these systems, subscribers can subscribe themselves to different combinations of events and are only notified by the event service if this combination of events is provided by the publishers.

In the case of content-based publish/subscribe, the event service provides the subscribe() method with an additional argument. This argument defines the content to be subscribed to and can be either a string or a template object. By using template objects, the subscriber provides an object that it is interested in and the event service notifies the subscriber only on occurrence of events of the same properties. The advantage over topic-based publish/subscribe is that events are no longer bound to fix criteria such as the topic name and that content-based publish/subscribe is highly expressive.

11.4.1.3 Type-Based Publish/Subscribe

In a type-based scheme, subscribers express their interest in events by subscribing to the type of an object. This scheme guarantees type safety at compile-time and encapsulation, which is an advantage in implementation simplicity over the other two schemes. By using public class members, type-based publish/subscribe would transform into content-based publish/subscribe. However, in type-based publish/-subscribe, private class members, which can only be accessed through public methods, are used.

11.4.2 Decoupling

A publish/subscribe system enables decoupling in the following dimensions:

- Space decoupling: Modules do not need to know where they and other modules are located in the network. This means that publishers do not hold any references to subscribers and vice versa. In a visual sensor network (VSN), a publisher of images does not need to care if they are delivered to one or more displays or other modules.
- Time decoupling: Publishers and subscribers do not need to participate in an interaction at the same time. The publisher might for instance publish an event while there is no subscriber connected. Publishers which start after a subscriber can still be matched to an earlier subscription request. In a VSN, cameras may not start up at the same time; still, they must form one distributed application.
- Synchronisation decoupling: Preparing events does not block the publishers, and subscribers can be notified of an event even though they are currently executing another activity. As an example, a publisher of images can capture the next image while the current image is still being delivered to subscribers.

11.4.3 Publish/Subscribe for SACS

Publish/subscribe enables self-awareness and self-expression, making this paradigm very suitable for SACS. As already described above, publish/subscribe supports an application in being stimulus-aware, time-aware and self-expressive.

Subscriptions of data consumers enable private and public self-awareness. Nodes perceive their environment and react properly by subscribing to specific information and data. In tracking, for instance, a camera recognises whenever the object is going to leave the FOV (private self-awareness). The camera may then decide to hand over tracking responsibility to a neighbouring camera (public self-awareness).

By publishing data, self-expression is achieved. Since the publisher forgets about the data as well as the publishing process after publishing, we call it fire-and-forget self-expression.

11.5 Ella: A Publish/Subscribe-Based Hybrid Middleware

In the previous sections we have seen how different middleware paradigms can support different aspects of SACS. Of course, an application does not need to miss self-awareness features which are not supported by the middleware. However, they need to be implemented on the application layer.

In this section, we present a middleware implementation called Ella [100] which is available as an open source project[1]. Ella is designed as a hybrid middleware system combining features of host-centric and content-centric paradigms in order to provide an improved basis for self-aware applications. Ella in its basics is a distributed publish/subscribe middleware system. However, additional features have been added to allow for interaction- and self-awareness. Ella in its core builds on decoupled modules which do not have direct references to each other. This modularised architecture enables application modules to be exchanged at runtime.

11.5.1 Architecture

In this section, we present the architecture of the Ella middleware system. While there are several paradigms for middleware systems, like RPC or OOM, we chose a data-driven publish/subscribe approach. This allows for a high degree of flexibility since it provides decoupling in space and time. In contrast to other publish/subscribe implementations such as [388], Ella is completely distributed without the need for any central coordination. The publish/subscribe implementation of Ella is explained in the following sections.

11.5.1.1 Subscribing

Ella uses type-based subscriptions, which means that a subscriber specifies a certain data type to subscribe to. Additionally, Ella provides the possibility to request a template object. The middleware system will then ask each publisher (which is matching in type) to generate such a template object and will hand it to the prospec-

---

[1] https://ella.codeplex.com/

tive subscriber. The subscriber can then decide whether this specific publisher is accepted or not.

In addition, upon subscribing, the subscriber can decide to exclude remote publishers to obtain only subscriptions from the local node. This can be useful whenever node-specific information is requested, such as the current resource allocation.

Optionally, a subscriber can provide a callback method which is called whenever it is subscribed to a new publisher. The subscriber will be provided with a handle object which can be used to distinguish between multiple publishers but without making them identifiable.

11.5.1.2 Publishing

A publisher publishes events of a certain data type, which is then matched to subscribers by Ella. Similarly to subscribers, publishers can also provide callback methods which are called by Ella for each new subscriber to a certain event. Publishers can use this information to publish certain information to only a subset of subscribers. As an example, some subscribers to images of a camera may send an application message to the publisher asking for a reduced frame rate. The publisher can then publish at a lower rate to those subscribers which have requested this.

11.5.1.3 Realisation of Publishers and Subscribers

Creating a publisher or a subscriber for Ella is a straightforward task. Instead of forcing developers to implement a specific class hierarchy (i.e., subclassing a base class or implementing an interface), Ella uses code annotations to declare a certain class to be a publisher or a subscriber. These code annotations can be reflected by Ella at runtime to detect publishers and subscribers in code libraries. This approach has several advantages. First, it makes it very easy to adapt existing code to run on top of Ella. It basically requires annotating the existing code and adapting the way of data passing to the mechanism provided by Ella. Second, it makes the development of modules more flexible because developers are not bound to any inheritance hierarchy and thus it is easier to integrate Ella into any software architecture. Third, Ella-based code is easily readable and maintainable because the annotations directly inform about what the specific module does.

11.5.1.4 Network Management and Remote Operation

To support a convenient way of developing and deploying software modules for Ella, the middleware system provides a transparent node discovery mechanism which is used to detect any running Ella instances on other nodes in the network. This relieves the developer of the need for managing other nodes in the network. As soon as an Ella instance is detected, it is registered as a known host and it will also be checked for suitable publishers of events requested by local subscribers. With this approach, it is much easier to scale an existing application without having to modify existing code. As soon as Ella detects other instances, it will include them in its operation.

On start-up, Ella tries to first discover other nodes in the network. By default discovery is realised with a UDP broadcast. This broadcast also contains connection information necessary to address this node in the network. However, this may be exchanged with any other suitable discovery provider (e.g., for non-IP compatible media, like ZigBee). Upon reception of a broadcast message, a node will send a unicast answer to the broadcasting node with its own connection information. Thus, each node keeps a local directory of known remote hosts. This directory is used when searching for matching publishers on other nodes.

Whenever a subscriber requests a new subscription, all remote hosts will be queried about matching publishers. If any matches are found, proxy objects at the remote node and stubs at the subscriber node will be created which act as transparent transport points for published event data. A proxy acts as subscriber at the remote node, serialises the event data and sends it to the stub. The stub deserialises it and publishes it as a local publisher for the original subscriber to receive.

The requested subscription types by each subscriber module are cached by the local Ella instance. Whenever a new node is discovered in the network which runs suitable publishers, they will start to deliver their events to the local node. In addition, if a publisher has delayed its start, the node sends out a notification to other nodes that a potential publisher has become available for their subscribers. These two functionalities enable soft time decoupling and do not require publishers and subscribers to start at the same time.

11.5.1.5 Communication

Ella instances on remote nodes use an efficient message structure to exchange data. A binary protocol is used to encode message types and to transport any necessary data. For any given message payload, only nine bytes of overhead are added, one byte for the message type, and four bytes each for the sender node ID and the message ID. For small networks this can be reduced by only using single bytes for the sender node ID.

Besides data communication, Ella provides also a unicast control channel between publishers and subscribers. This can be used to exchange application-specific messages as in RPC systems. Instead of transporting data, control commands can be sent between publishers and subscribers. As an example, an image processing module could instruct the image capturing module to adapt its frame rate. This unicast channel was introduced to enable support of interaction–awareness (see Section 11.2), for which nodes have to know each other in order to enable participation in interactions with specific neighbours.

11.5.1.6 Implementation Details

Ella has been developed in C#.Net. It is capable of running in the open source Mono[2] runtime and can thus be deployed on all major operating systems and many other platforms. Since it is only performing high-level tasks like I/O and management of subscriptions, its overhead compared to a native implementation is very low. In addition, it is easily possible to integrate native code components into any .Net application. Thus, performance-critical applications parts can be written in C++ and be integrated into Ella with low effort. Of course, pre-existing native code can be integrated as well.

Subscription Handling:
On each node, Ella keeps a list of all subscriptions relevant to this node, i.e., all subscriptions where modules of this node are publishers and/or subscribers (this is also true for subscriptions only on the local nodes). Whenever a publisher publishes a new event, all subscribers are found in this list. In the simplest case, this data is delivered to a local subscriber (which is on the same node). For remote subscribers, this list contains the proxy at the publisher node. A proxy serialises the event data and sends it to the receiver node. There, a stub reconstructs the data and publishes it locally for the intended subscriber to receive it. In cases where unreliable transport can be used to deliver data (i.e., where loss of data can be tolerated), a UDP multicast mechanism can be used in order to save communication costs.

Event Correlation:
In some cases, a user might want to indicate that two events are somehow correlated. For example, the image of a camera and the result of a tracking algorithm might correspond to each other. For this case, Ella provides a simple mechanism where a publisher can indicate such a correspondence. This is then delivered to all modules which subscribed to both events. This is also part of Ella's interaction-awareness support capabilities because it enables nodes to handle correlations of the stimuli they are emitting.

11.5.2 SACS-Specific Features in Ella

Ella provides several useful services to an application. It handles discovery and communication, provides a control channel and helps in modularising an application. It enables flexible reconfiguration of an application with its module-based architecture. The use of code annotations to declare Ella-specific code regions makes it very easy for developers to port their existing code. The transparent subscription mechanism of publish/subscribe enables decoupling in space and time and the synchronisation of modules.

---

[2] http://www.go-mono.org

Ella provides specific features for self-awareness and self-expression in various aspects described in Table 11.1. In an application which uses Ella, the nodes can communicate without any concerns on the communication channel. The middleware system cares about message and data delivery and additionally enables application-specific 1:1 messages.

Further, Ella enables context-awareness, reacting properly when the network is congested. It informs the appropriate publisher that it is congesting the network. The application developer can than decide what to do with this information. In the case of a VSN application, the cameras may reduce the resolution or the frame rate to achieve a continuous and reliable image stream without congesting the network.

Ella inherently deals with modularised applications since it connects individual modules by address-free communication mechanisms. Hence, a module does not need to have any reference to another module in order to exchange data. In addition, modules can start and terminate arbitrarily during application runtime. This not only increases robustness to module failures, it also enables the exchange of specific modules at runtime.

11.5.3 Ella in Practice

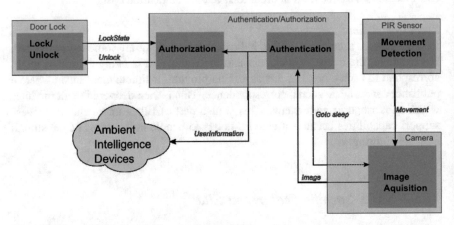

Fig. 11.4 The networked application consisting of PIR sensors, cameras, an authentication/authorisation module, door locks and ambient intelligence devices. Solid lines indicate published data, dotted lines indicate one-to-one messages.

Using the example introduced in Section 11.1.2 we now show how a self-aware application can be realised using Ella. We assume all components of the system to be on individual nodes. Figure 11.4 shows the architecture of the application.

PIR sensors act as pure publishers. Whenever a PIR sensor detects motion it will publish a new event of type *Movement*.

Cameras subscribe to this type of data. As soon as a camera receives this signal, it will start to record and publish the data type *Image*.

The authentication/authorisation module is subscribed to *Image* and performs face recognition. When it has received enough images from a camera, it sends a unicast message back to the corresponding camera telling it to go back into sleep mode and wait for the next *Movement*.

The lock in the door continuously publishes *LockState* information indicating if it is closed or open. The authentication module sends a control message to the lock to open it for an authorised user. Further, it publishes *UserInformation* to all devices which are concerned with adapting to specific user needs. Ambient intelligence devices like lighting controls, tablets, displays and others are subscribed to this data type and use the received events to change their state or display specific information.

Using Ella in this application brings various advantages. First, the application is very flexible in terms of data sources for stimuli input. As an example, also other sensor types may publish *Movement* events which activate cameras. This may be pressure mats on the floor or acoustic sensors. They can be integrated into the system without changing the application. The same holds true for bringing new subscribers into the system, a dedicated application part which consumes published data for archiving purposes.

Second, the possibility to use control messages in addition to publishing data enables feedback channels to publishers and interaction-aware behaviour of the application. The authentication/authorisation module can use this feature to disable a specific subset of cameras after an authentication fails and thus may record additional images of unauthorised persons.

Third, the subscription callback mechanism can be used by modules to dynamically adapt their behaviour. As an example, the authentication module may also subscribe to *Fingerprint* data. If it receives a subscription callback it knows that there is a fingerprint reader in the system. Upon identifying a person it can then wait for fingerprint information before authorising him.

11.6 Conclusion

This chapter provided an overview of distributed self-aware computing systems and and how middleware systems can support their development. Distributed SACS have additional requirements concerning communication, robustness and scalability as well as architectural primitives of self-aware applications.

A categorisation of middleware systems into host-centric and content-centric systems and their abilities to support SACS were presented. The publish/subscribe paradigm was discussed as a solid basis for a self-aware application since it enables decoupling of components along with scalability and robustness. However, it does not support all self-awareness aspects that can be emphasised by a middleware system.

We introduced Ella, a hybrid middleware system based on a distributed publish/-subscribe implementation and enhanced with additional concepts. Ella combines the best of different middleware paradigms in terms of self-awareness support. For our discussion, we used an example application to describe a possible implementation of a distributed SACS.

Part IV
Applications and Case Studies

Part IV demonstrates how self-awareness and self-expression are useful in the three widely different application domains of financial computation, multi-camera networks and active music systems, respectively. Chapter 12 demonstrates how complex financial models can be speeded up using reconfigurable hardware combined with optimisation algorithms. Object tracking in multi-camera networks is the topic of Chapter 13, where autonomous monitoring of each camera in a network is combined with learning mechanisms to adapt its behaviour to changing conditions. Finally, Chapter 14 illustrates how persons without musical skills can influence music in interactive music systems using nature and socially-inspired methods.

Chapter 12
Self-aware Hardware Acceleration of Financial Applications on a Heterogeneous Cluster

Maciej Kurek, Tobias Becker, Ce Guo, Stewart Denholm, Andreea-Ingrid Funie, Mark Salmon, Tim Todman, and Wayne Luk

Abstract This chapter describes self-awareness in four financial applications. We apply some of the design patterns of Chapter 5 and techniques of Chapter 7. We describe three applications briefly, highlighting the links to self-awareness and self-expression. The applications are (i) a hybrid genetic programming and particle swarm optimisation approach for high-frequency trading, with fitness function evaluation accelerated by FPGA; (ii) an adaptive point process model for currency trading, accelerated by FPGA hardware; (iii) an adaptive line arbitrator synthesising high-reliability and low-latency feeds from redundant data feeds (A/B feeds) using FPGA hardware. Finally, we describe in more detail a generic optimisation approach for reconfigurable designs automating design optimisation, using reconfigurable hardware to speed up the optimisation process, applied to applications including a quadrature-based financial application. In each application, the hardware-

Maciej Kurek
Imperial College London, UK, e-mail: mk306@imperial.ac.uk

Tobias Becker
Imperial College London, UK, e-mail: tobias.becker@imperial.ac.uk

Ce Guo
Imperial College London, UK, e-mail: ce.guo10@imperial.ac.uk

Stewart Denholm
Imperial College London, UK, e-mail: stewart.denholm10@imperial.ac.uk

Andreea-Ingrid Funie
Imperial College London, UK, e-mail: andreea.funie09@imperial.ac.uk

Mark Salmon
University of Cambridge, UK, e-mail: mhs39@cam.ac.uk

Tim Todman
Imperial College London, UK, e-mail: timothy.todman@imperial.ac.uk

Wayne Luk
Imperial College London, UK, e-mail: w.luk@imperial.ac.uk

accelerated self-aware approaches give significant benefits: up to $55\times$ speedup for hardware-accelerated design optimisation compared to software hill climbing.

12.1 Introduction

Financial applications often involve computationally complex models, such as numerical solvers. To cope with rapidly changing market conditions, such applications require frequent updating or tuning. In this chapter, we explore the acceleration of financial applications using reconfigurable hardware in heterogeneous clusters. Compared to homogeneous clusters, with possibly hundreds of software cores, heterogeneous clusters using smaller numbers of reconfigurable accelerators can significantly improve performance, power consumption, and energy efficiency.

Our work combines reconfigurability and self-optimisation: the reconfigurable hardware is customised to requirements. Optimising reconfigurable applications often means exploring a very large design space; we develop techniques to automatically explore this space, optimising the reconfigurable design. Many financial applications need to adapt to changing market conditions; we support such dynamic applications, allowing applications to adapt their behaviour using self-awareness and self-expression. We use learning algorithms to tune application algorithms, with fast learning enabled by reconfigurable hardware. We develop tools to automatically explore design spaces, allowing financial applications to self-optimise.

Self-optimisation is another form of self-expression based on self-awareness; self-optimisation is the ability of a system to optimise itself by improving metrics such as performance or power consumption.

The rest of this chapter is organised as follows: Section 12.1.1 gives an overview of our techniques and tools; Sections 12.2, 12.3 and 12.4 respectively describe rule-based currency trading, model-based algorithmic trading and market data feed arbitration, highlighting the link to self-awareness; Section 12.5 goes into more detail on the generic optimisation algorithm accelerated by reconfigurable hardware, applied to a quadrature-based financial application; Section 12.6 summarises and outlines current and future work.

12.1.1 Overview of Techniques and Tools

Research at Imperial College, London has led to multiple approaches to financial computation on heterogeneous clusters, including:

- Rule-based algorithmic trading (Section 12.2);
- Model-based algorithmic trading (Section 12.3);
- Market data feed arbitration (Section 12.4);
- Self-optimisation of reconfigurable designs (Section 12.5).

In each case, our self-adaptive approach gives significant benefits, up to 55 × speedup versus previous approaches. Table 12.1.1 summarises these applications, showing in each case the basis for self-awareness and the architecture for self-expression and the benefits of the approach.

Table 12.1 Overview of techniques and tools

Application	Basis for self-awareness	Architecture for self-expression	Benefits
Rule-based algorithmic trading	Genetic programming and particle swarm optimisation	Hardware pipeline for fitness evaluation	15.7 times speedup of fitness evaluation
Model-based algorithmic trading	Point process and classification	Hardware pipeline for point process encoding, classifier for training and prediction	Real-time performance enabled
Market data feed arbitration	Time-based and count-based windowing	Line arbitrator supporting configurable threshold	4.1 times lower latency for OPRA protocol
Self-optimisation of reconfigurable designs	Surrogate modelling with Expected Improvement metric	Parallel Monte Carlo evaluator	55 times speedup over software hill climbing approach

12.2 Rule-Based Algorithmic Trading

Algorithmic foreign exchange trading involves generating small profits by rapidly trading in currency pairs, e.g., (USD, GBP), but this needs to be timely to be effective, or any trading opportunities identified will have disappeared. Challenges include: the industry is already moving towards hardware acceleration, so naively implementing software approaches on hardware will not be effective; fixed strategies may be easier to compute, but have limited potential in changing market conditions; learning algorithms can adapt to changing conditions, but are complex. To address these challenges, our approach accelerates the complex learning part of the algorithm in hardware.

Figure 12.1 illustrates our approach using self-awareness by combining Genetic Algorithms (GA) and Particle Swarm Optimisation (PSO) and achieving self-expression using an FPGA hardware pipeline for fitness function evaluation (the computationally-intensive part of GA).

In this application, self-awareness allows adapting to changing markets by learning techniques, and self-expression allows rapid fitness function evaluation using an FPGA hardware pipeline; the hardware can adapt to updating trading rules. Our

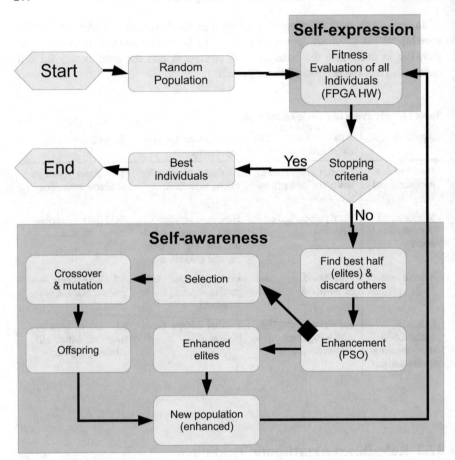

Fig. 12.1 Hybrid Genetic Algorithm (GA) and Particle Swarm Optimisation (PSO), highlighting self-awareness (combining GA and PSO) and self-expression (FPGA hardware pipeline for fitness evaluation), after [139]

approach combines genetic programming and particle swarm optimisation, using self-adaptive genetic operators, accelerating the fitness evaluation in hardware. The benefits of our approach include a $15.7\times$ speedup of fitness evaluation; testing on historical data shows improvement in profitability compared to static trading rules [139]. Our use of genetic programming and particle swarm optimisation can both be seen as examples of nature-inspired learning (Section 7.3).

12.3 Model-Based Algorithmic Trading

Computing near-term predictions of currency exchange rates needs to overcome several challenges: achieving high predictive power while addressing the high computational complexity needed. Our approach is an *adaptive point process model*, allowing fast prediction based on current events, which frequently self-updates the prediction model based on recent history. *Up* and *down* events are encoded as states. A mapping function is generated, enabling fast prediction, mapping states of the last 30 minutes to a 30-minute prediction window.

The approach self-adapts the mapping function by training on the last 2.5 hours of data. Our implementation uses hardware acceleration of both training and prediction, addressing the computational complexity challenge. The benefits are that the new point process is significantly faster than traditional point process models, enabling real-time performance, and supporting real-time adaptation [153, 154]. Our model-based learning can be seen as an example of an online learning algorithm (Section 7.2).

12.4 Market Data Feed Arbitration

Market data feed arbitration poses several challenges: providing very low latency data feeds to financial institutions *without retransmission*, using arbitration to provide some protection against packet loss, exploring the trade-off between latency and reliability. Our solution is hardware-accelerated A/B line arbitration: providing simultaneous low-latency and high-reliability streams, given two input streams, allowing three windowing modes. Our approach is self-adaptive: it can dynamically reconfigure the windowing mode, adapting to changes in requirements from downstream applications. The design is customisable and can be extended to support any protocol. Benefits include a thoughput of > 20 Gbps, with latencies 10× lower than commercial solutions.

Figure 12.2 shows the architecture used for self-expression in the market data feed arbitration. Given two identical market data streams A and B, the architecture synthesises both high-reliability and low-latency streams.

Architectures include a Virtex-5 solution tested with real market data over a network card (2 × 10 Gbps) and a Virtex-6 design implementing TotalView-ITCH and OPRA and ARCA protocols, achieving latencies down to 5.25 ns (low latency) and 36.75 ns (high reliability) [96, 97]. Our market data feed arbitration can also be seen as an example of online learning (Section 7.2).

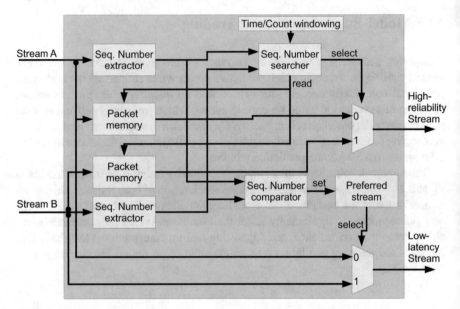

Fig. 12.2 Layout of A/B line arbitration design, used as the architecture for self-expression for our approach to market data feed arbitration, after [97]

12.5 In Detail: ARDEGO — Machine Learning-Based Optimisation of Reconfigurable Systems

We now present the last technique in Table 12.1.1—machine learning-based optimisation of reconfigurable designs—in more detail. Field-Programmable Gate Arrays (FPGAs), and other reconfigurable computing devices, can provide high computational performance but their productive adoption has been hampered due to the long hardware design cycle. Describing designs in low-level hardware description languages is more complex than software approaches and the hardware build process is also time-consuming: synthesising and implementing a single design can take several hours for large modern FPGAs. Advances have been made in high level FPGA design approaches but the problem of long hardware build time remains. In addition to specifying design functionality, a designer is often confronted with multiple non-functional design parameters such as degree of pipelining, memory transfer rates and clock frequency that have to be optimised for performance or power consumption. This optimisation requires tremendous effort in analysing the application to create models and benchmarks, which are then used in the parameter optimisation process. Analytical models have been used to tackle this problem for multi-FPGA systems [174]. Numerical representation is known to have significant impact on resource utilisation and the number of possible FPGA kernels, and therefore design throughput [393]. Optimisation of coefficients for constant multipliers can also yield improvement [202]. Determining optimal stencil configuration is known to be a difficult problem [292], which could highly benefit from automated optimisation.

Ideally, design parameters would be automatically optimised using a calibration-free algorithm; by *calibration-free*, we mean without needing manual setup or tuning, like exhaustive search or hill climbing. The designer would select the target parameters, and run the optimisation algorithm. However, this is impractical when optimising hardware designs as these schemes require hundreds of data points, while even a single test design can take hours to build. Furthermore, the combinatorial space of possible design options grows exponentially with any additional design parameter. Some optimisation techniques like mathematical programming or gradient descent make assumptions about the optimised system, like convexity or continuity of the underlying problem, which have to be verified by the designer; thus the *calibration-free* principle does not hold. To address these issues the parallel Efficient Global Optimisation (EGO) algorithm [198] is used as the base for our new methodology, updated with a novel sampling plan and accelerated with FPGAs to form Automatic Reconfigurable Design Efficient Global Optimisation (ARDEGO). This approach does not require the designer to tune the algorithm or to analytically study and model the underlying effects. Instead it automatically derives performance model of the design using Gaussian Process (GP) regression techniques [301], allowing to avoid unnecessary hardware generations. Furthermore, ARDEGO can be sped up by parallel evaluation of designs on multiple worker nodes. We also accelerate the computationally demanding parts of EGO using FPGAs. Thus, ARDEGO makes use of FPGA accelerators to speed up the optimisation of FPGA designs.

Summarising, our contributions are:

- A novel ARDEGO algorithm that offers *calibration-free* automatic optimisation of reconfigurable designs, based on the parallel EGO algorithm. See Section 12.5.2.
- Acceleration of the parallel EGO algorithm using an FPGA system. The circuit used for acceleration is customisable and portable, allowing for implementation on a range of platforms. See Section 12.5.3.
- An evaluation of ARDEGO using two case studies: a quadrature design for financial computation [393], and a Reverse Time Migration (RTM) design for seismic imaging with multiple parameters [292]. See Section 12.5.4.

12.5.1 Background

The design of reconfigurable designs is a time-consuming process, and automation is highly desired. The process of optimisation is illustrated in Figure 12.3. Such approaches have been explored in [174], [393], [292]. The designer starts with describing the design and coding of benchmarks. Benchmarks evaluate the reconfigurable design's parameters, which is a time-consuming process and often involves hardware generation and software execution. The output of a benchmark is a scalar performance measure: execution time, energy or any other target quality. This measure is called fitness in the optimisation literature. In the case of reconfigurable designs, the *fitness function f* represents the benchmark, and the vector **x** is the parameter setting within the design space $\mathcal{X}$ with D dimensions (parameters). The *fitness function* returns a scalar metric of *fitness*, $f(\mathbf{x}) = y$.

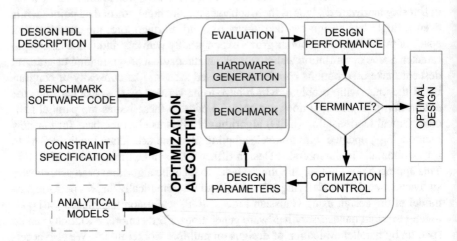

Fig. 12.3 Optimisation approach, after [229]

Once the benchmark code and the design description are ready, the designer specifies the design space and the design constraints. The space specifies the architecture and the physical settings of the FPGA design. If the design fails any of the constraints, the benchmark informs the designer about the error using an appropriate exit code t [230]. The exit codes can be used for prediction of design failure. A design may build successfully or violate one of several constraints such as area, timing or power. Analytical models are often constructed to predict the performance of a design, their development requires a high level of expertise [174], [292]. The optimisation is controlled either manually or by an application-specific algorithm developed to explore the design space.

The optimisation can also be carried by a generic design optimisation algorithm. Again, the designer codes the design, and specifies the design space and constraints. Instead of developing analytical models and application-specific optimisation algorithms an automated *calibration-free* algorithm controls the optimisation flow. The benefit comes from avoiding construction of analytical models and algorithms. Exhaustive search or hill climbing-based search can be applied to a wide range of problems; however, those algorithms are potentially time-consuming, involving multiple hardware generations lasting many days; hence, they are rarely applicable. Optimisation algorithms like mathematical programming or gradient descent rely on assumptions about the design space and the fitness function. If assumptions such as space continuity, convexity or others do not hold, the algorithm is not applicable.

The designer has to verify those assumptions, hence the *calibration free* principle is violated.

Surrogate model-based design optimisation. Using information feedback from design evaluation such as the performance metric y and the exit code t, a surrogate model can be constructed and used by an automatic optimisation algorithm. A *surrogate model* is an approximation technique that is used when the behaviour of the underlying problem is not known or too expensive to measure. Speeding up time-consuming high-level synthesis using surrogate modelling has been explored in [323] using fitness inheritance. Surrogate models are also used in Machine Learning Optimiser (MLO), developed to optimise parameters of reconfigurable designs [230]. Since these models are orders of magnitude faster to evaluate than hardware generation and benchmarks execution, they can substantially accelerate optimisation and enable an automated approach. One remaining challenge is that all of the reviewed algorithms require manual tuning of algorithm parameters, which control its behaviour (**Challenge 1**). Thus they are violating the *calibration-free* principle. In [323] the algorithm's sensitivity to its parameters is analysed; it is shown that they can have significant impact on the optimisation time. The evaluation of the algorithm is based on a set of benchmarks which take unrealistically small amount of time for synthesis. For a realistic evaluation, designs taking hours to synthesis are required. MLO sensitivity to its parameters is not thoroughly investigated in [230], although the evaluated designs are more representative of real problems.

ALGORITHM 1: Surrogate model optimisation algorithm, after [229]

```
/* Sampling                                                              */
Sample designs;
Build initial surrogate model;
while  not Termination condition do
    /* Infill                                                           */
    Collect result of the evaluated design;
    Update the surrogate model;
    Generate the next infill point and evaluate it;
end
```

A typical surrogate-based optimisation algorithm is presented in Algorithm 1. The algorithm starts with sampling of the design space. Sampling is necessary to construct an initial surrogate model. After evaluation of sampled designs has finished, construction of the surrogate model, assessment of the next design to be evaluated and, finally, evaluation of the design follows in an iterative fashion. The design which is evaluated during every iteration is called infill: It can be understood as filling in the gaps in the understanding of the problem; different possible infill criteria exist [204].

Often a random sampling plan will be chosen; however, other plans like the Latin hypercube plan [261] offer better space filling qualities which improve performance

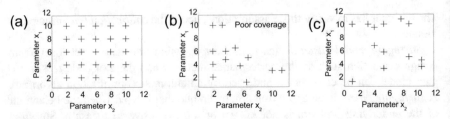

Fig. 12.4 Grid, random and Latin hypercube sampling, after [229]

of the optimisation algorithm [203]. Figure 12.4 presents three different sampling plans. A grid sampling plan would seem ideal, yet it struggles to deal with problems that exhibit periodic behaviour. A random sampling plan has a tendency to oversample areas of space, while undersampling others. This is visible in Figure 12.4 where Random sampling plan oversamples the top right corner and leaves a patch of empty space. The Latin hypercube sampling plan's space-filing properties are nearly as good as the grid plan's, but introduce randomness. All of the plans struggle if only a fraction of the design space produces designs satisfying all constraints (**Challenge 2**). In such a case, the majority of sampled designs are wasted as they violate some constraints.

The GP based surrogate models have been used for optimisation of expensive fitness functions by many researchers [204], [203]. GP is a supervised learning method capable of *regression* [333]. *Regression* is an approach to model dependency between variable changes and scalar output of a system. GPs are often used when a predictive distribution is required instead of a simple point estimate. The goal is to obtain the distribution of the function $\hat{f}$ at input $\mathbf{x}$ given a set of input vectors $\mathbb{X}$, the associated past observations $f(\mathbb{X}) = \mathbb{Y}$, and the *kernel* function $k(\mathbf{x}, \mathbf{x}')$. *Kernel functions* are used to transform the original space to a different space, possibly yielding better predictive power. The standard deviation estimate $\sigma(\mathbf{x})$ is often interpreted as a measure of the uncertainty of the prediction at point $\mathbf{x}$.

Usually, a *surrogate model* consists of a series of regressors, although for constrained optimisation problems they can include classifiers, Support Vector Machines (SVMs) in particular [230], [30]. In such a case, besides modelling the fitness function the *surrogate model* determines which, if any, constraints are likely to fail. The goal of classification is to construct a decision function d allowing prediction of a class label $d(\mathbf{x}) = t$ for an unseen input $\mathbf{x}$. The decision function is constructed based on the observed data $\mathbb{X}, \mathbb{T}$ and the kernel function $k(\mathbf{x}, \mathbf{x}')$. The target labels associated with observations at $\mathbb{X}$ are denoted $\mathbf{T}$. Extension of the surrogate model with a classifier allows for pruning of the design space: the classifier predicts which regions yield valid designs and thus prevents evaluation of unpromising designs, which fail in some constraints such as sufficient accuracy or low resources usage. This is crucial when the design space is large and designs take a long time to evaluate.

Expected Improvement: An Expected Improvement (EI) metric is defined in [204] using a GP surrogate model. The metric tells how much a design $\mathbf{x}$ is likely to improve over the best currently found design fitness $f_{best} = min(\mathbb{Y})$. Given the mean estimate $\hat{f}(\mathbf{x})$ and standard deviation $\sigma(\mathbf{x})$ by the surrogate model GP regression, an improvement $I(\mathbf{x})$ over f_{best} is defined as seen in Equation 12.1. Its expectation $E[I(\mathbf{x})]$ is used as the EI metric. $Y_{(\omega)}(\mathbf{x})$ is a Gaussian random number conditioned on the past observations $\mathbb{X}$, $\mathbb{Y}$, i.e., $Y_{(\omega)}(\mathbf{x}) \sim \mathcal{N}(\hat{f}(\mathbf{x}), \sigma(\mathbf{x}))$, the point distribution returned by the GP regression. The presented equations are defined for a minimisation problem, such as execution time optimisation. They can be easily modified for maximisation problems.

$$I(\mathbf{x}) = [f_{best} - Y_{(\omega)}(\mathbf{x})]^+, \text{where}[\bullet] = \max(0, \bullet) \qquad (12.1)$$

The EI concept is used in the EGO algorithms [198], [204], [30]. EGO's advantage is that the algorithms have no parameters of their own, and hence require no tuning. They have been defined for parallel systems of P worker nodes [198], based on a revised EI metric $E[I^{(\mu,\lambda)}]$. At any given time μ nodes are busy and λ are idle. Although parallelisation decreases optimisation time, the time-consuming $E[I^{(\mu,\lambda)}]$ metric calculation introduces a significant computational burden (**Challenge 3**).

12.5.2 ARDEGO Approach

The ARDEGO is used in the optimisation approach outlined in Figure 12.3. The algorithm is designed to offer automatic *calibration-free* design optimisation, requiring no knowledge of the nature of the optimisation problem. It is based on a surrogate model with an integrated SVM classifier. The key steps of the ARDEGO algorithm are illustrated in Figure 12.5. The algorithm starts to build the initial surrogate model with sampling of the design space, generating hardware for these design samples and evaluating their fitness. After the initial surrogate model is constructed, an iterative process follows where the model is refined with infill. The goal of infill is to find the designs that are most likely to improve over the currently best found design. Hardware is then built for them, their fitness is evaluated and the surrogate model is updated accordingly. The three challenges mentioned earlier are addressed by ARDEGO in the following ways:

1. ARDEGO is based on EGO and has no parameters of its own that require calibration. Hence, it can be used fully automatically;
2. A novel adaptive sampling plan addresses the issue of design spaces with a small number of valid designs. This behaviour is often found in reconfigurable designs;
3. Design evaluation is accelerated by parallelising it on multiple worker nodes. Furthermore, the EI metric which is the computationally demanding component of ARDEGO is accelerated using FPGAs.

Algorithm: The ARDEGO algorithm follows the pseudocode illustrated in Algorithm 2. The algorithm is based on parallel EGO [198], constrained EGO [30] and the new adaptive sampling plan. The optimisation procedure itself is a sequential process that is carried out by a control node. The control node invokes several worker nodes which can build and evaluate designs in parallel. Whenever a worker node finishes evaluation of a design, the control node searches for new infill designs. The goal during infill is to find the set of λ most promising designs $\mathbf{X}_{idle} = [\mathbf{x}_1, \mathbf{x}_2, \cdots \mathbf{x}_\lambda]$, with highest $E[I^{(\mu,\lambda)}(\mathbf{X}_{idle})]$. The infill search on the control node does not block the optimisation on the worker nodes, and multiple worker nodes can finish evaluation at similar times. This often happens during later stages of optimisation when two worker nodes finish evaluation while infill is performed by the control node. The termination condition is checked during infill whenever a worker node finishes evaluation.

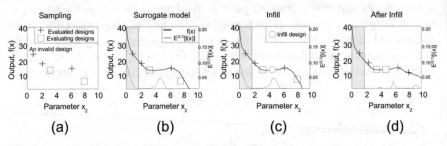

Fig. 12.5 ARDEGO iteration with three worker nodes ($P = 3$): the algorithm starts with sampling (a), after which it moves into the infill state. At first the *surrogate model* is constructed, with two designs being evaluated (b). With one worker node free, the current surrogate model is used to search for infill design (c). In (d) a worker has finished evaluation and a new infill has to be found; the process repeats. After [229]

There are two possible termination conditions: 1. setting a fixed time budget for optimisation, which likely leads to a longer exploitation phase with possibly better optimisation results, which might be desirable if compute time is acceptable; 2. terminating if no design offers minimum expected improvement value [204], for example, 1% or 10%, over f_{best}. The two termination criteria do not violate the *calibration-free* principle; they offer the user a simple choice depending on his compute time constraints.

The *surrogate model* consists of a GP regressor and an SVM classifier. The GP assesses the standard deviation estimate $\sigma(\mathbf{x})$ and the fitness $\hat{f}(\mathbf{x})$ of not evaluated parameter configurations x; the SVMs predicts the class $d(\mathbf{x}) = t$. Regressions are made using the training set obtained from hardware generation and benchmark execution $\mathbb{X}, \mathbb{Y}$, while classification is done using the training set $\mathbb{Y}, \mathbb{T}$. The Gaussian *kernel function* is used for the SVM classifier as it offers universal approximation properties and was shown effective in solving practical problems [33]. The GP regressor uses the anisotropic Gaussian kernel function, which assumes nearby designs fitness to be highly correlated, and the correlation drops off as a function of distance in the design space [301]. The anisotropic property means that different parameters have different impact on the design space.

ALGORITHM 2: ARDEGO, based on parallel EGO [198], after [229]. Suitable global optimisers include CMA-ES [157].

Evaluate the always valid design; /* Adaptive   */
Sample designs using Latin hypercube; /* Sampling   */
Sample random designs (do not evaluate designs predicted to fail constraints);
After every sample, update SVM model;
while *not Termination condition* **do**
  /* Infill   */
  [non-blocking] Collect results of evaluated designs;
  Update the surrogate model;
  Update λ and μ ;
  Generate next λ infill points by $\max E[I^{(\mu,\lambda)}(\mathbf{X}_{idle})]$ using global optimiser;
  Send them to worker nodes for evaluation;
end

The optimisation time is heavily based on the average hardware generation time. To decrease average design evaluation time, hardware generation is terminated when the design is predicted to overmap on resources. It is based on the simple built-in preliminary resource reports, available for both Altera and Xilinx. This yields substantial time savings, as it can be detected as early as 20 minutes after build process initialisation. Hardware is not generated for designs that are predicted to use over 95% of resources. It is unlikely for these designs to meet timing constraints and they are very likely to overmap. This feature does not violate the *calibration-free* principle, as to our best knowledge it is applicable to all modern FPGA platforms.

Adaptive sampling plan: The algorithm involves a two-stage adaptive sampling plan to allow good coverage of the valid space. Initially, if available, an always valid design is evaluated. It is a design representing the most basic configuration which the designer is certain will succeed to generate. This allows ARDEGO to localise the region of space which does not violate any constraints. This is especially important if the region is relatively small. The sampling plan has two stages, the Latin Hypercube and random sampling. Initial sampling is done using a Latin hypercube method with $5 \times D$ designs, which is followed by a random sampling of $5 \times D$ designs. This gives a total of ten designs per dimension, as recommended by [204]. The Latin hypercube sampling provides an initial estimate of the surrogate model while the subsequent random sampling is mainly used to improve the GP regression. This sampling methodology provides better regressions in cases when the valid region is small relative to the design space.

A visualisation of the adaptive sampling plan is presented in Figure 12.6. Initially the always valid design is evaluated (a), followed by a five-design Latin hypercube sampling (b). The subsequent random sampling plan (c) evaluates five randomly chosen designs from the valid area, tuning the shape of the valid region. Figure 12.7 contains the result of sampling and initial classification in the three other sampling plans. Although the identified valid area has a similar shape, the number of designs available for regression is severely limited.

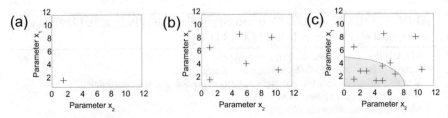

Fig. 12.6 Adaptive sampling plan; gray area indicates valid region. After [229]

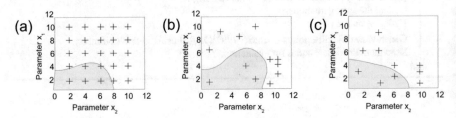

Fig. 12.7 Grid, random and Latin hypercube sampling plans, after [229]

Infill: This step consists of searching the design space to find λ designs that deliver the maximum expected improvement and of their subsequent evaluation on the idle worker nodes. Meanwhile, μ designs $\mathbf{X}_{busy} = [\mathbf{x}_\lambda, \mathbf{x}_{\lambda+1}, \cdots \mathbf{x}_{\lambda+\mu}]$ are being evaluated on the busy nodes. The $E[I^{(\mu,\lambda)}]$ is used as the EI metric. A closed form solution to the metric exists only for $E[I^{(0,1)}]$ and $E[I^{(0,2)}]$. In general, the metric has to be estimated using numerical procedures, in particular computationally demanding Monte Carlo methods [198]. It is a multi-dimensional integral calculation, where the number of dimensions is $\lambda \times D$, which is dependent on the number of idle nodes and the size of the design space. The search space increases being evaluated on the busy nodes. The $E[I^{(\mu,\lambda)}]$ is used as with λ and the calculation of it to discriminate between two designs with similar performance requires a high number of Monte Carlo simulations, further aggravating the problem. The Monte Carlo estimation of $E[I^{(\mu,\lambda)}]$ with *sims* simulations is presented in Algorithm 3. It consists of estimation of the fitness and uncertainty, which is subsequently used to calculate $E[I^{(\mu,\lambda)}]$ using a Monte Carlo technique.

ALGORITHM 3: $E_{MC}I^{(\mu,\lambda)}(\mathbf{X}_{idle})$. For notation purposes busy and idle designs are aggregated in $\mathbf{X} = [\mathbf{x}_1, \mathbf{x}_2 \cdots \mathbf{x}_{\lambda+\mu}]$. After [229]

$\hat{f}(\mathbf{X}), \omega(\mathbf{X}) = \text{surr\_model\_pred}(\mathbf{X})$;
sum=0 ;
for $i \in [0, 1, \cdots sims]$ **do**
$\quad \Big| \quad \mathbf{Y}^\mu_{(\omega)}, \mathbf{Y}^\lambda_{(\omega)} \sim \mathcal{N}(\hat{f}(\mathbf{X}), \sigma(\mathbf{X}));\ \text{sum} += [\min(f_{best}, \mathbf{Y}^\mu_{(\omega)}) - \min(\mathbf{Y}^\lambda_{(\omega)})]^+$;
end
sum = sum $\div$ sims ;

The SVM output is used to disqualify designs which violate constraints. The classification mechanism is incorporated in the surrogate model by returning a Gaussian distribution with f_{best} mean and standard deviation 0 for the designs which are predicted to violate some constraints. As those designs offer no improvement over f_{best}, they do not contribute towards $E[I^{(\mu,\lambda)}]$. Exhaustive search is used over the design space to find designs with maximum expected improvement. It can be a computationally expensive process and its burden depends on the design space size. If the design space becomes too large, a global optimiser is used. The $EI^{(\mu,\lambda)}$ is multi-modal and therefore gradient-based optimisation cannot be used [198], but global optimisers are applicable [157].

12.5.3 Acceleration of ARDEGO

The acceleration of ARDEGO relies on fast evaluation of Monte Carlo estimation of the $E_{MC}[I^{\mu,\lambda}]$. The throughput of software implementations of $E_{MC}[I^{\mu,\lambda}]$ is too low, often involving hours of computation. This is the motivation behind the FPGA acceleration of ARDEGO.

ALGORITHM 4: C-slowed $E_{MC}I^{(\mu,\,\lambda)}([\mathbf{X}^1_{idle}, \mathbf{X}^2_{idle}, \cdots \mathbf{X}^C_{idle}])$, after [229]

```
/* Software                                                                      */
for c ∈ [0, 1, ··· C] do
    |   f̂(Xᶜ) = surr_model_pred(Xᶜ) ;
end
sum = [0, 0, ··· 0] ;
/* Hardware                                                                      */
for core ∈ [0, 1, ··· cores] do
    |   for ii ∈ [0, 1; ··· sims = (C × unroll × cores)] do
    |   |   for c ∈ [0, 1, ··· C] do
    |   |   |   Yᵘ₍ω₎, Yˡ₍ω₎ ~ 𝒩(f̂(Xᶜ), σ(Xᶜ)) ;
    |   |   |   sum += [min(f_best, Yᵘ₍ω₎) − min(Yˡ₍ω₎)]⁺ ;
    |   |   |   ··· ;
    |   |   |   Yᵘ₍ω₎Yˡ₍ω₎ ~ 𝒩(f̂(Xᶜ), σ(Xᶜ)) ;
    |   |   |   sum += [min(f_best, Yᵘ₍ω₎) − min(Yˡ₍ω₎)] ;
    |   |   end
    |   end
end
/* Software                                                                      */
for c ∈ [0, 1, ···C] do
    |   sum = sum + (sum[c] ÷ sims) ;
end
```

The most computationally demanding component of ARDEGO is the Monte Carlo estimates of $E[I^{(\mu,\lambda)}]$. Most reconfigurable designs have between a few thousand and several million possible design parameter settings, meaning that during the exhaustive search $E[I^{(\mu,\lambda)}]$ has to be evaluated for that many points. Even when global optimisers like CMA-ES [157] are used instead of exhaustive search with as little as few hundred Monte Carlo simulations per point, millions of simulations have to be conducted [198]. This means that lines 3 to 5 from Algorithm 3 have to be executed billions of times during each infill. This makes the ARDEGO algorithm extremely computationally demanding. The problem has no data hazards and easily maps onto a streaming computing model. The code from Figure 8 is rewritten, making it suitable for the streaming computing model, as shown in Algorithm 4. The simulation loop from lines 3 to 6 is split into multiple nested loops which can be computed in parallel. The loop is also unrolled, as seen in lines 6 to 15. For maximum throughput, the design is heavily pipelined and the C-slowing technique is applied. The engine presented in Figure 12.8 computes lines lines 2 to 13 of the code listed in Algorithm 4. The surrogate model prediction and the summation from lines 16 to 17 are not accelerated; it is more beneficial to calculate them in software. The surrogate model prediction and accumulation and division of sum are not computationally intensive and thus not accelerated.

The hardware implementation (Figure 12.8) consists of multiple parallel data-paths. Memory (elements) store the data required to generate Gaussian random numbers $Y^{\mu}_{(\omega)}$ and $Y^{\lambda}_{(\omega)}$.

The data is the surrogate estimation of $\hat{f}$ and ω for every point involved in the $E_{MC}[I^{\mu,\lambda}]$ calculation. Two Block RAMs (BRAMs) labelled with "BRAM μ" or "BRAM λ_i" store the mean and standard deviation of the associated $\mathbf{Y}^{\mu}_{(\omega)}$ or $\mathbf{Y}^{\lambda}_{(\omega)}$. The memory element also contains a register storing the f_{best} value. Each data-path generates a Gaussian random number [384] (labelled "GNRG" on the circuit diagram) for every $\mathbf{Y}^{\mu}_{(\omega)}$ and $\mathbf{Y}^{\mu}_{(\lambda)}$. The rest of the data-path performs the simple calculation as defined in $E_{MC}[I^{\mu,\lambda}]$. The results from each data-path are aggregated in the C-slowing buffer and returned from the engine after a programmed number of sims. The maximum μ and λ circuit limit is set equal to P, allowing ARDEGO to support up to P nodes. The design has a customisable number of *data-paths* per number of *cores*, offering throughput and resource trade-off.

12.5.4 Evaluation

The evaluation has two aspects: the acceleration of the algorithm is evaluated and the optimisation time using ARDEGO is compared with alternative algorithms. The accelerated ARDEGO is compared to its software implementation in C. Two case application studies are used to evaluate ARDEGO optimisation time: a quadrature-based financial design with customisable precision [393] and a high performance RTM design with seven parameters [292]. Both involve complex design choices, and have non-trivial constraints. Especially in the RTM case the number of parameters poses an optimisation challenge. ARDEGO is compared with MLO [230] and

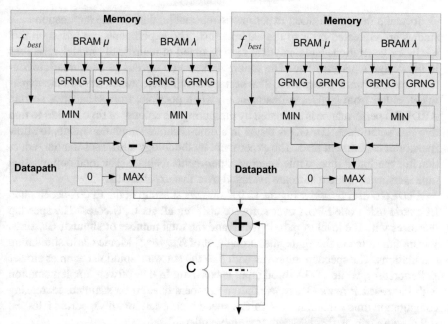

Fig. 12.8 A visualisation of a one core circuit with a loop unrolling factor of 2. In the image the λ and μ values are limited to 2. Each data-path has an associated memory component. After [229]

hill climbing. Hill climbing is a trivial optimisation algorithm which can deal with mixed continuous and discrete design spaces, and makes no assumptions about the fitness function. With multiple restarts, it has a good chance of converging to the optimal design. All algorithms terminate hardware generation if predicted to overuse resources.

Setup: A software implementation of the $E_{MC}[I^{\mu,\lambda}]$ is run on a high performance Intel Xeon x5650 (32 nm, six cores, 2.67 GHz) CPU. The application was compiled using Intel icc compiler with "-O3" and "-fast" optimisation flags. The Abramowitz and Stegun method [2] is used for software Gaussian random number generation, although Marsaglia [256] and others were investigated. The method offers good performance and reasonable quality random numbers.

The ARDEGO FPGA engine is implemented on a Maxeler Max 4 system with an Altera Stratix V GS 5SGSD8 FPGA. On this FPGA, we implement two cores each with an unrolling factor of 5 and one memory unit. For testing purpose the maximum μ and λ are set to 6 to allow different levels of ARDEGO parallelism without regeneration of the circuit. Double precision arithmetic is used throughout the circuit. The frequency of the system is set at 100 MHz. The Cslowing buffer depth was set to 50. Each point estimate of $E_{MC}[I^{\mu,\lambda}]$ uses 200,000 Monte Carlo simulations; this ensures high quality of the estimate. The optimised application designs are targeting a Maxeler Max 3 system with a Xilinx Virtex-6 XC6VSX475T FPGA.

To make the optimisation experiments repeatable, data on the application case studies are collected prior to the experiments. The experiments are based on real hardware generations and analytical models tuned to realistically model variations in the design's performance. The tuning is based on dozens of benchmark executions and hardware generations. The best solution is determined prior to the experiment, either from analytical treatment or, when possible, from exhaustive search. ARDEGO performance is evaluated by measuring the amount of time it takes to find this best solution. ARDEGO is tested in a time-accurate simulation mode, to allow for shorter experiment time. The experiment methodology allows substantial reduction in experiment time, while keeping experiments realistic. The total optimisation time is measured; the results are averaged over five experiments.

ARDEGO acceleration: Figure 12.9 shows the speedup of the FPGA-accelerated EI metric using one FPGA over software utilising all six CPU cores. The speedup increases with the level of parallelism P and the total number of simulations, sims. We are interested in the aggregate throughput of $E_{MC}[I^{\mu,\lambda}]$ Monte Carlo simulation calculations. The speedup improves with P as the software solution becomes slower as P increases, while FPGA throughput is constant. In the software implementation with increased P more Gaussian random numbers have to be sampled, increasing computation time per simulation. FPGA speed is constant for all supported P levels; if P is submaximal, the data-path is simply underutilised.

The speedup also varies with the total number of simulations, sims. For a small number of simulations, instantiation of the FPGA on-chip memory unit takes a relatively large amount of time, yet becomes insignificant as sims increases. With a high number of simulations, the accelerated $E_{MC}[I^{\mu,\lambda}]$ engine offers very high speedup. The difference can be as big as 104.6 million simulations per second for software vs. 1.98 billion for hardware, a 19× speedup. By using bitstreams specialised to specific λ and μ values, the 19× speedup could be achieved for lower P values.

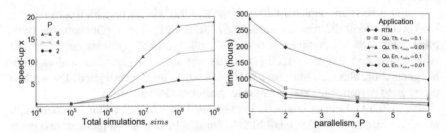

Fig. 12.9 Speedup over software implementation of $E[I^{\mu;\lambda}]$, after [229]

Fig. 12.10 Impact of P on ARDEGO optimisation time, after [229]

12.5.4.1 ARDEGO Optimisation Performance

1) Quadrature-based financial design. In [393] the authors present a precision optimisation methodology for a quadrature-based numerical integration solver for financial option pricing on reconfigurable devices. The design has two benchmarks measuring the throughput and energy consumption of a given configuration. The goal is to find the design offering the highest throughput or the lowest energy consumption given a required minimum accuracy ε_{rms} by optimising the three parameters. The three parameters are mantissa width m_w of the floating point operators, the number of computational cores *cores*, and the density factor df which specifies the density of quadratures used for integral estimation. ARDEGO is evaluated for three different ε_{rms}, which influence the ratio of valid to invalid designs. The design space $\mathscr{X}$ spans 18,000 configurations, with an average hardware generation time of around two hours. The total optimisation time using the application-specific optimisation methodology, including hardware generation and benchmark execution, is 198 hours. Depending on ε_{rms}, the optimal design can be between 10 to 100 times faster and up to 200 times more power efficient than the most basic design.

As seen in Table 12.2, in all of the cases P offers significant speedup in optimisation time, yet even when P = 1 ARDEGO is substantially faster than MLO. The hill climbing algorithm offers very poor performance; in the worst case it takes twice the amount of time needed by the manual approach (502 vs. 198 hours). The advantage of the manual approach is certainty in finding the optimal design, although with high effort [393]. Although ARDEGO does not guarantee that the optimal configuration is found, it can substantially reduce the optimisation time. Seen in Figure 12.10, P offers logarithmic speedup, consistent with [198].

The total ARDEGO computational overhead is around an hour for any ε_{rms}, including infill search and surrogate model training. The search space is small and infill search for the FPGA-accelerated ARDEGO takes around three minutes. Software optimisation takes up to an hour. Compared to the total design optimisation times it is insignificant. The FPGA acceleration becomes relevant for designs with larger numbers of dimensions, as shown in the next example.

2) Reverse Time Migration design: We have also applied ARDEGO to a non-financial design, showing its generality. In [292] the designer faces a problem of optimising seven parameters of a high performance Reverse Time Migration (RTM) design, and there are nearly 27 million possible parameter combinations. The RTM design is used for seismic imaging to detect terrain images of geological structures and involves stencil computation, and most of the parameters are related to balancing communication and computation ratios as well as controlling the internal architectural settings such as parallelism and numerical precision to find an optimal design. Table 12.2 shows hill climbing is unrealistically slow, but ARDEGO can find optimal designs within practical time limits. Multiple cores (P) result in additional speedup of 41%.

Table 12.2 Total design optimisation time given in hours. HC stands for hill climbing. MLO follows [230]. QEn is an abbreviation for Quadrature-based Financial Design energy benchmark, QTh is the corresponding throughput benchmark; after [229]

Design	MLO [230]	ARDEGO P				HC
		1	2	4	6	
QEn (ε_{rms} 0.01)	113	81	53	30	19	226
QEn (ε_{rms} 0.1)	139	120	54	35	27	273
QTh (ε_{rms} 0.01)	122	109	43	30	26	328
QTh (ε_{rms} 0.1)	174	131	72	41	39	502
RTM	*	285	197	123	97	5350

12.6 Conclusion

This chapter presented multiple self-aware financial applications accelerated on heterogeneous clusters involving reconfigurable hardware. We applied several self-awareness and self-expression concepts, including online learning for self-optimisation and rule-based algorithmic trading using nature-inspired learning. The applications include: (i) rule-based algorithmic trading for foreign exchange, combining genetic programming and particle swarm optimisation, using hardware to speedup fitness evaluation by $15\times$ compared to software; (ii) model-based algorithmic trading by using an adaptive point process model, where hardware acceleration enables real-time performance; (iii) adaptive low-latency arbitration, providing very low-latency data feeds to financial institutions, using hardware acceleration to provide simultaneous low-latency and high-reliability output streams, achieving latencies up to $10\times$ lower than commercial solutions; and (iv) automatic optimisation of design parameters, applied to applications including a quadrature-based financial application, achieving up to $55\times$ speedup compared to software hill climbing. Current and future work on ARDEGO include combining surrogate models with embedded platform-specific knowledge, similarly to the early hardware termination principle. Such an algorithm could be used for anything starting with higher-level synthesis and ending with soft processor optimisation.

Chapter 13
Self-aware Object Tracking in Multi-Camera Networks

Lukas Esterle, Jennifer Simonjan, Georg Nebehay, Roman Pflugfelder, Gustavo Fernández Domínguez, and Bernhard Rinner

Abstract This chapter discusses another example of self-aware and self-expressive systems: a multi-camera network for object tracking. It provides a detailed description of how the concepts of self-awareness and self-expression can be implemented in a real network of smart cameras. In contrast to traditional cameras, smart cameras are able to perform image analysis on-board and collaborate with other cameras in order to analyse the dynamic behaviour of objects in partly unknown environments. Self-aware and self-expressive smart cameras are even able to reason about their current state and to adapt their algorithms in response to changes in their environment and the network. Self-awareness and self-expression allow them to manage the trade-off among performance, flexibility, resources and reliability during runtime. Due to the uncertainties and dynamics in the network a fixed configuration of the cameras is infeasible. We adopt the concepts of self-awareness and self-expression for autonomous monitoring of the state and progress of each camera in the network and adapt its behaviour to changing conditions. In this chapter we focus on describing the building blocks for self-aware camera networks and demonstrate the key characteristics in a multi-camera object tracking application both in simulation and in a real camera network. The proposed application implements the goal sharing

Lukas Esterle
Alpen-Adria-Universität Klagenfurt, Austria, e-mail: lukas.esterle@aau.at

Jennifer Simonjan
Alpen-Adria-Universität Klagenfurt, Austria, e-mail: jennifer.simonjan@aau.at

Georg Nebehay
Austrian Institute of Technology, Austria, e-mail: gnebehay@gmail.com

Roman Pflugfelder
Austrian Institute of Technology, Austria, e-mail: roman.pflugfelder@ait.ac.at

Gustavo Fernández Domínguez
Austrian Institute of Technology, Austria, e-mail: gustavo.fernandez@ait.ac.at

Bernhard Rinner
Alpen-Adria-Universität Klagenfurt, Austria, e-mail: bernhard.rinner@aau.at

with time-awareness capability pattern, including meta-self-awareness capabilities as discussed in Chapter 5. Furthermore, the distributed camera network employs the middleware system described in Chapter 11 to facilitate distributed coordination of tracking responsibilities. Moreover, the application uses socially inspired techniques and mechanisms discussed in Chapter 7.

13.1 Smart Camera Networks

Recent advances in technology make cameras almost omnipresent in our everyday life. Cameras are widely used for applications in security, disaster response, environmental monitoring and smart environments, among others. *Smart cameras* have emerged recently by bringing together advances in computer vision, embedded computing, image sensors and networks [415, 341]. They provide image sensing, processing, storage and communication capabilities onboard an embedded device. Smart cameras gained acceptance due to various reasons, including the low system costs, the ability to avoid network loads and the wide range of possible application scenarios. Soon enough, single smart cameras were connected to distributed smart camera networks. They are real-time, distributed, embedded systems that perform computer vision tasks using multiple cameras [45, 342, 336]. Compared to a network of traditional cameras, smart cameras offer the benefit that raw data do not have to be transmitted via the network. These raw data are processed on the sensor platform and necessary results are transmitted. As each camera has reasonable computing and communication capabilities, such a network of smart cameras can be treated as a distributed system for image processing.

In order to allow useful in-network processing of captured imagery, smart camera networks have to deal with various challenges. These challenges vary from highly dynamic behaviour of objects over partially unknown environments to required cooperation with neighbouring cameras on demand. To enable future cameras to deal with these challenges without human interaction, they are required to achieve advanced levels of autonomous behaviour to adapt themselves at runtime and learn appropriate behaviours for changing conditions. A particular challenge is to manage the trade-off of conflicting objectives such as high performance, low resource consumption and high reliability. Not knowing possible changes due to dynamics in the objects' behaviour, the environment or the network itself does not allow a fixed configuration [348]. An adaptive approach only allows a system to change based on predefined options and again only based on expected and foreseen trade-offs.

In this chapter we adopt the concepts of *self-awareness* and *self-expression* as a successful alternative to fixed configurations. We translate these concepts to computational analogies and apply them to smart camera networks [340]. As introduced in Chapter 2, self-awareness refers to the ability of a system to obtain and maintain knowledge about its state, behaviour and progress, enabling self-expression, the generation of autonomous behaviour based on such self-awareness. Combining both self-awareness and self-expression allows for adaptation of a camera's behaviour to changing conditions in an effective and autonomous manner. We therefore implement the reference architecture using the goal sharing pattern. We enhance this pattern with meta-self-aware capabilities to improve our application even further. The fundamental building blocks to achieve self-awareness and self-expression in a camera are effective sensing of the environment, learning models of the camera's state and context during runtime as well as decentralised decision making. The building blocks and their interactions are explained in this chapter in order to build a computationally self-aware and self-expressive camera network. Its features and capabilities are demonstrated with a distributed multi-camera tracking application as an example.

13.2 Object Tracking

Object tracking is an important topic and an extensively investigated subject within the field of computer vision. Research of object tracking algorithms has generated great interest in the computer vision community due to the many fields of application such as automated surveillance, security, object indexing and retrieval, human computer interaction, transportation and activity recognition. Object tracking can be classified as an intermediate-level computer vision task. Low-level information, such as edge segments or corner points, is used to build the desired trajectory in order to provide it for high-level tasks such as object retrieval or activity recognition. Basically, the goal of a tracking task is to recover the motion paths or trajectories of objects using detected object locations, such that each recovered trajectory represents the motion of a single object. Given an object i in frame t (noted as O_i^t) and a set of 'n' candidate objects in frame $(t+1)$ O_k^{t+1} (where $k = 1, ..., n$), the tracking problem consists of selecting an object j in frame $(t+1)$, i.e., O_j^{t+1}, from among all 'n' objects which best matches with object O_i^t. The term object refers to image objects and the best matching is specified by some distance measure.

Unfortunately, the captured visual data is usually contaminated with noise, and missing observations increase the complexity of the tracking task. Objects can appear in different orientations, rotations and shapes depending on how they are oriented towards the camera. Occlusions can occur any time while reliable detections are still needed. Changing lighting conditions may appear in many environments and add another challenge to the tracking algorithms. Therefore, advanced techniques for data association and state estimation are necessary to provide robustness in the generation of the objects' trajectories, i.e., to succeed in the tracking task. Tracking algorithms can be classified according to different criteria, including the type of information the algorithms extract and use, the extraction method, and the matching approach (e.g., deterministic or stochastic). We refer the interested reader to [421] for a survey of tracking algorithms, to [392] for video tracking and to [184] for visual surveillance tracking.

In a multi-camera network the goal is to detect, localise and track moving objects such as pedestrians or vehicles within the fields of view (FOVs) of all cameras throughout the whole network. Besides the aforementioned difficulties, multi-camera tracking poses additional challenges due to dynamics of the environment, uncertainties of camera pose and network topology. In comparison to tracking objects in a single camera, multi-camera tracking requires the cooperation of the cameras to delegate tracking responsibilities. The use of epipolar geometry to fuse the object locations is useful when cameras have overlapping FOVs [209, 319]. However, in many scenarios and applications, cameras do not have overlapping FOVs. In such situations, assumptions about the path followed by the object or its speed [185, 214, 200], use of a motion model [320], assumptions about geometry of the scene [253] or a combination of learning the spatial links between cameras, movement of the object and colour information [298] are useful to track objects successfully.

13.3 Multi-camera Tracking Coordination

Transferring tracking from single cameras to a network of multiple cameras requires coordination of the tracking responsibility. Coordinating this responsibility for tracking an object among multiple cameras is a fundamental issue in online multi-camera tracking. A particular challenge is maintaining the association of objects when they move among cameras, i.e., to re-identify tracked objects among multiple cameras [381]. Once a desired object to be tracked is identified, the camera network has to decide by itself how to track this object through the network whenever this object is present in the observed scenario.

In centralised coordination, the cameras send the traces of the objects within their FOV to a central node which then selects the "best" trace. Various approaches have been proposed to coordinate tracking responsibilities in camera networks [199, 77, 245]. A central component for coordination of tracking responsibilities introduces benefits and drawbacks. On the downside, gathering all information on a single entity adds significant communication overhead and computational load on a single component. Furthermore, a centralised approach limits scalability and introduces a single point of failure, reducing the applicability of this approach in large camera networks. On the upside, a centralised approach may achieve a better tracking performance due to the availability of complete tracking and state information from all sensors in a single node. A comprehensive analysis of the state of the art is given in [322, 280].

In distributed coordination each camera decides on its own when and to whom to hand over the tracking responsibility. This distributes the computational load to the cameras in the network and reduces the communication overhead by avoiding transmitting full state information of all cameras. This makes the network not only highly scalable but also quite robust to failure of single cameras or even to changes in network topology caused by dynamically adding cameras. Various approaches for distributed coordination without centralised control have been presented in the literature [118, 328, 128]. Deriving the handover decision based on possibly incomplete, local information and with a camera's limited resources is still a fundamental challenge in distributed coordination.

In the following we summarise how self-aware and self-expressive approaches are used in order to enable a network of smart cameras to coordinate tracking responsibilities autonomously and efficiently among each other.

13.4 Self-aware and Self-expressive Building Blocks

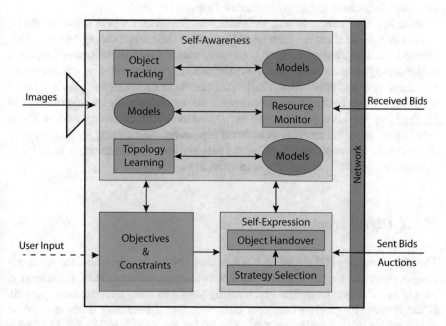

Fig. 13.1 The architecture of a self-aware and self-expressive camera node composed by six building blocks

To endow a camera network with self-awareness and self-expression, dedicated self-aware and self-expressive building blocks are implemented in the individual camera nodes. The different building blocks are able to interact with blocks locally on each camera as well as with other blocks in the network on remote cameras. The blocks depicted in Figure 13.1 aim for a generic and flexible design and implement architectural self-aware and self-expressive patterns as discussed in Chapter 5. Self-aware building blocks, such as *object tracking*, *resource monitoring*, and *topology learning*, are able to monitor its state and behaviour. Utilising online learning techniques allows them to maintain models of their states and the respective behaviour rather than rely on predefined knowledge and rules. The generated models are then used in the self-expressive building blocks (i.e., *object handover* and *strategy selection*) to steer the behaviour of the entire system. The *objectives & constraints* block represents the camera's goals and resource constraints. Both highly influence the other blocks and hence the behaviour and interaction of each individual camera in the network.

The aggregation of individual camera nodes allows the composition of a truly self-aware and self-expressive decentralised camera network. As our employed embedded camera platforms are rather resource-constrained, we focussed the design of each individual building block on resource awareness. Thus, while each building block can be utilising a diverse number of algorithms from computer vision, online learning, distributed coordination and decision making for their implementation, all building blocks have to be able to execute in real time.

In an iterative design process, we initially implemented the interaction awareness pattern in order to allow the cameras to learn about their neighbouring cameras. This allowed the individual cameras to coordinate tracking responsibilities among local neighbouring cameras rather than the entire network. In the next step, we introduced time awareness and enabled each individual camera to also "unlearn" previous learnt information. This becomes important in case the network changes during runtime. In the final step for this application, we introduced meta-self-aware capabilities. These capabilities allow the cameras to trade off exploration, identifying local changes in the network, and exploitation, using the previously learnt information in order to optimise coordination of tracking responsibilities.

13.4.1 Object Tracking

The *object tracking* (OT) building block of each camera is responsible for acquiring images, detecting objects and tracking them within the camera's FOV. Additionally, the OT block transmits images and tracking results to other interested components in the system (for example, the user interface), and if necessary it can update the model of the object during runtime. Identifying the model within the FOV of a camera relates to *private self-awareness* whereas creating and adapting the model of an object corresponds to the *model* in our reference architecture (cf. Chapter 3). The tracking process is described as semi-automatic because the user has to select the desired object to be tracked. After initialising the process, the computer vision tasks run automatically.

The implementation of this self-aware and self-expressive camera network application has been performed in multiple iterations. While the initial version was limited to tracking a single object within the camera network, the final version was capable of simultaneously tracking three people on a network of six smart cameras. The camera network employs appearance-based tracking without using temporal information from previous frames. As soon as the object has been identified within the FOV of the camera, the OT starts tracking the object. By doing so, the OT can adapt the visual representation of the object and improve the internal model if this is desired. This simple approach achieves the necessary computational resource and real-time requirements as well as acceptable accuracy and robustness against dropped frames, occlusions and disappearance of objects.

The approach employs general assumptions about the cameras and visual discriminability of objects. The final implementation of the camera network [362] exploited the static camera assumption and built a detailed model of the background. The static background model enables us to implement a fast and reliable foreground object detector. To perform the association of foreground objects with the desired object to be tracked, colour histograms are compared using appropriate distance metrics. The method is illustrated in Figure 13.2. In a first step, foreground pixels in each camera are identified by comparing the camera image to the background image learned by each camera individually. Foreground constitutes moving objects that comprise objects of interest for tracking. These foreground pixels are then grouped into candidate objects based on their connectedness. In a second step, an association is performed between these candidate objects and a template database that contains the objects of interest (Figure 13.2(d)). To this end, a measure of similarity is employed according to [82] which is interpreted as the confidence of the validity of the association. This is depicted by different colours in Figure 13.2(c).

The association between templates and candidates is established by interpreting the problem of associating templates and candidates as a transportation problem, where the distances between the respective feature vectors are the transportation costs and the goal is to minimise all transportation costs. This problem can be solved optimally by employing the well-known Hungarian algorithm [227]. Additionally, the reciprocal of the transportation cost for a successful association is reported to other components as a confidence value of the current object. In each frame, the *object tracking* block searches for an assignment minimising the overall transportation cost of the system. In this way, a satisfying tracking performance is achieved even in difficult scenarios.

13.4.2 Object Handover

The *object handover* block coordinates the object tracking responsibility in the camera network. We apply a novel market-based handover approach [125]. A more detailed description of this approach is presented in Section 7.4. In this artificial market, the cameras act as traders and treat object tracking responsibilities as goods. For trading purposes, an artificial currency is used. This currency is provided by each object tracking responsibility as some utility over time. This makes these responsibilities worthwhile for cameras to own. The cameras can decide in a self-expressive manner on their own when to "sell" tracking responsibilities to other cameras using single sealed-bid auctions. Employing the Vickrey auction mechanism, which sells the good to the highest bidder for the second highest price, makes truthful bidding the dominant strategy among the participating cameras. Whenever a camera decides to sell an object, it initiates an auction for this particular object by transferring an

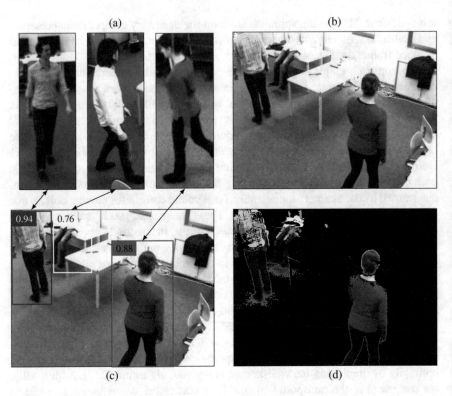

Fig. 13.2 Object tracking approach. (a) Template database images, (b) original image, (c) matching results between foreground objects and objects templates, (d) foreground image

object description to the other cameras. The receiving cameras request the state of the searched object from the *object tracking* block. This means, the *object tracking* block searches within its own FOV for the object and values the object based on detection confidence and visibility. The *object handover* blocks of the cameras return their valuation as a bid. The auctioneering camera selects the highest bidder and transfers the tracking responsibility. An important question for the selling camera is to whom to send the auction invitations. Without any a priori knowledge about the network topology, invitations could be broadcasted to all cameras. By following this policy the "best" camera for taking over the tracking responsibility will receive an invitation (and may respond with the highest bid). However, the BROADCAST policy causes a significant communication and computation load because each camera has to perform an object detection after receiving the invitation.

13.4.3 Topology Learning

If the auctioneering cameras are aware of the potentially "best" cameras in their neighbourhood, this knowledge can be exploited to significantly reduce the overhead. Such topological information can be initially assigned to the cameras or computed by means of multi-camera calibration during the deployment of the camera network. However, in a self-aware manner we learn the network topology by observing the bidding behaviour of cameras over time. Each camera keeps track of its local neighbours and uses artificial pheromones to express the likelihood of a handover to that camera. Whenever a handover has taken place, the artificial pheromone to the succeeding camera is strengthened. If no trading occurs, pheromones evaporate over time. This mechanism enables each camera to deal with network uncertainties and to adapt to changes in its neighbourhood topology caused by addition, removal or failure of cameras or changes in the movement pattern of the objects.

We exploit the learnt neighbourhood topology through three different communication policies for the handover: broadcast auctions to all cameras (BROADCAST), a smooth probabilistic multicast (SMOOTH) and a threshold-based probabilistic multicast (STEP) [125]. The SMOOTH policy sends auction invitations to all neighbours, with probability normalised to the current pheromone level. The STEP policy sends invitations to all neighbours with pheromone level above a certain threshold and to neighbours below the threshold with some (low) probability. Details and formal definitions to these different policies are also given in Section 7.4.1.1. Additionally, it is not only important to whom to advertise object tracking responsibilities, but also when to initiate such auctions. We distinguish between sending out invitations at regular intervals (ACTIVE) or only when the object is about to leave the FOV (PASSIVE). While a PASSIVE schedule ensures we keep track of each object continuously, in comparison the ACTIVE schedule achieves higher network-wide tracking utility as cameras assign tracking responsibilities to the camera "seeing" the object best at all times. Combining the variations in which cameras to invite and when to send out the invitations results in six different self-expressive handover strategies, each of which obviously influences tracking utility as well as communication and computational overhead.

13.4.4 Strategy Selection

While the six handover strategies allow us trade off communication overhead for tracking utility and hence influence the behaviour of the network, selecting a strategy is a difficult decision. The performance of each strategy strongly depends on factors such as the placement of the cameras, the movement patterns of the objects and the object tracking algorithm. In principle, we can follow three approaches for strategy selection: (i) a homogeneous assignment, where all cameras employ the same strategy from deployment time on, (ii) a heterogeneous assignment, where at least two cameras in the network use different strategies from deployment time on, and

(iii) a dynamic selection, where each camera can select its strategy autonomously during runtime.

As discussed in Section 7.4.1.3, we use online learning algorithms, specifically multi-armed bandit problem solvers, within each camera to learn the appropriate strategy for each node during runtime to trade off communication overhead for achieved tracking utility. The bandit solvers balance the exploitation behaviour, where a camera achieves high performance by using its currently best known strategy, with exploration, where the camera explores the effect of using other strategies to build up its knowledge [239, 238]. Dynamic strategy selection leads to a meta-self-aware behaviour of the individual camera nodes and by extension of the entire camera network. This allows the network as a whole to achieve a more Pareto efficient global performance than with any static strategy assignment at deployment time.

13.4.5 Resource Monitoring

Resource monitoring is an important aspect of computational self-awareness, and its main objective is to observe the available resources on the camera nodes. The monitored data is further used to build up models of resource consumption for each task a camera is capable of performing. The knowledge generated by the self-aware *resource monitoring* block is provided to the individual self-aware and self-expressive blocks on the camera. *Object handover* can use this information on the one hand to reason about submitting bids for a new object tracking responsibility and on the other hand to factor in available resources at the time of bidding for its valuation of the object. The *strategy selection* block uses the information from the resource monitor to reason not only about the performance of each task but also about its respective resource consumption. In our network, we currently monitor required processing power, available and allocated memory, and network traffic.

13.4.6 Constraints and Objectives

In order for a camera to become self-aware, it not only requires constraints, objectives and goals but also has to be aware of them. In our system every camera has its own constraints and objectives which it needs to consider for its self-aware and self-expressive operation. Constraints specify some limitations of the available resources (processing, memory and networking). These constraints help us decide on whether to bid for an object, but also help us evaluate the performance of the different strategies. In contrast to hardware-defined constraints, objectives are defined by the user during runtime or the designer of the system before deployment. Objects in our system drive the behaviour of the cameras and specify, for example, some quality of service parameters or certain tasks the cameras should achieve.

13.5 Camera Network Case Study

13.5.1 Camera Network Setup

For our experimental study, we set up a smart camera network in our laboratory at Alpen-Adria-Universität Klagenfurt. The network consists of six cameras, four of them in the laboratory room with overlapping fields of view and two more in the corridor and lounge area, respectively. An illustration of the camera layout is given in Fig. 13.3. Figure 13.4 shows snapshots of the six cameras' FOVs. This heterogeneous network is composed of different hardware platforms. Cameras 1 to 4 are equipped with Atom processors and connected via wired Ethernet. Cameras 5 and 6 are based on Pandaboards equipped with ARM processors and use WiFi for communication. All cameras run standard Linux and our distributed publish-subscribe middleware system to provide a flexible software platform for software development (cf. Chapter 11). The building blocks have been implemented in C++ and C# using the middleware services for communication and control.

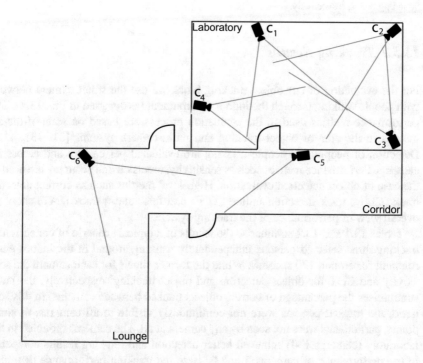

Fig. 13.3 The smart camera network composed of six cameras deployed in an indoor environment. The cameras are depicted by a black camera symbol and their FOVs are indicated by orange lines.

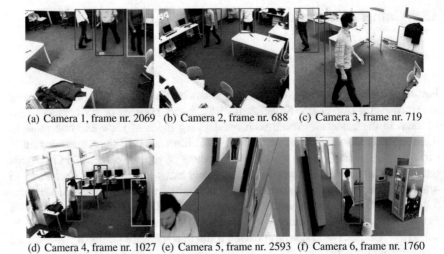

(a) Camera 1, frame nr. 2069   (b) Camera 2, frame nr. 688   (c) Camera 3, frame nr. 719

(d) Camera 4, frame nr. 1027   (e) Camera 5, frame nr. 2593   (f) Camera 6, frame nr. 1760

Fig. 13.4 Snapshots (a) to (f) of cameras 1–6 at different times. The images show the status of object tracking over time; the system tracks three people who are marked by red, blue and green bounding boxes, respectively.

13.5.2 Tracking Results

For the evaluation of our object tracking block we use the smart camera network with people walking through the indoor environment (as depicted in Fig. 13.4). The performance metrics used in the evaluation process are based on state-of-the-art metrics in the area of object tracking and multi-camera systems [31, 431, 422]. Detection of people and people tracking are evaluated per camera and across all cameras. Performance metrics such as sensitivity, precision and accuracy are used in the case of object detection evaluation. High-level metrics such as correct detected track (CDT), track detection failure (TDF) and false alarm track (FAT) show an overall view of performance of the tracking system.

Tables 13.1 and 13.2 summarise the results of a typical scenario of concurrently tracking three selected persons independently walking around in the indoor environment for around 120 seconds. While the former shows for each camera the sensitivity and CDT for object detection and object tracking, respectively, the latter summarises the percentage of correct objects tracked between cameras. In this scenario, the tracked persons were not continuously visible to all cameras; at some points, participants were not seen by any camera at all. The cameras mounted in the laboratory (cameras 1–4) achieved better detection and tracking results compared to the performance of cameras 5 and 6. Here the tracking performance degraded slightly due to changes in lighting and object appearance.

Table 13.1 Performance of object detection and object tracking: single camera evaluation

Camera number	Object detection: Sensitivity	Object tracking: CDT
1	0.78	3
2	0.73	2
3	0.87	3
4	0.76	3
5	0.48	2
6	0.54	2

Table 13.2 Performance of correctly tracked objects between cameras

Camera pair	(1,2)	(1,3)	(1,4)	(1,5)	(2,3)	(2,4)	(2,5)	(3,4)	(3,6)	(4,5)	(5,6)
Percentage	66.67	100.00	100.00	66.67	66.67	66.67	33.33	100.00	66.67	66.67	66.67

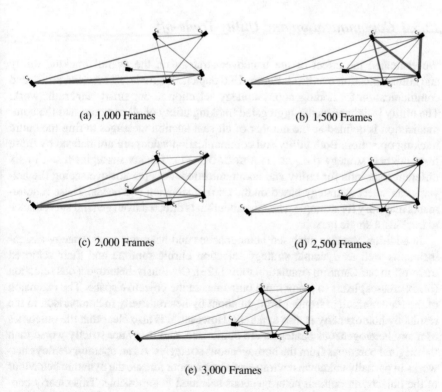

(a) 1,000 Frames

(b) 1,500 Frames

(c) 2,000 Frames

(d) 2,500 Frames

(e) 3,000 Frames

Fig. 13.5 The graph shows the learnt topology at various times of the experiment exploiting the trading behaviour. The thickness of the red line indicates the pheromone level of the link.

13.5.3 Topology Learning

Our self-aware topology learning block builds up a neighbourhood relationship graph locally on each camera. Figure 13.5 shows snapshots of the graph for the entire network at various time steps in a typical test run. The thickness of the red lines indicates the strength of the artificial pheromone deposit on this link, and corresponds to the probability of an object transiting between the connected cameras. Initially, links are created between the camera in the lounge (camera 6) and those in the laboratory (cameras 1–4) due to misdetections of camera 5. Nevertheless, due to the evaporation of the artificial pheromones, these inaccurate links are "forgotten" over time. Furthermore, cameras can deal not only with errors induced by the tracking block but also with changes in the topology due to hardware errors, vandalism, or maintenance when cameras are being removed, added, or moved to a different location. Over time invalid links evaporate and a qualitatively correct neighbourhood graph emerges again.

13.5.4 Communication and Utility Trade-off

We evaluated the effect of the handover strategy on the overall tracking utility and communication overhead. Figure 13.6 depicts this trade-off between utility and communication for homogeneous strategy selection in our smart camera network. The utility is defined as the aggregated tracking utility of all cameras, and the communication is defined as the number of all sent auction messages during the entire tracking operation. Both utility and communication values are normalised by those from the best strategy (i.e., ACTIVE BROADCAST). The six strategies result in six different trade-offs for utility and communication. An operator overseeing the network can select a strategy based on the current situation and needs. These requirements may vary for example when the attention is directed from general surveillance to tracking a single person.

In addition, we also analysed homogeneous and heterogeneous strategy assignments, as well as dynamic strategy selection during runtime and their achieved trade-off in our CamSim simulation tool[1] [123]. Obviously, heterogeneous selection (black crosses) leads to many more outcomes in the objective space. The extension of the Pareto efficient frontier brought about by heterogeneity in comparison to the results by homogeneity is also apparent. However, it is also clear that the outcomes of many heterogeneous strategies are dominated, and many are strictly worse than the original outcomes from the homogeneous strategies. As an operator deploys networks in partially unknown environments and cannot foresee the dynamic behaviour of the objects, an optimal heterogeneous selection is impossible. This clearly benefits the self-expressive behaviour of our cameras, facilitating a dynamic strategy

---

[1] http://www.epics-project.eu/CamSim/

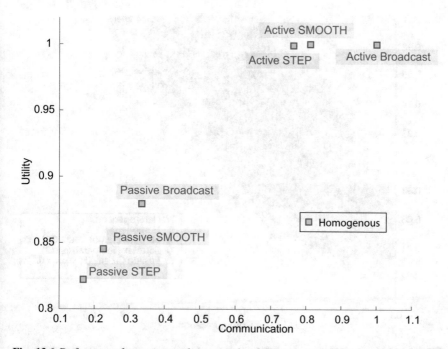

Fig. 13.6 Performance for two exemplary scenarios from our smart camera network showing homogeneous (red and blue squares), heterogeneous (black crosses), and dynamically learned strategy assignments (coloured symbols). The results have been normalised by the maximum value of the ACTIVE BROADCAST strategy and are averages over 30 runs with $1,000$ time steps each [239, 238].

selection. Dynamic strategy selection (coloured symbols) is able to outperform the static (homogeneous and heterogeneous) strategies and to extend the Pareto front.

13.6 Conclusion and Outlook

This chapter presented how self-awareness and self-expression were implemented in a real smart camera network for a person tracking application. By enabling each camera to learn about its environment, its topology, and its performance, the entire network was able to perform continuously well to achieve a common goal. The facilitated building blocks allowed the camera network to face various challenges such as the limitation of available resources of the cameras, the continuous changes in a real scenario and the intrinsic problem of robustness in people tracking. These six different building blocks, encapsulating the entire processing, were embedded into resource-limited smart camera nodes and aggregated into a completely decentralised and thus scalable network.

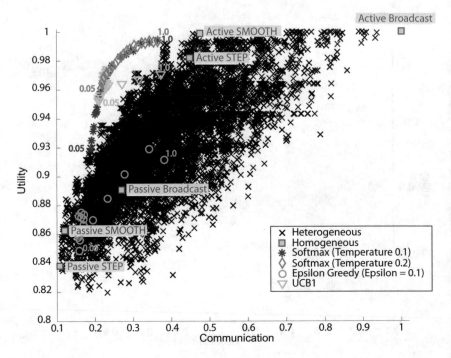

Fig. 13.7 Performance for two exemplary scenarios from our simulation environment showing homogeneous (red and blue squares), heterogeneous (black crosses), and dynamically learned strategy assignments (coloured symbols). The results have been normalised by the maximum value of the ACTIVE BROADCAST strategy and are averages over 30 runs with 1,000 time steps each [239, 238].

Nevertheless, the different blocks are implemented at different levels of self-awareness and self-expression. While the *object tracking* block is only *stimulus-aware*, other blocks such as the *object handover*, *topology learning*, or *strategy selection* are *interaction-aware*, *time-aware*, and *meta-self-aware*, respectively. Merging the different blocks in every single camera allows each camera, and by extension the entire network, to achieve higher levels of self-awareness and self-expression. However, one could still introduce additional blocks or refine existing ones in order to improve the overall performance. An example would be the *object handover* block which currently combines multiple strategies, each one consisting of an auction schedule and a communication policy. In the presented version, the auction schedules are static for all cameras, and communicate either at regular intervals or when the object is at a specified position within the FOV of the camera. In contrast, a camera could learn the best timing for a handover during runtime. In such a setting, the camera could start with an active approach and refine the timing based on the received bid in the advertised auctions.

The concepts of computational self-awareness and self-expression are not limited to camera networks alone. The previous chapter presented another application and the next chapter introduces a third one using real-time interaction between humans playing music. In fact, we are confident that self-awareness and self-expression could serve as an enabling technology for future systems and networks, meeting a multitude of requirements with respect to functionality, flexibility, performance, resource usage, costs, reliability and safety.

Chapter 14
Self-awareness in Active Music Systems

Kristian Nymoen, Arjun Chandra, and Jim Tørresen

Abstract Self-aware and self-expressive technologies may be used to improve user experience in interactive music systems. This chapter presents how the concepts and techniques from the first parts of the book may be exploited to develop better technologies for active music. Nature-inspired and socially-inspired methods, as introduced in Chapter 7, are used to allow music listeners to influence high-level parameters of the music, such as mood or tempo, without requiring the skill of a professional musician. Several of the examples presented in this chapter utilise the same algorithms as presented for the multi-camera networks in Chapter 13, thus demonstrating the broad application domain of these algorithms. The chapter is organised as a discussion of three example systems. First, a mechanism for conflict resolution in a distributed active music system is presented. The second and third example presented take inspiration from the pheromone mechanism used in Ant Colony Optimisation. The approach is first used for continuous classification of the movement patterns of listeners to incorporate adaptive mapping between sensor data and musical output. In the third example the same mechanism enables a system to remember the preferences of a user when navigating in a musical space.

14.1 Introduction

Picture an axis displaying the amount of control possible in different musical devices. At one end of the axis music boxes, media players, software like iTunes or VLC, or online services like Soundcloud or Spotify, are examples of technologies

Kristian Nymoen
University of Oslo, Norway, e-mail: kristian.nymoen@imv.uio.no

Arjun Chandra
Studix, Oslo, Norway, e-mail: arjun@studix.com

Jim Tørresen
University of Oslo, Norway, e-mail: jimtoer@ifi.uio.no

where a user has little control over the musical output. For the music box, control is limited to starting the music. Some of the other technologies allow the user to fast-forward, skip songs, change the volume, etc. Still, the control possibilities are quite limited when compared to the other end of this axis. The opposite extreme includes musical instruments that provide skilled musicians with a wide range of options. Not only the selection of tones and their timing, but an innumerable range of expressive qualities such as the many dimensions of timbre.

The axis in the above example illustrates how musical devices enable varying degrees of control. In between the two extremes, as presented in Figure 14.1, are media players that offer slightly more control over the music, and musical instruments that do not require the hard-earned skills of a professional musician and at the same time do not offer the same range and resolution in terms of control possibilities. Toward the centre of the axis are what we call *active music technologies* that enable users to change the music to which they listen and at the same time allow users to be *listeners* as opposed to *performers*. The process has previously been labelled *active music listening* [145].

Fig. 14.1 Active music allows greater control than traditional media players, but more constraints than a traditional musical instrument.

A range of examples of such technologies have been developed, both commercial products like apps for remixing songs, and music games such as Guitar Hero. Other examples are more experimental research outcomes that enable automatic song selection [37, 117, 257] or time stretching of ongoing music [169, 274] based on the walking tempo of the user.

Higher levels of self-awareness are not a prerequisite for implementing an active music system. However, levels beyond stimulus-awareness allow the system to respond better to the control actions of the user. Take as an example a very simple active music system, where a pre-defined melody is being played, and a user has some interface to control timbral parameters. Let us first consider the scenario in Figure 14.2, where the melody would remain the same without any alterations of the pitch or the duration of the tones. The control actions of the user change variables such as loudness, brightness, roughness, etc. While this too could constitute an interesting active music system, there is no self-awareness or self-expression in the melody playback mechanism (and only stimulus-awareness is present in the filtering process).

In order to implement self-awareness and self-expression in the melody-making mechanism, the system would have to reason about its own musical output and how the output matches the control actions of the user. For instance, if the control input

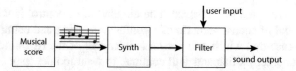

Fig. 14.2 A simple interactive system where the musical score remains the same regardless of the control actions of the user. Self-awareness or self-expression is not exhibited by the melody playback mechanism.

by the user conveys a sad or mellow mood, the melody could be adapted to play fewer tones and a minor scale. Figure 14.3 shows how the short phrase in Figure 14.2 can be altered while retaining some qualities from the original score.

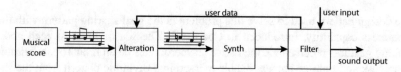

Fig. 14.3 Goal-awareness in the melody playback mechanism may be implemented by allowing the user's data to alter the score, in this scenario, by changing the phrase to lower activity and a minor scale, with the goal of adapting to the activity level of the user.

Three problem domains in active music are presented in this chapter. The framework presented earlier in this book provides solutions to difficult problems. First, in Section 14.2, the challenge of decentralised allocation and circulation of control in a musical band will be presented. Second, in Section 14.3 the problem of adapting the mapping between sensor data and musical parameters to movement patterns of a user is addressed. In Section 14.4, we introduce how flexible pieces of music may be navigated within by a user of an active music system, and how an artificial agent may learn (and subsequently cater for) the preferences of the user. Finally, we conclude and present pointers for future research in this field in Section 14.5.

14.2 Decentralised Circulation of Musical Control

The first example in this chapter considers a group of people playing together, each controlling an active music node. Only one person is in control of the rhythm of the music at any given time. We call this person the leader. However, over time, the responsibility circulates across the group. The crucial point in the circulation of control is for this circulation to happen in a decentralised manner. If many people want to control the same musical feature (e.g., the rhythm), which can be expected when people with little or no musical training participate in a band, a conflict arises. Only one will end up being responsible for controlling the given feature once the conflict gets resolved. Over time however, the responsibility may circulate, as fu-

ture conflicts arise and get resolved. The circulation of control is what we see as
a stylised model of such a scenario of a conflict occurring and being resolved. We
consider the case that whilst one person is in charge of the rhythmical patterns of
the music, the other participants will continue to want to take responsibility of the
rhythm. We find it impractical for such conflicts, which may occur at arbitrary times,
to be resolved centrally. Thus, the decentralised element within a group forms the
crux of the music system we call *SoloJam*. The group is composed of nodes which
have to act in a decentralised fashion. We take inspiration from economics in order
to orchestrate this decentralisation.

14.2.1 SoloJam Algorithmic Details

The design pattern used to solve this problem is the goal sharing pattern with time-
awareness capability, introduced in Chapter 5. The sensor input to each node is
shaking gestures received via the native accelerometer on an iPod Touch. Stimulus-
awareness and time-awareness enable extracting a rhythmic pattern from the shaking gestures. At each point in time, one node will be the *leader* of the group. The
current leader gets to play its rhythmic pattern during the current measure (bar). At
every measure, interaction occurs as the leader also broadcasts an *auction* in which
all other nodes can *bid* in order to become the next leader. We design the system
such that every node can evaluate the *utility* of its rhythmic patterns. The nodes use
the utilities of their respective rhythmic pattern at any time step as their respective
bids. In this manner, at any given time, the node with the highest utility is the leader.
The leader and bidders can also change their respective rhythmic patterns at every time step. The transfer of leadership responsibility happens when a bidder wins
the auction. This happens when the utility of the rhythmic pattern that the leader
is currently playing is lower than at least one of the bids it receives. For the sake
of simplicity in the algorithm, rather than working with complicated rhythmic patterns, each pattern is represented as a bit string parsed from left to right, where a 1
indicates a 'downbeat' and a 0 represents 'not a downbeat'. Algorithm 5 shows the
SoloJam procedure.

The following equation specifies part of the utility function that a node uses to
evaluate its rhythmic pattern:

$$u_i = \frac{c}{(1+aD_l)(1+bT_l)}. \tag{14.1}$$

Here, D_l is the Hamming distance of a node's current rhythmic pattern with respect to the leader's current rhythmic pattern, T_l is the length of time a node has been
playing the solo, i.e., the number of time steps (bars) a node has played rhythmic
patterns as a leader, the coefficient a is a weighting given to D_l, the coefficient b is
a weighting given to T_l, and c is a normalisation constant. In addition to this, two
clauses complete the utility function:

ALGORITHM 5: The SoloJam procedure

---

1. Randomly initialise the rhythmic pattern (ρ_i^t) for the first bar ($t = 1$) for each node i.
2. Randomly choose the leader (ℓ) from within the participating nodes.
3. For each bar (t) until the maximum duration of play:

    a. If $t > 1$, set ℓ as the winner of auction from bar $t - 1$.
    b. For each node:

       • As the leader ℓ in bar t:
         i. Play ρ_ℓ^t.
         ii. Compute utility of ρ_ℓ^t using Equation 14.1 and the associated clauses.
         iii. Generate (by mutation as agent or by shaking iPod as human user) ρ_ℓ^{t+1} (optional in case of agent interaction).
         iv. Broadcast the auction for taking responsibility next (to start in bar $t + 1$).
         v. Accept bids from other nodes.
         vi. If highest bid > leader's utility at bar t:
            A. Set winner of auction as either the highest bidder or a randomly chosen bidder in case of a tie.

       • As a bidder j in bar t:
         i. If $t = 1$:
            A. Compute utility of ρ_j^1 using Equation 14.1 and the associated clauses.
            B. Set bid as utility of ρ_j^1.
         ii. If $t > 1$:
            A. Recompute utility of ρ_j^t (using Equation 14.1 and the associated clauses) if a transfer of responsibility happened in bar t.
            B. Generate (by mutation as agent or by shaking iPod as human user) candidate rhythmic pattern for bar $t + 1$ ($\rho_j^{\prime t+1}$).
            C. Compute utility of $\rho_j^{\prime t+1}$ using Equation 14.1 and the associated clauses.
            D. If utility of $\rho_j^{\prime t+1}$ > utility of ρ_j^t, set $\rho_j^{t+1} = \rho_j^{\prime t+1}$, and set bid as utility of $\rho_j^{\prime t+1}$.
         iii. Bid with the set bid.

---

1. The utility is zero for a bidder node if D_l goes below $\varepsilon \lambda$, where ε is a small percentage of the length of the rhythmic pattern (λ).
2. The utility is *zero* for a bidder node if the node has handed over control to a new leader node in the previous time step.

The first clause penalises perpetual repetition of the same rhythmic pattern by all the nodes of SoloJam, which would be monotonous. The second clause allows for a node to not take over the leadership responsibility directly after it has released it, which may happen otherwise, since the node's rhythmic pattern would already be a slight variation of the new leader that took over the responsibility from this node. Not considering this clause may thus reduce the variations that may occur in the music in the global sense.

The longer a node plays the same rhythmic pattern as a leader, the less its value becomes, this indicating boredom or fatigue, enabled by time-awareness. The node possesses interaction-awareness to enable direct communication with the current leader, and goal-awareness in terms of the Hamming distance between its own and the leader's respective rhythmic patterns. The closer a node can match its rhythmic pattern against the leader's rhythmic pattern, the higher the node values its rhythmic pattern. This remains true as long as the match does not get closer than or equal to $\varepsilon\lambda$.

The leader holds a sealed-bid auction, in particular, the Vickrey auction [400]. The reason for this design choice is that in Vickrey auctions, truthful bidding is the dominant bidding strategy. This means that a bidder can do no better than bidding with the true utility of its rhythmic pattern. Ties in bids, when the bids are higher than the leader's utility, are broken randomly in our system. The sealed-bid nature of the auction requires that the bids should not be public and known only to the bidder and leader.

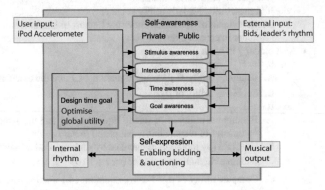

Fig. 14.4 The SoloJam system within the reference architecture from Chapter 4 (see Figure 4.2)

14.2.2 SoloJam Implementation

Figure 14.5 outlines the schematic of the implementation of SoloJam. The current SoloJam scenario has been implemented on a Macintosh computer, in conjunction with iOS devices for human interaction within the scenario. The setup can be broken down into four modules: the Computation module, the Interaction module, the Sound interfacing module, and the Sound synthesis module.

The Computation module is implemented in Python and simulates the auction mechanism and solo utility computation, with a thread representing each node. These threads interface with the Interaction module as well as the Sound interfacing module. The Interaction module can function in two ways. If an artificial agent is to be part of the node, the thread in the Computation module representing this node is made to implement the functionality of the agent in terms of the manner in which this agent comes up with rhythmic patterns. If a human user is to be part of the nodes, mobile devices are used for sensing human motion, and specifically for SoloJam, sensing the shaking of the device (using the built-in inertial sensors of an iPod Touch). The signals from shaking are sent as Open Sound Control (OSC) [416] messages to a thread in the Computation module associated with the device, which are then converted into rhythmic patterns within this thread. The bit strings representing rhythmic patterns are further sent as OSC messages to the Sound interfacing module, together with the utilities/bids (computed within the Computation module) of leader/bidder node rhythmic patterns at every bar.

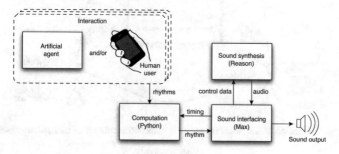

Fig. 14.5 A schematic of the SoloJam implementation

Figure 14.6 illustrates the SoloJam scenario within the context of the aforementioned implementation. It shows three agents or human users participating in the scenario. The rhythmic patterns associated with each participant at various bars are shown. These rhythmic patterns are fed into our auction-based approach for effective participation simulated by the Computation module. As per the rhythmic patterns shown, one possibility for the transfers of responsibility of playing solos is indicated in the figure. A video of SoloJam with three agent-controlled nodes is available online.[1]

The Sound interfacing module is implemented in the graphical programming environment Max [260]. It serves as a control module for the SoloJam scenario, accepting strings of rhythmic patterns, synchronising and converting them to control signals for the Sound synthesis module. The beats received from the nodes are interpreted as musical downbeats, while heuristics are used to determine the intermediate rhythmic patterns. A downbeat is directly mapped to a bass drum. If the downbeats are three quarters of a measure or further apart, a snare drum will be played in between (e.g., if there is a downbeat at the first beat of a four-by-four groove, a snare drum will be put at the third beat). In addition, percussive elements are selected for playback at each 16th timestep with some probability which is predefined but also adjustable by the user. All in all, this results in a diverse musical output, and allows a high degree of dynamic control by the users. An example of a rhythmic pattern generated from a bit string is shown in Figure 14.7. The Sound synthesis module is currently instantiated as a virtual sound module rack in Reason. A drum kit synthesiser module is used for each node. Reason is controlled by the Sound interfacing module through the ReWire interface. MIDI signals are sent to the synthesisers, and

---

[1] http://vimeo.com/67205603

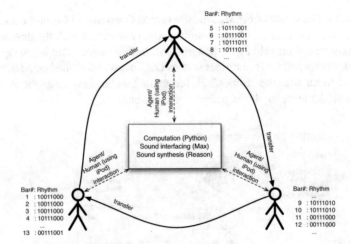

Fig. 14.6 An illustration of the SoloJam scenario within the context of its implementation

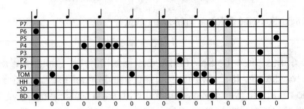

Fig. 14.7 Example of two measures generated by the SoloJam heuristics from the downbeat bit-string [1 0 0 0 0 0 0 0 0 1 0 1 0 0 1 0]. The drums name in the left column are: Bass Drum (BD), Snare Drum (SD), Hi-hat (HH), Rack Tom (Tom), and various bright percussive electronic sounds (P1-7). Each column corresponds to one 16th note.

the audio streams are output back to the Sound interfacing module where mixing and effects are applied.

14.3 Adaptive Mapping in Active Music Systems

One of the many possibilities offered by active music technologies is to enable users to control music through *indirect* control. Indirect control is best understood when compared to *direct* control, which involves making intentional decisions to change specific parameters of the ongoing music. While direct control actions typically use control interfaces such as push-buttons, touch screens, or extraction of *discrete* control actions from, indirect control involves a continuous monitoring of the current state of the user. Typical examples used of such monitoring is measuring heart rate or galvanic skin response in order to estimate the mood of the user. A self-aware active music system may use such information to adapt the ongoing music without the need of direct (intentional) control actions.

In research on new digital musical instruments, great effort has been spent on defining a terminology for *mapping*. Most such devices consist of a separate *gestural controller* and *sound generation unit* [270]. Mapping is concerned with defining a strategy for connecting the outputs of the gestural controller (control variables) to the inputs of the sound generation unit (sound parameters). A wide range of complexity exists for mapping strategies. In its simplest form, one control variable is connected directly to a sound parameter (typically also filtered and scaled) in a *one-to-one* mapping. Other strategies may let a single sound parameter be controlled by a range of control variables (*many-to-one*), or let a single control variable influence many sound parameters (*one-to-many*). Even more complicated *many-to-many* mappings are used. Hunt et al. [189] performed user studies with a variety of mappings, and found that people preferred the more complicated mapping strategies, even when they required some training to master fully. Many-to-many mapping strategies using neural networks have been investigated since the 1990s (e.g., [129, 273]), but also in later work [134]. However, mapping in active music technologies aimed at controlling an entire piece of active music using only a limited set of sensors has not been a major research topic.

This section explores the use of indirect control actions through on-body sensors. More specifically, the hardware in our prototype consists of a sole made from rubber foam, with three Interlink 402 force sensitive resistors (FSR) attached with duct tape; see Figure 14.8. One sensor is placed below the heel, and the two other in the front of the sole on each side, to capture sideways tilting of the foot. The FSRs are connected by cable to our General Purpose Sensor Platform (GPSP) [389]. Through the use of a novel gait recognition algorithm, an active music system is able to distinguish between variations in the gait pattern of the user and adapt the music accordingly. Our algorithm combines concepts from research on mapping, gesture recognition, and active music, and performs a complex mapping between sensor data and individual control parameters. In this way both the musical output and the user's control of it are adapted to the current activity of the user.

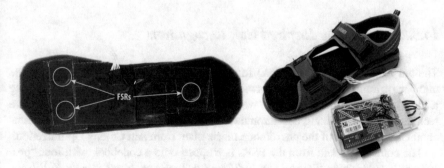

Fig. 14.8 The sensor sole with three force-sensitive resistors attached with duct tape (left) and a sandal connected to the GPSP interface with exposed electronics (right)

14.3.1 Gesture Recognition in Active Music Systems

Hidden Markov Models (HMMs) are among the most popular models for gesture recognition today. HMM has amongst other scenarios been applied to wearable sensing technologies, such as accelerometer-based gesture recognition systems [36, 211, 327, 352]. Pylvänäinen used continuous HMMs to recognise gestures [327], demonstrating effects of data quality on recognition performance, which included both sampling rate and vector quantization. Schlömer et al. [352], using a Wii-controller as input device, estimated the performance of both left-to-right and ergodic HMM on five simple gestures, given the classic recognition pipeline.

HMM has proved to be effective for gesture recognition. However, it still has several shortcomings. HMM-based methods suffer from large time complexity for training and inference [329, 287], and their performance depends highly on the number of training instances [423]. Consequently, using HMM-based systems can be quite difficult in several scenarios. For example, it requires quite a long time and extensive physical work to build a gesture library with the necessary amount of gestures. As a result, user fatigue can greatly reduce the quality of training gesture instances, and lengthy training can be very tedious as well. Moreover, such systems could be weak in cases like real-time gesture recognition on low-end devices.

A variant of HMM, namely segmental HMMs, was applied by Caramiaux et al. in [57] for segmenting and classifying motion capture data from a clarinet player. The method requires a dictionary of short base motion segments and a sequential order of these, and segmentation and classification of continuous motion data is performed by comparing to this predefined dictionary.

We note that dynamic time warping is a very effective technique for recognizing accelerometer-based gestures [287, 247] and has extensively been studied recently. It requires low computational overhead and few training samples, which can be performed on a mobile device with promising performance. However, to the best of our knowledge, both the length and number of templates increase the classification time. Compared with dynamic time warping, we have previously shown that it is possible to reduce the classification time using our approach [367].

14.3.2 Pheromone-Inspired Gait Recognition

The ant learning algorithm (ALA) for gait recognition aims at reducing both the number of training instances and computational overhead, while still maintaining a high recognition accuracy. Our work has proven that ALA is a very effective technique for recognizing accelerometer-based gestures [367]. ALA demonstrates another exploitation of the pheromone mechanism from Ant Colony Optimization.

The continuous data from the FSRs is mapped onto a codebook with four "protostates", as shown in Figure 14.9: *FullRelease, ToePress, HeelPress,* or *FullPress*. Each state is identified by a nearest neighbour algorithm compared against a prerecorded segment of each of the four protostates.

Fig. 14.9 The data is quantised into four states: 1-FullRelease, 2-ToePress, 3-HeelPress, 4-FullPress

Next, the sequence of codebook entries are accumulated in a *pheromone table*. Separate pheromone tables are trained for each type of movement that is to be recognised. Figure 14.10 shows the process for a (very short) *walking* sequence. For every two successive frames, a corresponding increment is found in the pheromone table below. When one state is followed by the same state, the corresponding value along the diagonal is incremented; when the state is different from the previous one; one of the other cells are incremented. A separate pheromone table is trained for each gait to be recognised.

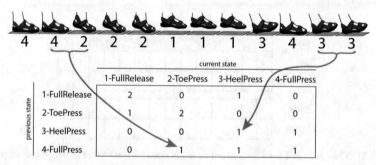

Fig. 14.10 A simplified example of a short walking sequence made from 12 successive states. Below is the corresponding pheromone table.

Recognition of the movement pattern is done by calculating the Euclidean distance between the currently received pheromone table and the set of trained pheromone tables. To allow movement that is not part of the training data (such as standing still), a confidence level has been defined. This level is defined by segmenting the training data (e.g., a 30-second recording of walking data is segmented into five six-second segments), and the distances between each of the pheromone tables of these segments are calculated. Calculating the mean and standard deviation of the inter-table distances, a confidence interval is defined as:

$$\text{confidence interval} = \left[\overline{dist} - \alpha \sigma_{dist} \quad , \quad \overline{dist} + \alpha \sigma_{dist} \right], \qquad (14.2)$$

where $\overline{dist}$ is the mean and σ_{dist} the standard deviation of all distances in the training data, and α is the confidence level, defining the threshold within which a pheromone table must be in order to belong to a given class.

14.3.3 Music Synthesis in Funky Sole Music

The gait recognition system described above has been included in the active music system called *Funky Sole Music* [297]. The sound generation unit of our system has been implemented using Max [260] and Reason [334]. The current setup involves a flexible musical composition with a hierarchy of loops. Figure 14.11 shows how Part A is the main overarching structure, covering 12 measures, and determining the chord progressions in a 12-measure blues scheme for the bass and Wurlitzer instruments. The voicing of the Wurlitzer repeats every two measures, and the most prominent bass tones repeat every measure. The guitar and drum loops also repeat every measure, but these remain unaffected by the chord progressions. In Part B, all the loops change, and repeat every measure. Only one chord (the subdominant) is used in this part.

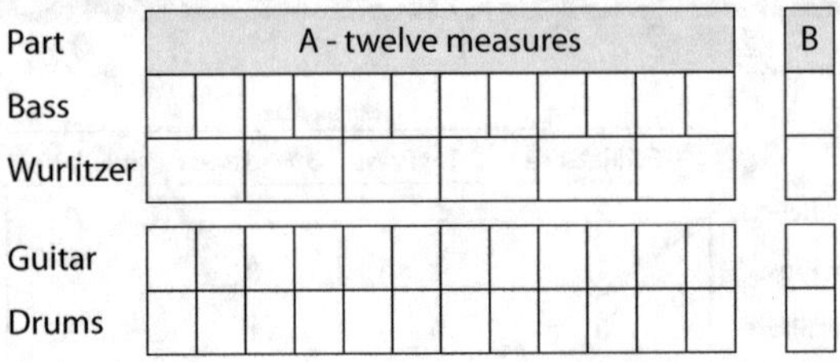

Fig. 14.11 The music in Funky Sole Music is made from a hierarchy of loops. Please refer to the main text for details.

In addition to the main loops, various heuristics determine the note selection of the wurlitzer in Part B, and of the bass instrument in the whole piece, similarly to the *Pheromusic* application described in Section 14.4.

14.3.4 Adaptive Mapping

A traditional way of implementing mapping between a gestural controller and a sound engine is through a direct mapping between the available control variables and sound synthesis parameters, often through a set of layers to create a complex many-to-many relationship. In our system, we want the user to have control over many aspects of the musical output, and a direct mapping between the three force sensors and the musical engine would run the risk of being too simple, even with complex many-to-many mappings. The classification results of the ALA classifier are utilised as indirect control information to change between different states in the system. Such states can involve both states in the music (e.g., changing between different sections of the music) and changing the mapping between sensor data and sound synthesis parameters. In the prototype, three different mapping strategies between sensor data and musical output are available. The strategies are selected by the system based on the current state of the user.

Walking: While walking, the tempo of the music follows the footsteps of the user. All other parameters are predefined and the twelve-measure blues is played according to the approach described in Section 14.3.3.

Tapping back: When tapping the heel to the floor, the tempo control is disabled. A wah-wah effect is enabled and controlled by the data from the force sensors. Also, the bass activity level increases with increased foot activity.

Tapping front: Tapping with the front of the foot takes the music to part B. In this part, a Wurlitzer solo is enabled when the front of the foot touches the floor.

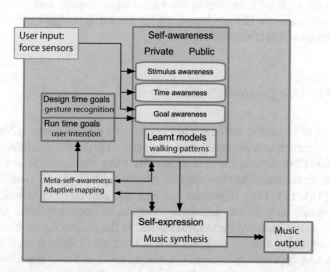

Fig. 14.12 The Funky Sole Music system within the reference architecture from Chapter 4. Meta-self-awareness is realised by changing the mapping strategy when a different input pattern from the user is detected.

Figure 14.12 puts the Funky Sole Music system within the reference architecture laid out in Chapter 4. The system consists of a single node which adapts its internal state according to user input. Thus no interaction awareness is present. The design time goal of the system is to perform correct classifications of the input gesture, and the run time goal is to adapt the music to the user intention by changing the mapping strategy, e.g., to accommodate for tempo changes when walking.

The Funky Sole Music prototype employing the sensor sandal has not yet been subject to quantitative tests, however, qualitative evaluations have been made during the implementation of the system. A presentation video is available online,[2] which clearly shows that the various walking patterns are satisfactorily recognised. Occasional misclassifications as "no pattern" do not cause any harm, as only the predefined patterns cause state changes in the system.

14.4 Pheromone Trails in a Musical Space

Many of the existing interactive music systems use pre-recorded sound files (labelled "samples") in their sound generation. Each sound sample may represent a single sonic object (e.g., a drum stroke) or a large piece of music, such as an entire song. Triggering of individual sounds require skills in both selecting the appropriate sample and to time the onset correctly. Long sound samples on the other hand offer a limited number of options for manipulation. In order to allow non-expert (in a musical sense) users to use an active music system with a certain amount of options for manipulation, flexible musical systems must be developed. This section presents the system *Pheromusic* [296], where users are invited to navigate within a musical space composed of flexible musical scenes.

14.4.1 Flexible Musical Scenes

The concept of generative music has been explored widely since the 1950s, with everything from computer-produced musical scores to real-time algorithmically generated music [79]. Among the approaches taken to generate music in real time in later years are the use of Markov models, genetic algorithms and so-called feature-feedback [114, 115, 176]. Eigenfeldt and Pasquier applied these techniques to develop rhythmic, melodic and harmonic structures in electronic dance music. This approach was not intended for non-musicians, as the control inputs from users are limited to algorithm-specific variables such as selection of individuals in the genetic algorithm. Our approach to generative music is novel in that it is interactive, pop-oriented, and targeted at non-musicians. This means that quite simple control inputs by a user (e.g., tilting an iPhone or walking) must be interpreted by the self-aware system and translated to enjoyable music (self-expression).

---

[2] http://vimeo.com/74219398

We identify three layers encompassing the mechanisms necessary to develop flexible, pop-oriented, interactive music, all three of which are present in the application *PheroMusic*:

The upper layer: Involves overarching structures in the music. Examples are mechanisms for navigation between different parts of the music and mechanisms for controlling the overall dynamics in a piece.

The middle layer: Involves generation of shorter, often repeating, patterns in the music. Some examples are melodic, harmonic, and rhythmic patterns.

The lower layer: Involves the sound generation itself. While mechanisms in the middle layer defines, for instance, which tone is to be played, lower layer mechanisms could be selection of a sound sample or a particular sound synthesis technique.

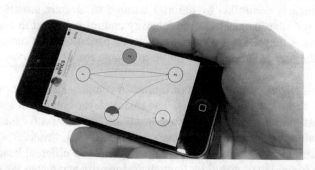

Fig. 14.13 A picture of the *PheroMusic* app for iOS

In PheroMusic, which was first prototyped and described in [296] and later released on the iTunes App store in August 2014 [222] (see Figure 14.13), the lower layer is composed of a set of basic building blocks available in the sound programming environments Max [260] and Pure Data [326]. Examples of the building blocks are wave generators such as sine, triangle and sawtooth, noise generators, filters, and envelopes. All the sounds in our example are made using the same basic principle: A white noise signal, multiplied with an envelope, passed through one or more resonant filters. Figure 14.14 shows this, both as a basic principle and with screen captures from the bass drum and melody implementations in Max.

A range of parameters are defined for each musical scene. Some examples are:

- filter gains, bandwidths, and centre frequencies
- envelope attack and decay slopes
- thresholds for triggering percussive/melodic tones
- a range of available tones to be played
- harmonic progressions
- tempo
- number of subdivisions in the triggering mechanism

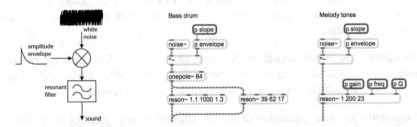

Fig. 14.14 The basic principle and two examples of the triggering tone mechanism in PheroMusic. The red Max objects indicate parameters varied according to user input and/or heuristics and include slope parameters for the amplitude envelope, as well as filter gain, centre frequency and bandwidth.

Among these parameters some have fixed values, while a range of possible values, ultimately controlled by the user through an abstract *intensity* parameter, is defined for other parameters. The intensity control is limited to a simple tilt control which is mapped to a wide range of sound parameters. As such, the framework allows the user to influence the music to a large extent using only simple control actions.

Our implementation consists of four main components: percussion, bass drone, melody, and digital audio effects. The middle layer contains a heuristic that defines possible sequences of the above-mentioned parameters for these components. The result is different harmonic progressions, melodies, "trimbral" qualities, etc. A probabilistic approach is used for triggering tones on different beats, and the user controls triggering probabilities through the intensity parameter. We provide a video example[3] which shows an implementation with five musical scenes. The implementation in the video is on a laptop, using a mobile device for control gestures.

14.4.2 Pheromone Mechanism

As mentioned in the previous section, the upper layer is concerned with overarching structures in the music. In our system these correspond to transitions between musical scenes, connected through a *transition graph*. An example of such a graph is shown in Figure 14.15. It is a directed graph with loops, where nodes in the graph represent scenes. Each node in the graph connects with every other node and itself, via a link. The links in the graph represent transitions from one scene to the other. These links have an associated strength value, the thickness indicating this strength.

The graph is updated and traversed at runtime using in a method inspired by the *pheromone mechanism* of Ant Colony Optimization (ACO) [81, 107]. ACO takes inspiration from the foraging behaviour of ants in nature. As ants in nature search for

---

[3] http://vimeo.com/89418486

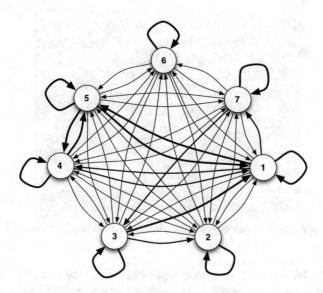

Fig. 14.15 Nodes and links on a transition graph in a musical space with seven scenes

food they lay down pheromone trails on their respective paths, and these pheromone trails evaporate gradually. A path is more attractive to ants if it has larger amounts of pheromone, the shortest (or optimal) path thus becoming the most attractive. In ACO, the artificial ants deposit pheromones on their paths while moving, and a probabilistic model is used to decide future movements of these ants. In our implementation of this mechanism, the probability p_{ij} of the agent choosing a scene j after i is given by,

$$p_{ij} = \frac{\tau_{ij}}{\sum_{k=1}^{N} \tau_{ik}}$$

where τ_{ik} denotes the strength of the link between scenes i and k, and N is the total number of scenes the transition graph is composed of.

Every time a user makes a selection in the navigation of a sequence of scenes, the pheromone graph gets updated in the form of strengthening of the links from one scene to the next. Moreover, each time a transition is made, a fraction of the pheromone on every link on the graph necessarily evaporates. This link strength update mechanism can be written as

$$\tau_{ij} = \begin{cases} (1-\rho)\tau_{ij} & \text{if link not selected by user} \\ (1-\rho)\tau_{ij} + \Delta & \text{if link selected by user} \end{cases}$$

Thus, the link that is traversed has a positive value $\Delta \in \mathbb{R}$ added to it. And, the links between the nodes necessarily have their strength decreased over time at a rate given by $\rho \in [0.0, 1.0]$, known as the *evaporation rate*.

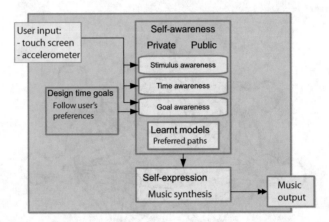

Fig. 14.16 The Pheromusic system within the reference architecture from Chapter 4

Figure 14.16 puts the Pheromusic application within the reference architecture. The system consists of a single node which learns the preferred paths selected by the user.

14.5 Conclusion

In this chapter we introduced a number of different active music systems allowing for varying degrees of user interaction with the music. This includes both direct and indirect control using a variety of sensors which are available as smartphone-integrated or external sensors. The systems presented display different levels of control over the music. Our system spans a wide range—from controlling the overarching structure of the selection of musical patterns to be combined down to what sounds to generate.

SoloJam provides a mechanism for controlling both rhythm and harmonic patterns for two or more people. This is mostly using direct control input, and includes a mechanism for resolving the conflict that arises when multiple people want to have control of the musical output. The second example, Funky Sole Music, uses external force sensors in a sole worn by the user to control a variety of levels in the music being synthesised. It represents the example with the largest degree of indirect control of the three introduced. In the third example, PheroMusic, most levels of control are applied and provide agent-assisted navigation within a musical space. This example has been made publicly available through an iOS app available in the iTunes store.

The examples of active music systems presented in this chapter demonstrate the many possible directions for active music systems. Thus, there are a number of systems that can be developed with extensions and improvements.

Chapter 15
Conclusions and Outlook

Peter R. Lewis, Marco Platzner, Bernhard Rinner, Jim Tørresen, and Xin Yao

In this book we have argued that in order to deal with the complexities of future computing systems, including size, decentralisation, uncertainty, dynamics and heterogeneity, greater levels of self-awareness on the part of such systems will be important. While this has long been agreed in principle, only now have we begun to establish a principled approach to the design of self-aware systems.

As this book highlights, this new approach will require a rethinking of current design and operation principles in order to handle the scale, variety and changing nature of requirements and optimisation goals efficiently. Novel design and operation principles and methods are needed, possibly those that radically break with the static ones we have built into our systems and the fixed abstraction layers we have learned to cherish over the years.

Taking inspiration from self-awareness in humans, in this book we introduced the new notion of *computational self-awareness* as a fundamental concept for designing and operating computing systems. The basic ability of such self-aware computing systems is to collect information about their state and progress, building and maintaining knowledge that enables them to reason about their behaviour (self-awareness). Further, self-aware computing systems will have the ability to utilise this knowledge to effectively and autonomously adapt and explain their behaviour, in changing conditions (self-expression). This book addresses these fundamental

Peter R. Lewis
Aston University, UK, e-mail: p.lewis@aston.ac.uk

Marco Platzner
University of Paderborn, Germany, e-mail: platzner@upb.de

Bernhard Rinner
Alpen-Adria-Universität Klagenfurt, Austria, e-mail: bernhard.rinner@aau.at

Jim Tørresen
University of Oslo, Norway, e-mail: jimtoer@ifi.uio.no

Xin Yao
University of Birmingham, UK, e-mail: x.yao@cs.bham.ac.uk

concepts from an engineering perspective, aiming at developing *primitives for building* systems and applications. To this end, we introduced a theory of computational self-awareness and an associated reference architecture for describing self-aware and self-expressive computing systems. We developed patterns for designing systems and applications, provided guidelines for knowledge representation and common techniques, and demonstrated the effectiveness of the primitives for developing self-aware nodes and networks as well as for three selected applications. This, along with associated self-expression, behaviour based on self-awareness, will comprise various capabilities which can provide computing systems with advanced levels of autonomous behaviour. This enables them to adapt themselves at run time, learning behaviours to manage complex trade-offs in changing conditions.

15.1 Computational Self-awareness

In order to develop a theory of *computational self-awareness*, we argued that it is beneficial to first consider self-awareness in humans. Part I of the book began building our new notion of computational self-awareness based on three key concepts from psychology. First, as with human self-awareness, systems possessing computational self-awareness should be concerned not only with knowledge of their own internals (*private self-awareness*), but also with their experiences of, impact on, and role within the world (*public self-awareness*). A second concept is the existence of different *levels of self-awareness*, and since self-aware systems may vary a great deal in their complexity, Neisser's [284] broad set of levels of self-awareness was chosen as a source of inspiration. Our developed *levels of computational self-awareness* [236] are inspired by Neisser's levels for humans but translated appropriately for describing the capabilities of computer systems. While "full-stack" computational self-awareness may often be beneficial, with several processes responsible for one or more levels of self-awareness, there are also cases where a more minimal approach is appropriate, and these levels provide a common language for considering this design aspect. A third concept is that self-awareness can be a property of collective systems, even when there is no single component with a global awareness of the whole system [272]. This challenges an important assumption regarding the design of self-aware systems: it highlights that any choice to include a component with a global view (an awareness of the entire system) is a choice, not a requirement. We encapsulated the concepts of computational self-awareness and self-expression into a reference architecture, which is used to instantiate architectural patterns for specific systems. We draw attention to a methodology outlined in full in our handbook [70] for the design of self-aware systems, refined based on our experience of applying computational self-awareness concepts to a wide range of applications over recent years. Compared to other architectures, the presented framework facilitates principled design decisions concerning whether, how, and to what extent to include self-awareness and self-expression capabilities.

The notion of *computational self-awareness* intentionally includes many existing and prior systems which have not been previously described as self-aware. Nevertheless these systems use capabilities which according to our framework fulfill some self-awareness aspect, often due to their benefit in a complex environment. Based on our experience, we also believe that further self-awareness will be necessary to deal with increasing complexity. Our choice to use Neisser's broad set of levels, rather than others presented in the psychology literature, is based on our intention to explicitly account for a full spectrum of existing and future systems, and not be concerned only with what one might typically consider highly advanced artificial intelligence.

These primitives have been successfully applied in building self-aware and self-expressive nodes and networks as well as in developing three distinct and diverse case studies, which have provided tangible contributions and benefits not just as case studies for self-aware computing, but in their own respective domains. Nevertheless, there are many challenges which need to be faced in both developing and applying self-aware systems, and a comprehensive theory on self-aware comping systems is still an ongoing research endeavour.

15.2 Challenges and Research Questions

During our work on self-aware computing, we have identified a number of new research questions. Some of the most pressing of these are:

- *Holistic system model*
  Our reference architecture represents a preliminary formal representation of common components and interactions of a self-aware and self-expressive computing system. This architecture allows us to describe the system's structure and functionality, but also being able predict the system's behaviour is important. Developing a holistic system model that enables us to reason about structure and functionality and also about the behaviour of such a dynamic system is a formidable challenge.
- *System performance and properties*
  For engineering a self-aware computing system we must be able to estimate— or at least to provide bounds on—the system's performance and properties such as resource usage, cost, reliability and safety. The effect of different levels of self-awareness and self-expression on these properties represents an important research question. Moreover, self-awareness incurs overheads and the corresponding trade-offs between these overheads and the system capabilities need to be explored and quantified.
- *Claims and expectations despite uncertainty*
  For general acceptance of self-awareness methods, an important question is how to formulate claims about what we can expect from self-aware systems deployed in uncertain and dynamic environments. To address this challenge we might

need to bridge progress in holistic system models and techniques to estimate system properties with system theoretical considerations.

- *Design processes and tools*
  In this book we have made a first and essential step towards engineering self-aware computing systems. Our design patterns and common techniques for self-aware and self-expressive computing help engineers create real-world applications. Yet, this line of research needs to be extended to provide engineers with useful and efficient design processes and corresponding tools to support them in their task.

Despite the progress made in recent years, there is still much to understand about how to incorporate self-awareness properties into computing systems.

As shown in this book, self-awareness and self-expression are fundamental concepts. In fact, we are confident that self-awareness and self-expression could serve as enabling technologies for future systems and networks meeting a multitude of requirements with respect to functionality, flexibility, performance, resource usage, costs, reliability and safety. Even now, as research communities are converging around questions concerned with self-awareness and self-expression in computing systems, in the long-term, self-aware computing might well develop into a dedicated research field fostering an interdisciplinary community of academics and professionals and creating a new breed of systems and applications.

References

1. Aberdeen, D., Baxter, J.: Emmerald: a fast matrix-matrix multiply using Intel's SSE instructions. Concurrency and Computation: Practice and Experience **13**(2), 103–119 (2001)
2. Abramowitz, M., Stegun, I.: Handbook of Mathematical Functions. Dover Publications (1965)
3. Agarwal, A., Harrod, B.: Organic computing. Tech. Rep. White paper, MIT and DARPA (2006)
4. Agarwal, A., Miller, J., Eastep, J., Wentziaff, D., Kasture, H.: Self-aware computing. Tech. Rep. AFRL-RI-RS-TR-2009-161, MIT (2009)
5. Agne, A., Hangmann, H., Happe, M., Platzner, M., Plessl, C.: Seven recipes for setting your FPGA on fire – a cookbook on heat generators. Microprocessors and Microsystems **38**(8), 911–919 (2014). DOI 10.1016/j.micpro.2013.12.001
6. Agne, A., Happe, M., Keller, A., Lübbers, E., Plattner, B., Platzner, M., Plessl, C.: ReconOS: An Operating System Approach for Reconfigurable Computing. IEEE Micro **34**(1), 60–71 (2014). DOI 10.1109/MM.2013.110
7. Agne, A., Platzner, M., Lübbers, E.: Memory virtualization for multithreaded reconfigurable hardware. In: Proceedings of the International Conference on Field Programmable Logic and Applications (FPL), pp. 185–188. IEEE Computer Society (2011). DOI 10.1109/FPL.2011.42
8. Ahuja, S., Carriero, N., Gelernter, D.: Linda and friends. IEEE Computer **19**(8), 26–34 (1986). DOI 10.1109/MC.1986.1663305
9. Al-Naeem, T., Gorton, I., Babar, M.A., Rabhi, F., Benatallah, B.: A quality-driven systematic approach for architecting distributed software applications. In: Proceedings of the 27th International Conference on Software Engineering, pp. 244–253. ACM (2005). DOI 10.1145/1062455.1062508. URL http://doi.acm.org/10.1145/1062455.1062508
10. Ali, H.A., Desouky, A.I.E., Saleh, A.I.: Studying and Analysis of a Vertical Web Page Classifier Based on Continuous Learning Naive Bayes (CLNB) Algorithm, pp. 210–254. Information Science (2009)
11. Alippi, C., Boracchi, G., Roveri, M.: Just-in-time classifiers for recurrent concepts. IEEE Transactions on Neural Networks and Learning Systems **24**(4), 620–634 (2013)
12. Amir, E., Anderson, M.L., Chaudhri, V.K.: Report on DARPA workshop on self-aware computer systems. Tech. Rep. UIUCDCS-R-2007-2810, UIUC Comp. Sci. (2007)
13. ANA: Autonomic Network Architecture. URL www.ana-project.org. (accessed March 8, 2016)
14. Angelov, P.: Nature-inspired methods for knowledge generation from data in real-time (2006). URL http://www.nisis.risk-technologies.com/popup/Mallorca2006_Papers/A333_13774_Nature-inspiredmethodsforKnowledgeGeneration_Angelov.pdf

15. Apache: Hadoop. http://hadoop.apache.org/docs/r1.2.1/mapred_tutorial.html. (Accessed March 8, 2016)
16. Araya-Polo, M., Cabezas, J., Hanzich, M., Pericàs, M., Rubio, F., Gelado, I., Shafiq, M., Morancho, E., Navarro, N., Ayguadé, E., Cela, J.M., Valero, M.: Assessing accelerator-based HPC reverse time migration. IEEE Transactions on Parallel and Distributed Systems 22(1), 147–162 (2011)
17. Asanovic, K., Bodik, R., Catanzaro, B.C., Gebis, J.J., Husbands, P., Keutzer, K., Patterson, D.A., Plishker, W.L., Shalf, J., Williams, S.W., Yelick, K.A.: The landscape of parallel computing research: A view from Berkeley. Tech. Rep. UCB/EECS-2006-183, EECS Department, University of California, Berkeley (2006)
18. Asendorpf, J.B., Warkentin, V., Baudonnière, P.M.: Self-awareness and other-awareness. II: Mirror self-recognition, social contingency awareness, and synchronic imitation. Developmental Psychology 32(2), 313 (1996)
19. Athan, T.W., Papalambros, P.Y.: A note on weighted criteria methods for compromise solutions in multi-objective optimization. Engineering Optimization 27(2), 155–176 (1996)
20. Auer, P., Cesa-Bianchi, N., Fischer, P.: Finite-time analysis of the multiarmed bandit problem. Machine Learning 47(2–3), 235–256 (2002)
21. Babaoglu, O., Binci, T., Jelasity, M., Montresor, A.: Firefly-inspired heartbeat synchronization in overlay networks. In: First International Conference on Self-Adaptive and Self-Organizing Systems (SASO), pp. 77–86 (2007)
22. Babenko, B., Yang, M.H., Belongie, S.: Robust object tracking with online multiple instance learning. IEEE Transactions on Pattern Analysis and Machine Intelligence 33(8), 1619–1632 (2011)
23. Bader, J., Zitzler, E.: HypE: an algorithm for fast hypervolume-based many-objective optimization. Tech. Rep. TIK 286, Computer Engineering and Networks Laboratory, ETH Zurich, Zurich (2008)
24. Baena-García, M., Campo-Ávila, J.D., Fidalgo, R., Bifet, A.: Early drift detection method. In: Procedings of the 4th ECML PKDD International Workshop on Knowledge Discovery From Data Streams (IWKDDS), pp. 77–86. Berlin, Germany (2006)
25. Baker, S.: The identification of the self. Psyc. Rev. 4(3), 272–284 (1897)
26. Banks, A., Gupta, R.: MQTT Version 3.1.1. http://docs.oasis-open.org/mqtt/mqtt/v3.1.1/os/mqtt-v3.1.1-os.html (2014)
27. Bartolini, D.B., Sironi, F., Maggio, M., Cattaneo, R., Sciuto, D., Santambrogio, M.D.: A Framework for Thermal and Performance Management. In: Proceedings of the Workshop on Managing Systems Automatically and Dynamically (MAD) (2012)
28. Basheer, I.A., Hajmeer, M.: Artificial neural networks: fundamentals, computing, design, and application. Journal of Microbiological Methods 43(1), 3–31 (2000)
29. Basseur, M., Zitzler, E.: Handling uncertainty in indicator-based multiobjective optimization. International Journal of Computational Intelligence Research 2(3), 255–272 (2006)
30. Basudhar, A., Dribusch, C., Lacaze, S., Missoum, S.: Constrained efficient global optimization with support vector machines. Structural and Multidisciplinary Optimization 46(2), 201–221 (2012)
31. Baumann, A., Boltz, M., Ebling, J., Koenig, M., Loos, H.S., Merkel, M., Niem, W., Warzelhan, J.K., Yu, J.: A review and comparison of measures for automatic video surveillance systems. EURASIP Journal on Image and Video Processing 2008(4) (2008). DOI 10.1155/2008/824726
32. Becker, T., Agne, A., Lewis, P.R., Bahsoon, R., Faniyi, F., Esterle, L., Keller, A., Chandra, A., Jensenius, A.R., Stilkerich, S.C.: EPiCS: Engineering proprioception in computing systems. In: Proceedings of the International Conference on Computational Science and Engineering (CSE), pp. 353–360. IEEE Computer Society (2012)
33. Ben-Hur, A., Weston, J.: A user's guide to support vector machines. Data Mining Techniques for the Life Sciences 609, 223–239 (2010)
34. Betts, A., Chong, N., Donaldson, A.F., Qadeer, S., Thompson, P.: GPUVerify: a verifier for GPU kernels. In: Proceedings of the ACM International Conference on object-oriented Programming Systems Languages and Applications (OOPSLA) (2012)

35. Beume, N., Naujoks, B., Emmerich, M.: SMS-EMOA: Multiobjective selection based on dominated hypervolume. European Journal on Operational Research **181**(3), 1653–1669 (2007)
36. Bevilacqua, F., Zamborlin, B., Sypniewski, A., Schnell, N., Guédy, F., Rasamimanana, N.: Continuous realtime gesture following and recognition. In: Gesture in embodied communication and human-computer interaction, pp. 73–84. Springer (2010)
37. Biehl, J.T., Adamczyk, P.D., Bailey, B.P.: Djogger: A mobile dynamic music device. In: Proceedings of CHI '06 Extended Abstracts on Human Factors in Computing Systems, pp. 556–561. ACM (2006)
38. Bishop, C.M.: Neural Networks for Pattern Recognition. Oxford University Press, United Kingdom (2005)
39. Bojic, I., Lipic, T., Podobnik, V.: Bio-inspired clustering and data diffusion in machine social networks. In: Computational Social Networks, pp. 51–79. Springer (2012)
40. Bongard, J., Lipson, H.: Evolved machines shed light on robustness and resilience. Proceedings of the IEEE **102**(5), 899–914 (2014)
41. Bongard, J., Zykov, V., Lipson, H.: Resilient machines through continuous self-modeling. Science **314**(5802), 1118–1121 (2006)
42. Borkar, S.: Designing Reliable Systems from Unreliable Components: The Challenges of Transistor Variability and Degradation. IEEE Micro pp. 10–16 (2005)
43. Bouabene, G., Jelger, C., Tschudin, C., Schmid, S., Keller, A., May, M.: The Autonomic Network Architecture (ANA). IEEE Journal on Selected Areas in Communications **28**(1), 4–14 (2010). DOI 10.1109/JSAC.2010.100102
44. Boyd, J.: The Essence of Winning and Losing. http://dnipogo.org/john-r-boyd/ (1996). (Accessed March 8, 2016)
45. Bramberger, M., Doblander, A., Maier, A., Rinner, B., Schwabach, H.: Distributed Embedded Smart Cameras for Surveillance Applications. IEEE Computer **39**(2), 68–75 (2006)
46. Brdiczka, O., Crowley, J.L., Reignier, P.: Learning situation models in a smart home. IEEE Transactions on Systems, Man, and Cybernetics, Part B **39**, 56–63 (2009)
47. Breiman, L.: Bagging predictors. Machine Learning **24**(2), 123–140 (1996)
48. Breiman, L.: Random forests. Machine Learning **45**(1), 5–32 (2001)
49. Brockhoff, D., Zitzler, E.: Improving hypervolume-based multiobjective evolutionary algorithms by using objective reduction methods. In: Proceedings of the 2007 IEEE Congress on Evolutionary Computation, pp. 2086–2093 (2007)
50. Buchanan, J.T.: A naive approach for solving MCDM problems: The GUESS method. Journal of the Operational Research Society **48**(2), 202–206 (1997)
51. Buck, J.: Synchronous rhythmic flashing of fireflies. The Quarterly Review of Biology **13**(3), 301–314 (1938)
52. Buck, J.: Synchronous rhythmic flashing of fireflies II. The Quarterly Review of Biology **63**(3), 265–289 (1988)
53. Burke, E.K., Gendreau, M., Hyde, M., Kendall, G., Ochoa, G., Ozcan, E., Qu, R.: Hyper-heuristics: A survey of the state of the art. Journal of the Operational Research Society **206**(1), 241–264 (2013)
54. Buschmann, F., Henney, K., Douglas, S.C.: Pattern-oriented software architecture: On patterns and pattern languages. John Wiley and Sons (2007)
55. Buss, A.H.: Self-consciousness and social anxiety. W. H. Freeman, San Fransisco, CA, USA (1980)
56. Calinescu, R., Ghezzi, C., Kwiatkowska, M., Mirandola, R.: Self-adaptive software needs quantitative verication at runtime. Communications of the ACM **55**(9), 69–77 (2012)
57. Caramiaux, B., Wanderley, M.M., Bevilacqua, F.: Segmenting and parsing instrumentalists' gestures. Journal of New Music Research **41**(1), 13–29 (2012)
58. Carver, C.S., Scheier, M.: Attention and Self-Regulation: A Control-Theory Approach to Human Behavior. Springer (1981)
59. de Castro, L.N.: Fundamentals of natural computing: basic concepts, algorithms, and applications. Chapman & Hall/CRC Computer and Information Sciences (2006)

60. Chandra, A.: A methodical framework for engineering co-evolution for simulating socio-economic game playing agents. Ph.D. thesis, The University of Birmingham (2011)

61. Chandra, A., Nymoen, K., Volsund, A., Jensenius, A.R., Glette, K., Tørresen, J.: Enabling participants to play rhythmic solos within a group via auctions. In: Proceedings of the International Symposium on Computer Music Modeling and Retrieval (CMMR), pp. 674–689 (2012)

62. Chandra, A., Yao, X.: Ensemble learning using multi-objective evolutionary algorithms. Journal of Mathematical Modelling and Algorithms **5**(4), 417–445 (2006)

63. Chang, C., Wawrzynek, J., Brodersen, R.W.: BEE2: a high-end reconfigurable computing system. IEEE Transactions on Design & Test of Computer **22**(2), 114–125 (2005)

64. Chen, J., John, L.K.: Efficient program scheduling for heterogeneous multi-core processors. In: Proceedings of the Design Automation Conference (DAC). ACM (2009)

65. Chen, R., Lewis, P.R., Yao, X.: Temperature management for heterogeneous multi-core FPGAs using adaptive evolutionary multi-objective approaches. In: Proceedings of the International Conference on Evolvable Systems (ICES), pp. 101–108. IEEE (2014)

66. Chen, S., Langner, C.A., Mendoza-Denton, R.: When dispositional and role power fit: implications for self-expression and self-other congruence. Journal of Personality and Social Psychology **96**(3), 710–27 (2009)

67. Chen, T., Bahsoon, R.: Self-adaptive and Sensitivity-aware QoS Modeling for the Cloud. In: Proceedings of the 8th International Symposium on Software Engineering for Adaptive and Self-Managing Systems (SEAMS), pp. 43–52. IEEE (2013). URL http://dl.acm.org/citation.cfm?id=2487336.2487346

68. Chen, T., Bahsoon, R.: Symbiotic and Sensitivity-aware Architecture for Globally-optimal Benefit in Self-adaptive Cloud. In: Proceedings of the 9th International Symposium on Software Engineering for Adaptive and Self-Managing Systems (SEAMS), pp. 85–94. ACM (2014). DOI 10.1145/2593929.2593931. URL http://doi.acm.org/10.1145/2593929.2593931

69. Chen, T., Bahsoon, R., Yao, X.: Online QoS Modeling in the Cloud: A Hybrid and Adaptive Multi-learners Approach. In: 2014 IEEE/ACM 7th International Conference on Utility and Cloud Computing (UCC), pp. 327–336 (2014)

70. Chen, T., Faniyi, F., Bahsoon, R., Lewis, P.R., Yao, X., Minku, L.L., Esterle, L.: The handbook of engineering self-aware and self-expressive systems. Tech. rep., EPiCS EU FP7 project consortium (2014). URL http://arxiv.org/abs/1409.1793. Available via EPiCS website and arXiv

71. Chen, X., Li, X., Wu, H., Qiu, T.: Real-time Object Tracking via CamShift-based Robust Framework. In: Proceedings of the International Conference on Information Science and Technology (ICIST). IEEE (2012)

72. Chow, G.C.T., Grigoras, P., Burovskiy, P., Luk, W.: An efficient sparse conjugate gradient solver using a Beneš permutation network. In: Proceedings of the 24th International Conference on Field Programmable Logic and Applications, pp. 1–7 (2014)

73. Chow, G.C.T., Tse, A.H.T., Jin, Q., Luk, W., Leong, P.H.W., Thomas, D.B.: A mixed precision Monte Carlo methodology for reconfigurable accelerator systems. In: Proceedings of the ACM/SIGDA 20th International Symposium on Field Programmable Gate Arrays, FPGA 2012, Monterey, California, USA, February 22-24, 2012, pp. 57–66 (2012)

74. Christensen, A.L., O'Grady, R., Dorigo, M.: From fireflies to fault-tolerant swarms of robots. IEEE Transactions on Evolutionary Computation **13**(4), 754–766 (2009)

75. Christensen, E., Curbera, F., Meredith, G., Weerawarana, S.: Web Services Description Language (WSDL) 1.1. World Wide Web Consortium (2001)

76. Chu, F., Zaniolo, C.: Fast and light boosting for adaptive mining of data streams. In: Proceedings of the Eighth Pacific-Asia Knowledge Discovery and Data Mining Conference (PAKDD), pp. 282–292. Sydney (2004)

77. Cichowski, A., Madden, C., Detmold, H., Dick, A., Van den Hengel, A., Hill, R.: Tracking Hand-off in Large Surveillance Networks. In: Proceedings of the International Conference Image and Vision Computing, pp. 276–281. IEEE Computer Society Press (2009). DOI 10.1109/IVCNZ.2009.5378396

78. Claus, C., Boutilier, C.: The Dynamics of Reinforcement Learning in Cooperative Multiagent Systems. In: Proceedings of the Conference on Artificial Intelligence/Innovative Applications of Artificial Intelligence, pp. 746–752. American Association for Artificial Intelligence (1998)

79. Collins, N.: The analysis of generative music programs. Organised Sound **13**, 237–248 (2008)

80. Collins, R.T., Liu, Y., Leordeanu, M.: Online selection of discriminative tracking features. IEEE Transactions on Pattern Analysis and Machine Intelligence **27**(10), 1631–1643 (2005). DOI 10.1109/tpami.2005.205

81. Colorni, A., Dorigo, M., Maniezzo, V., et al.: Distributed optimization by ant colonies. In: Proceedings of the first European conference on artificial life, vol. 142, pp. 134–142. Elsevier (1991)

82. Comaniciu, D., Ramesh, V., Meer, P.: Kernel-based object tracking. IEEE Transactions on Pattern Analysis and Machine Intelligence **25**(5) (2003). DOI 10.1109/tpami.2003.1195991

83. Connors, K.: Chemical kinetics: the study of reaction rates in solution. VCH Publishers (1990)

84. Cox, M.: Metacognition in computation: A selected research review. Artificial Intelligence **169**(2), 104–141 (2005)

85. Cramer, T., Schmidl, D., Klemm, M., an Mey, D.: OpenMP Programming on Intel Xeon Phi Coprocessors: An Early Performance Comparison. In: Proceedings of the Many-core Applications Research Community (MARC) Symposium, pp. 38–44. Aachen, Germany (2012)

86. Crockford, D.: The application/json Media Type for JavaScript Object Notation (JSON). RFC 7159, RFC Editor (2014). URL http://tools.ietf.org/pdf/rfc7159.pdf

87. Curreri, J., Stitt, G., George, A.D.: High-level synthesis of in-circuit assertions for verification, debugging, and timing analysis. International Journal of Reconfigurable Computing **2011**, 1–17 (2011). DOI http://dx.doi.org/10.1155/2011/406857

88. Czajkowski, T.S., Aydonat, U., Denisenko, D., Freeman, J., Kinsner, M., Neto, D., Wong, J., Yiannacouras, P., Singh, D.P.: From OpenCL to high-performance hardware on FPGAs. In: Proceedings of the 22nd International Conference on Field Programmable Logic and Applications (FPL), pp. 531–534. Oslo, Norway (2012)

89. Datta, K., Murphy, M., Volkov, V., Williams, S., Carter, J., Oliker, L., Patterson, D., Shalf, J., Yelick, K.: Stencil computation optimization and auto-tuning on state-of-the-art multi-core architectures. In: Proceedings of the International Conference for High Performance Computing, Networking, Storage and Analysis (SC 2008)., p. 4. IEEE (2008)

90. Davidson, A.A., Owens, J.D.: Toward techniques for auto-tuning GPU algorithms. In: Proceedings of the 10th International Conference on Applied Parallel and Scientific Computing (PARA), Revised Selected Papers, Part II, pp. 110–119. Reykjavík (2010)

91. Day, J.: Patterns in Network Architecture: A Return to Fundamentals. Prentice Hall International (2008)

92. Day, J., Matta, I., Mattar, K.: Networking is IPC: A Guiding Principle to a Better Internet. In: Proceedings of the 2008 ACM CoNEXT Conference, pp. 67:1–67:6 (2008). DOI 10.1145/1544012.1544079. URL http://doi.acm.org/10.1145/1544012.1544079

93. Dean, J., Ghemawat, S.: MapReduce: Simplified Data Processing on Large Clusters. In: Proceedings of the 6th Symposium on Operating System Design and Implementation (OSDI), pp. 137–150. San Francisco, California, USA (2004)

94. Deb, K.: Multi-objective optimization using evolutionary algorithms, vol. 16. John Wiley & Sons, England (2001)

95. Deb, K., Pratap, A., Agarwal, S., Meyarivan, T.: A fast and elitist multiobjective genetic algorithm: NSGA-II. IEEE Transactions on Evolutionary Computation **6**(2), 182–197 (2002)

96. Denholm, S., Inoue, H., Takenaka, T., Luk, W.: Application-specific customisation of market data feed arbitration. In: Proceedings of the International Conference on Field Programmable Technology (ICFPT), pp. 322–325. IEEE (2013)

97. Denholm, S., Inouey, H., Takenakay, T., Becker, T., Luk, W.: Low latency FPGA acceleration of market data feed arbitration. In: Proceedings of the International Conference

on Application-Specific Systems, Architectures, and Processors (ASAP), pp. 36–40. IEEE (2014). DOI 10.1109/ASAP.2014.6868628

98. Dennett, D.C.: Consciousness Explained. Penguin Science (1993)
99. Dennis, J.B., Misunas, D.: A preliminary architecture for a basic data flow processor. In: Proceedings of the 2nd Annual Symposium on Computer Architecture, pp. 126–132 (1974)
100. Dieber, B., Simonjan, J., Esterle, L., Rinner, B., Nebehay, G., Pflugfelder, R., Fernandez, G.J.: Ella: Middleware for multi-camera surveillance in heterogeneous visual sensor networks. In: Proceedings of the International Conference on Distributed Smart Cameras (ICDSC) (2013). DOI 10.1109/ICDSC.2013.6778223
101. Dietterich, T.G.: Ensemble methods in machine learning. In: Proceedings of the First International Workshop on Multiple Classifier Systems, Lecture Notes in Computer Science, pp. 1–15. Springer-Verlag (2000)
102. Diguet, J.P., Eustache, Y., Gogniat, G.: Closed-loop–based Self-adaptive Hardware/Software-Embedded Systems: Design Methodology and Smart Cam Case Study. ACM Transactions on Embedded Computing Systems 10(3), 1–28 (2011)
103. Dinh, M.N., Abramson, D., J. Chao, D.K., Gontarek, A., Moench, B., DeRose, L.: Debugging scientific applications with statistical assertions. Procedia Computer Science 9(0), 1940––1949 (2012)
104. Dobson, S., Denazis, S., Fernández, A., Gaïti, D., Gelenbe, E., Massacci, F., Nixon, P., Saffre, F., Schmidt, N., Zambonelli, F.: A survey of autonomic communications. ACM Transactions on Autonomous and Adaptive Systems 1(2), 223–259 (2006)
105. Dobson, S., Sterritt, R., Nixon, P., Hinchey, M.: Fulfilling the vision of autonomic computing. IEEE Computer 43(1), 35 –41 (2010)
106. Dobzhansky, T., Hecht, M., Steere, W.: On some fundamental concepts of evolutionary biology. Evolutionary Biology 2, 1–34 (1968)
107. Dorigo, M.: Optimization, learning and natural algorithms. Ph.D. thesis, Politecnico di Milano (1992)
108. Dorigo, M., Blum, C.: Ant colony optimization theory: A survey. Theoretical computer science 344(2), 243–278 (2005)
109. Dorigo, M., Maniezzo, V., Colorni, A.: Ant system: optimization by a colony of cooperating agents. IEEE Transactions on Systems, Man, and Cybernetics, Part B: Cybernetics 26(1), 29–41 (1996)
110. Dutta, R., Rouskas, G., Baldine, I., Bragg, A., Stevenson, D.: The SILO Architecture for Services Integration, controL, and Optimization for the Future Internet. In: Proceedings of the IEEE International Conference on Communications (ICC), pp. 1899–1904 (2007). DOI 10.1109/ICC.2007.316
111. Duval, S., Wicklund, R.A.: A theory of objective self awareness. Academic Press (1972)
112. Ehrgott, M.: Other Methods for Pareto Optimality. In: Multicriteria Optimization, Lecture Notes in Economics and Mathematical Systems, vol. 491, pp. 77–102. Springer (2000)
113. Eiben, A.E., Smith, J.E.: Introduction to evolutionary computing. Springer (2003)
114. Eigenfeldt, A., Pasquier, P.: Considering vertical and horizontal context in corpus-based generative electronic dance music. In: Proceedings of the Fourth International Conference on Computational Creativity, p. 72 (2013)
115. Eigenfeldt, A., Pasquier, P.: Evolving structures for electronic dance music. In: Proceedings of the 15th Annual Conference on Genetic and Evolutionary Computation (GECCO), pp. 319–326. ACM (2013)
116. Elkhodary, A., Esfahani, N., Malek, S.: FUSION: a framework for engineering self-tuning self-adaptive software systems. In: Proceedings of the eighteenth ACM SIGSOFT International Symposium on Foundations of Software Engineering, pp. 7–16. ACM (2010). DOI 10.1145/1882291.1882296. URL http://doi.acm.org/10.1145/1882291.1882296
117. Elliott, G.T., Tomlinson, B.: PersonalSoundtrack: context-aware playlists that adapt to user pace. In: Proceedings of CHI'06 Extended Abstracts on Human Factors in Computing Systems, pp. 736–741. ACM (2006)

118. Ellis, T., Makris, D., Black, J.: Learning a Multi-camera Topology. In: Proceedings of the Joint International Workshop on Visual Surveillance and Performance Evaluation of Tracking and Surveillance, pp. 165–171. IEEE Computer Society Press (2003)
119. Elwell, R., Polikar, R.: Incremental learning of concept drift in nonstationary environments. IEEE Transactions on Neural Networks **22**, 1517–1531 (2011)
120. Endo, T., Matsuoka, S.: Massive supercomputing coping with heterogeneity of modern accelerators. In: Proceedings of the 22nd IEEE International Symposium on Parallel and Distributed Processing (IPDPS), pp. 1–10 (2008)
121. Erdem, U.M., Sclaroff, S.: Look there! Predicting Where to Look for Motion in an Active Camera Network. In: Proceedings of the IEEE Conference on Advanced Video and Signal-based Surveillance, pp. 105–110. Como, Italy (2005)
122. Esterle, L., Lewis, P.R., Bogdanski, M., Rinner, B., Yao, X.: A socio-economic approach to online vision graph generation and handover in distributed smart camera networks. In: Proceedings of the International Conference on Distributed Smart Cameras (ICDSC), pp. 1–6. IEEE (2011). DOI 10.1109/ICDSC.2011.6042902
123. Esterle, L., Lewis, P.R., Caine, H., Yao, X., Rinner, B.: CamSim: A distributed smart camera network simulator. In: Proceedings of the International Conference on Self-Adaptive and Self-Organizing Systems Workshops, pp. 19–20. IEEE Computer Society Press (2013). DOI 10.1109/SASOW.2013.11
124. Esterle, L., Lewis, P.R., Rinner, B., Yao, X.: Improved adaptivity and robustness in decentralised multi-camera networks. In: Proceedings of the International Conference on Distributed Smart Cameras, pp. 1–6. ACM (2012)
125. Esterle, L., Lewis, P.R., Yao, X., Rinner, B.: Socio-economic vision graph generation and handover in distributed smart camera networks. ACM Transactions on Sensor Networks **10**(2), 20:1–20:24 (2014). DOI 10.1145/2530001
126. Eugster, P.T., Felber, P.A., Guerraoui, R., Kermarrec, A.M.: The Many Faces of Publish/Subscribe. ACM Computing Surveys **35**(2), 114–131 (2003)
127. Faniyi, F., Lewis, P.R., Bahsoon, R., Xao, X.: Architecting self-aware software systems. In: Proceedings of the IEEE/IFIP Conference on Software Architecture (WICSA), pp. 91–94. IEEE (2014)
128. Farrell, R., Davis, L.S.: Decentralized discovery of camera network topology. In: Proceedings of the International Conference on Distributed Smart Cameras, pp. 1–10. IEEE Computer Society Press (2008). DOI 10.1109/ICDSC.2008.4635696
129. Fels, S., Hinton, G.: Glove-talk: A neural network interface between a data-glove and a speech synthesizer. IEEE Transactiona on Neural Networks **4**(1), 2–8 (1993)
130. Feng, W.: Making a case for efficient supercomputing. ACM Queue **1**(7), 54–64 (2003)
131. Fenigstein, A., Scheier, M.F., Buss, A.H.: Public and private self-consciousness: Assessment and theory. Journal of Consulting and Clinical Psychology **43**(4), 522–527 (1975)
132. Fern, A., Givan, R.: Online ensemble learning: An empirical study. Machine Learning **53**(1–2), 71–109 (2003)
133. Fette, B.: Cognitive radio technology. Academic Press (2009)
134. Fiebrink, R., Trueman, D., Cook, P.R.: A meta-instrument for interactive, on-the-fly machine learning. In: Proceedings of the International Conference on New Interfaces for Musical Expression. Pittsburgh (2009)
135. Fielding, R.T., Taylor, R.N.: Principled design of the modern web architecture. ACM Transactions on Internet Technology **2**(2), 115–150 (2002). DOI 10.1145/514183.514185. URL http://doi.acm.org/10.1145/514183.514185
136. Freund, Y., Schapire, R.E.: Experiments with a new boosting algorithm. In: Proceedings of the 13th International Conference on Machine Learning, pp. 148–156 (1996)
137. Froming, W.J., Walker, G.R., Lopyan, K.J.: Public and private self-awareness: When personal attitudes conflict with societal expectations. Journal of Experimental Social Psychology **18**(5), 476 – 487 (1982). DOI 10.1016/0022-1031(82)90067-1
138. Fu, H., Sendhoff, B., Tang, K., Yao, X.: Finding robust solutions to dynamic optimization problems. In: Proceedings of the 16th European conference on Applications of Evolutionary Computation (EvoApplications), pp. 616–625 (2013)

139. Funie, A., Salmon, M., Luk, W.: A hybrid genetic-programming swarm-optimisation approach for examining the nature and stability of high frequency trading strategies. In: Proceedings of the 13th International Conference on Machine Learning and Applications (ICMLA), pp. 29–34. Detroit, USA (2014). DOI 10.1109/ICMLA.2014.11. URL http://dx.doi.org/10.1109/ICMLA.2014.11

140. Gallup, G.G.: Chimpanzees: self-recognition. Science (1970)

141. Gama, J., Medas, P., Castillo, G., Rodrigues, P.: Learning with drift detection. In: Proceedings of the 7th Brazilian Symposium on Artificial Intelligence (SBIA) - Lecture Notes in Computer Science, vol. 3171, pp. 286–295. Springer, São Luiz do Maranhão, Brazil (2004)

142. Gao, J., Fan, W., Han, J.: On appropriate assumptions to mine data streams: Analysis and practice. In: Proceedings of the Seventh IEEE International Conference on Data Mining (ICDM), pp. 143–152 (2007)

143. Garlan, D., Cheng, S.W., Huang, A.C., Schmerl, B., Steenkiste, P.: Rainbow: architecture-based self-adaptation with reusable infrastructure. IEEE Computer 37(10), 46–54 (2004)

144. Gelenbe, E., Loukas, G.: A self-aware approach to denial of service defence. Computer Networks 51(5), 1299–1314 (2007)

145. Goto, M.: Active music listening interfaces based on signal processing. In: Proceedings of the IEEE International Conference on Acoustics, Speech and Signal Processing, vol. 4, pp. 1441–1444 (2007)

146. Gouin-Vallerand, C., Abdulrazak, B., Giroux, S., Mokhtari, M.: Toward autonomic pervasive computing. In: Proceedings of the 10th International Conference on Information Integration and Web-based Applications & Services, iiWAS '08, pp. 673–676. ACM, New York, NY, USA (2008)

147. Goukens, C., Dewitte, S., Warlop, L.: Me, myself, and my choices: The influence of private self-awareness on preference-behavior consistency. Tech. rep., Katholieke Universiteit Leuven (2007)

148. Grabner, H., Bischof, H.: On-line Boosting and Vision. In: Proceedings of the IEEE Computer Society Conference on Computer Vision and Pattern Recognition, pp. 260–267 (2006)

149. Grabner, H., Leistner, C., Bischof, H.: Semi-supervised on-line Boosting for Robust Tracking. In: Proceedings of the European Conference on Computer Vision, Lecture Notes in Computer Science, vol. 5302, pp. 234–247 (2008)

150. Group, K.: The OpenCL specification, version: 1.1. http://www.khronos.org/registry/cl/specs/opencl-1.1.pdf. (Accessed March 8, 2016)

151. Gudger, E.W.: A historical note on the synchronous flashing of fireflies. Science 50(1286), 188–190 (1919)

152. Gudgin, M., Hadley, M., Mendelsohn, N., Moreau, J.J., Nielsen, H.F., Karmarkar, A., Lafon, Y.: SOAP Version 1.2. World Wide Web Consortium (2007)

153. Guo, C., Luk, W.: Accelerating Maximum Likelihood Estimation for Hawkes Point Processes. In: Proceedings of the International Conference on Field Programmable Logic and Applications (FPL), pp. 1–6. IEEE (2013)

154. Guo, C., Luk, W.: Accelerating parameter estimation for multivariate self-exciting point processes. In: Proceedings of the International Symposium on Field-Programmable Gate Arrays (FPGA), pp. 181–184. ACM (2014). DOI 10.1145/2554688.2554765

155. Haikonen, P.O.: Reflections of consciousness: The mirror test. In: Proceedings of the AAAI Fall Symposium on Consciousness and Artificial Intelligence, pp. 67–71 (2007)

156. Hamid, R., Maddi, S., Johnson, A., Bobick, A., Essa, I., Isbell, C.: A novel sequence representation for unsupervised analysis of human activities. Artificial Intelligence 173(14), 1221–1244 (2009). DOI DOI: 10.1016/j.artint.2009.05.002

157. Hansen, N.: The CMA evolution strategy: A comparing review. In: J. Lozano, P. Larrañaga, I. Inza, E. Bengoetxea (eds.) Towards a New Evolutionary Computation, Studies in Fuzziness and Soft Computing, vol. 192, pp. 75–102. Springer Berlin Heidelberg (2006)

158. Happe, M., Agne, A., Plessl, C.: Measuring and Predicting Temperature Distributions on FPGAs at Run-Time. In: Proceedings of the International Conference on Reconfigurable Computing and FPGAs (ReConFig), pp. 55–60. IEEE Computer Society (2011). DOI 10.1109/ReConFig.2011.59

159. Happe, M., Huang, Y., Keller, A.: Dynamic Protocol Stacks in Smart Camera Networks. In: Proceedings of the International Conference on Reconfigurable Computing and FPGAs (ReConFig), pp. 1–6. IEEE (2014)

160. Happe, M., Traber, A., Keller, A.: Preemptive Hardware Multitasking in ReconOS. In: Proceedings of the International Symposium on Applied Reconfigurable Computing (ARC), Springer (2015)

161. Hart, J.W., Scassellati, B.: Robotic self-modeling. In: J. Pitt (ed.) The Computer After Me, pp. 207–218. Imperial College Press / World Scientific Book (2014)

162. Heath, D., Jarrow, R., Morton, A.: Bond pricing and the term structure of interest rates: A new methodology for contingent claims valuation. Econometrica **60**(1), 77–105 (1992)

163. Hernandez, H., Blum, C.: Distributed graph coloring in wireless ad hoc networks: A lightweight algorithm based on Japanese tree frogs' calling behaviour. In: Proceedings of the 4th Joint IFIP Wireless and Mobile Networking Conference (WMNC), pp. 1–7 (2011)

164. Herzen, B.V.: Signal Processing at 250 MHz Using High-Performance FPGAs. In: Proceedings of the ACM Fifth International Symposium on Field-programmable Gate Arrays, pp. 62–68 (1997)

165. Ho, T.K.: The Random Subspace Method for Constructing Decision Forests. IEEE Transactions on Pattern Analysis and Machine Intelligence **20**(8), 832–844 (1998)

166. Ho, T.K., Hull, J.J., Srihari, S.N.: Decision Combination in Multiple Classifier Systems. IEEE Transactions on Pattern Analysis and Machine Intelligence **16**(1), 66–75 (1994)

167. Ho, T.S.Y., Lee, S.B.: Term Structure Movements and Pricing Interest Rate Contingent Claims. Journal of Finance **41**(5), 1011–1029 (1986)

168. Hoare, C.A.R.: An axiomatic basis for computer programming. Communications of the ACM **12**(10), 576–580 (1969)

169. Hockman, J.A., Wanderley, M.M., Fujinaga, I.: Real-time phase vocoder manipulation by runner's pace. In: Proceedings of the International Conference on New Interfaces for Musical Expression (2009)

170. Hoffmann, H., Eastep, J., Santambrogio, M., Miller, J., Agarwal, A.: Application heartbeats for software performance and health. In: ACM SIGPLAN Notices, vol. 45, pp. 347–348. ACM (2010)

171. Hoffmann, H., Eastep, J., Santambrogio, M.D., Miller, J.E., Agarwal, A.: Application Heartbeats: A Generic Interface for Specifying Program Performance and Goals in Autonomous Computing Environments. In: Proceedings of the International Conference on Autonomic Computing (ICAC) (2010)

172. Hoffmann, H., Holt, J., Kurian, G., Lau, E., Maggio, M., Miller, J.E., Neuman, S.M., Sinangil, M., Sinangil, Y., Agarwal, A., Chandrakasan, A.P., Devadas, S.: Self-aware computing in the Angstrom processor. In: Proceedings of the 49th Annual Design Automation Conference, DAC '12, pp. 259–264. ACM, New York, NY, USA (2012)

173. Hoffmann, H., Maggio, M., Santambrogio, M.D., Leva, A., Agarwal, A.: SEEC: A general and extensible framework for self-aware computing. Tech. Rep. MIT-CSAIL-TR-2011-046, Computer Science and Artificial Intelligence Laboratory, Massachusetts Institute of Technology (2011)

174. Holland, B., George, A.D., Lam, H., Smith, M.C.: An analytical model for multilevel performance prediction of Multi-FPGA systems. ACM Transactions on Reconfigurable Technology and Systems **4**(3), 27–28 (2011)

175. Holland, O., Goodman, R.B.: Robots with internal models: A route to machine consciousness? Journal of Consciousness Studies **10**(4), 77–109 (2003)

176. Holopainen, R.: Self-organised sound with autonomous instruments: Aesthetics and experiments. Ph.D. thesis, University of Oslo (2012)

177. Hölzl, M., Wirsing, M.: Towards a system model for ensembles. In: Formal Modeling: Actors, Open Systems, Biological Systems, pp. 241–261. Springer (2011)

178. Hölzl, M., Wirsing, M.: Issues in engineering self-aware and self-expressive ensembles. In: J. Pitt (ed.) The Computer After Me, pp. 37–54. Imperial College Press/World Scientific Book (2014)

179. Horn, J., Nafpliotis, N., Goldberg, D.E.: A niched Pareto genetic algorithm for multiobjective optimization. In: Proceedings of the 1st IEEE Conference on Evolutionary Computation, IEEE World Congress on Computational Intelligence, pp. 82–87 (1994)

180. Horn, P.: Autonomic computing: IBM's perspective on the state of information technology. Armonk, NY, USA. International Business Machines Corporation. (2001)

181. Hosseini, M.J., Ahmadi, Z., Beigy, H.: Using a classifier pool in accuracy based tracking of recurring concepts in data stream classification. Evolving Systems **4**(1), 43–60 (2013)

182. Hsu, C.H., Feng, W.C.: Reducing overheating-induced failures via performance-aware CPU power management. In: Proceedings of the 6th International Conference on Linux Clusters: The HPC Revolution (2005)

183. Hu, F., Evans, J.J.: Power and environment aware control of Beowulf clusters. Cluster Computing **12**, 299–308 (2009)

184. Hu, W., Tan, T., Wang, L., Maybank, S.: A Survey on Visual Surveillance of Object Motion and Behaviors. IEEE Transactions on Systems, Man and Cybernetics, Part C **34**(3), 334–352 (2004)

185. Huang, T., Russell, S.: Object Identification in a Bayesian Context. In: Proceedings of the International Joint Conference on Artificial Intelligence, pp. 1276–1283 (1997)

186. Huebscher, M., McCann, J.: Simulation Model for Self-Adaptive Applications in Pervasive Computing. In: Proceedings of the 15th International Workshop on Database and Expert Systems Applications, pp. 694–698. IEEE Computer Society (2004)

187. Hume, D.: A Treatise of Human Nature. Gutenberg eBook (1739). URL http://www.gutenberg.org/ebooks/4705. (Accessed March 8, 2016)

188. Hunkeler, U., Truong, H.L., Stanford-Clark, A.: MQTT-S–A publish/subscribe protocol for Wireless Sensor Networks. In: Proceedings of the Third International Conference on Communication Systems Software and Middleware and Workshops (COMSWARE), pp. 791–798. IEEE (2008)

189. Hunt, A., Wanderley, M.M., Paradis, M.: The importance of parameter mapping in electronic instrument design. In: Proceedings of the International Conference on New Interfaces for Musical Expression, pp. 1–6. National University of Singapore (2002)

190. IBM: An architectural blueprint for autonomic computing (2003). URL http://www-03.ibm.com/autonomic/pdfs/AC Blueprint White Paper V7.pdf. (Accessed March 8, 2016)

191. Iglesia, D.: MobMuPlat (iOS application). Iglesia Intermedia (2013)

192. Intel: Sophisticated library for vector parallelism. http://software.intel.com/en-us/articles/intel-array-building-blocks/. (Accessed March 8, 2016)

193. Investigating RINA as an Alternative to TCP/IP. URL http://irati.eu. (Accessed March 8, 2016)

194. Ishibuchi, H., Murata, T.: A multiobjective genetic local search algorithm and its application to flowshop scheduling. IEEE Transactions on Systems, Man, and Cybernetics, Part C: Applications and Reviews **28**(3), 392–403 (1998)

195. Ishibuchi, H., Tsukamoto, N., Nojima, Y.: Iterative approach to indicator-based multiobjective optimization. In: Proceedings of the IEEE Congress on Evolutionary Computation, pp. 3967–3974 (2007)

196. Ishibuchi, H., Tsukamoto, N., Nojima, Y.: Evolutionary many-objective optimization. In: Proceedings of the 3rd International Workshop on Genetic and Evolving Systems (GEFS), pp. 47–52. IEEE (2008)

197. James, W.: The principles of psychology. Henry Holt & Co. (1890)

198. Janusevskis, J., Riche, R.L., Ginsbourger, D., Girdziusas, R.: Expected Improvements for the Asynchronous Parallel Global Optimization of Expensive Functions: Potentials and Challenges. In: Y. Hamadi, M. Schoenauer (eds.) Learning and Intelligent Optimization, pp. 413–418. Springer (2012)

199. Javed, O., Khan, S., Rasheed, Z., Shah, M.: Camera Handoff: Tracking in Multiple Uncalibrated Stationary Cameras. In: Proceedings of the Workshop on Human Motion, pp. 113–118. IEEE Computer Society Press (2000). DOI 10.1109/HUMO.2000.897380

200. Javed, O., Rasheed, Z., Shafique, K., Shah, M.: Tracking across Multiple Cameras Disjoint Views. In: Proceedings of IEEE International Conference on Computer Vision, p. 952–957 (2003)

201. Jia, J., Veeravalli, B., Ghose, D.: Adaptive load distribution strategies for divisible load processing on resource unaware multilevel tree networks. IEEE Transactions on Computers 56(7), 999–1005 (2007)

202. Jin, Q., Becker, T., Luk, W., Thomas, D.: Optimising explicit finite difference option pricing for dynamic constant reconfiguration. In: Proceedings of the International Conference on Field Programmable Logic and Applications (FPL), pp. 165–172 (2012)

203. Jin, Y., Olhofer, M., Sendhoff, B.: A framework for evolutionary optimization with approximate fitness functions. IEEE Transactions on Evolutionary Computation 6(5), 481–494 (2002)

204. Jones, D.R., Schonlau, M., Welch, W.J.: Efficient global optimization of expensive black-box functions. Journal of Global Optimization 13(4), 455–492 (1998)

205. Jones, P., Cho, Y., Lockwood, J.: Dynamically optimizing FPGA applications by monitoring temperature and workloads. In: Proceedings of the International Conference on VLSI Design (VLSID). IEEE (2007)

206. Kalal, Z., Mikolajczyk, K., Matas, J.: Tracking-Learning-Detection. IEEE Transactions on Pattern Analysis and Machine Intelligence 34(7), 1409–1422 (2012)

207. Kalman, R.E.: A New Approach to Linear Filtering and Prediction Problems. Journal of Fluids Engineering 82(1), 35–45 (1960)

208. Kamil, S., Chan, C., Oliker, L., Shalf, J., Williams, S.: An auto-tuning framework for parallel multicore stencil computations. In: Proceedings of the IEEE International Symposium on Parallel & Distributed Processing (IPDPS), pp. 1–12 (2010)

209. Kang, J., Cohen, I., Medioni, G.: Continuous Tracking within and across Camera Streams. In: Proceedings of IEEE Conference on Computer Vision and Pattern Recognition, pp. 267–272 (2003)

210. Kant, I.: The critique of pure reason. Gutenberg eBook (1781). URL http://www.gutenberg.org/ebooks/4280. Digital edition 2003 (Accessed March 8, 2016)

211. Kela, J., Korpipää, P., Mäntyjärvi, J., Kallio, S., Savino, G., Jozzo, L., Marca, D.: Accelerometer-based gesture control for a design environment. Personal and Ubiquitous Computing 10(5), 285–299 (2006)

212. Keller, A., Borkmann, D., Neuhaus, S., Happe, M.: Self-Awareness in Computer Networks. International Journal of Reconfigurable Computing pp. 1–10 (2014). DOI 10.1155/2014/692076

213. Kephart, J.O., Chess, D.M.: The Vision of Autonomic Computing. IEEE Computer 36(1), 41–50 (2003)

214. Kettnaker, V., Zabith, R.: Bayesian Multi-Camera Surveillance. In: Proceedings of the International Conference on Computer Vision and Pattern Recognition, pp. 117–123 (1999)

215. Khan, M.I., Rinner, B.: Energy-aware task scheduling in wireless sensor networks based on cooperative reinforcement learning. In: Proceedings of the International Conference on Communications Workshops (ICCW). IEEE (2014). DOI 10.1109/ICCW.2014.6881310

216. Khare, V., Yao, X., Deb, K.: Performance scaling of multi-objective evolutionary algorithms. In: Evolutionary Multi-Criterion Optimization, Lecture Notes in Computer Science, vol. 2632, pp. 376–390. Springer (2003)

217. Kim, H.S., Sherman, D.K.: Express yourself: Culture and the effect of self-expression on choice. Journal of Personality and Social Psychology 92(1), 1–11 (2007). DOI 10.1037/0022-3514.92.1.1

218. Kim, J., Seo, S., Lee, J., Nah, J., Jo, G., Lee, J.: OpenCL as a unified programming model for heterogeneous CPU/GPU clusters. In: Proceedings of the 17th ACM SIGPLAN Symposium on Principles and Practice of Parallel Programming (PPOPP), pp. 299–300 (2012)

219. Kittler, J., Hatef, M., Duin, R.P., Matas, J.: On combining classifiers. IEEE Transactions on Pattern Analysis and Machine Intelligence 20(3), 226–239 (1998)

220. Klinglmayr, J., Bettstetter, C.: Self-organizing synchronization with inhibitory-coupled os-
     cillators: Convergence and robustness. ACM Transactions on Autonomous and Adaptive
     Systems **7**(3), 30:1–30:22 (2012)
221. Klinglmayr, J., Kirst, C., Bettstetter, C., Timme, M.: Guaranteeing global synchronization in
     networks with stochastic interactions. New Journal of Physics **14**(7), 1–13 (2012)
222. Knutzen, H., Nymoen, K., Tørresen, J.: PheroMusic [iOS application].   URL
     http://itunes.apple.com/app/pheromusic/id910100415
223. Kolter, J.Z., Maloof, M.A.: Dynamic weighted majority: An ensemble method for drifting
     concepts. Journal of Machine Learning Research **8**, 2755–2790 (2007)
224. Koski, J., Silvennoinen, R.: Norm methods and partial weighting in multicriterion optimiza-
     tion of structures. International Journal for Numerical Methods in Engineering **24**(6), 1101–
     1121 (1987)
225. Kramer, J., Magee, J.: Self-managed systems: an architectural challenge. In: Future of Soft-
     ware Engineering (FoSE), pp. 259–268. IEEE (2007)
226. Krishnamoorthy, S., Baskaran, M., Bondhugula, U., Ramanujam, J., Rountev, A., Sadayap-
     pan, P.: Effective automatic parallelization of stencil computations. In: Proceedings of the
     28th ACM SIGPLAN Conference on Programming Language Design and Implementation,
     pp. 235–244 (2007)
227. Kuhn, H.W., Yaw, B.: The Hungarian Method for the Assignment Problem. Naval Research
     Logistics Quarterly pp. 83–97 (1955)
228. Kuncheva, L.I.: A theoretical study on six classifier fusion strategies. IEEE Transactions on
     Pattern Analysis and Machine Intelligence **24**(2), 281–286 (2002)
229. Kurek, M., Becker, T., Chau, T.C., Luk, W.: Automating Optimization of Reconfigurable
     Designs. In: Proceedings of the International Symposium on Field-Programmable Custom
     Computing Machines (FCCM), pp. 210–213. IEEE (2014). DOI 10.1109/FCCM.2014.65
230. Kurek, M., Becker, T., Luk, W.: Parametric Optimization of Reconfigurable Designs Using
     Machine Learning. In: Proceedings of the International Conference on Reconfigurable Com-
     puting: Architectures, Tools and Applications (ARC), *Lecture Notes in Computer Science*,
     vol. 7806, pp. 134–145. Springer (2013)
231. Legrain, L., Cleeremans, A., Destrebecqz, A.: Distinguishing three levels in explicit self-
     awareness. Consciousness and Cognition **20**, 578–585 (2011)
232. Legrand, D.: Pre-reflective self-as-subject from experiential and empirical perspectives. Con-
     sciousness and Cognition **16**(3), 583–599 (2007)
233. Leidenfrost, R., Elmenreich, W.: Firefly clock synchronization in an 802.15. 4 wireless net-
     work. EURASIP Journal on Embedded Systems **2009**, 7:1–7:17 (2009)
234. Leland, W., Taqqu, M., Willinger, W., Wilson, D.: On the self-similar nature of ethernet
     traffic (extended version). IEEE/ACM Transactions on Networking **2**(1), 1–15 (1994). DOI
     10.1109/90.282603
235. Leutenegger, S., Chli, M., Siegwart, R.Y.: BRISK: Binary robust invariant scalable keypoints.
     In: Proceedings of the International Conference on Computer Vision, pp. 2548–2555. IEEE
     (2011). DOI 10.1109/iccv.2011.6126542
236. Lewis, P.R., Chandra, A., Faniyi, F., Glette, K., Chen, T., Bahsoon, R., Tørresen, J., Yao, X.:
     Architectural aspects of self-aware and self-expressive computing systems: From psychology
     to engineering. IEEE Computer **48**(8), 62–70 (2015)
237. Lewis, P.R., Chandra, A., Parsons, S., Robinson, E., Glette, K., Bahsoon, R., Tørresen, J.,
     Yao, X.: A Survey of Self-Awareness and Its Application in Computing Systems. In: Pro-
     ceedings of the International Conference on Self-Adaptive and Self-Organizing Systems
     Workshops (SASOW), pp. 102–107. IEEE Computer Society, Ann Arbor, MI, USA (2011)
238. Lewis, P.R., Esterle, L., Chandra, A., Rinner, B., Tørresen, J., Yao, X.: Static, Dynamic,
     and Adaptive Heterogeneity in Distributed Smart Camera Networks. ACM Transactions on
     Autonomous and Adaptive Systems **10**(2), 8:1–8:30 (2015). DOI 10.1145/2764460
239. Lewis, P.R., Esterle, L., Chandra, A., Rinner, B., Yao, X.: Learning to be Different: Het-
     erogeneity and Efficiency in Distributed Smart Camera Networks. In: Proceedings of the
     International Conference on Self-Adaptive and Self-Organizing Systems (SASO), pp. 209–
     218. IEEE Computer Society Press (2013). DOI 10.1109/SASO.2013.20

240. Lewis, P.R., Marrow, P., Yao, X.: Resource Allocation in Decentralised Computational Systems: An Evolutionary Market Based Approach. Autonomous Agents and Multi-Agent Systems **21**(2), 143–171 (2010)

241. Lewis, P.R., Platzner, M., Yao, X.: An outlook for self-awareness in computing systems. Awareness Magazine (2012). DOI 10.2417/3201203.004093

242. Li, B., Li, J., Tang, K., Yao, X.: An improved Two Archive Algorithm for Many-Objective Optimization. In: Proceedings of the IEEE Congress on Evolutionary Computation (CEC), pp. 2869–2876 (2014)

243. Li, G., Gopalakrishnan, G.: Scaleable SMT-based verification of GPU kernel functions. In: Proceedings of the Eighteenth International Symposium on the Foundations of Software Engineering (FSE-18) (2010)

244. Li, H., Zhang, Q.: Multiobjective optimization problems with complicated Pareto sets, MOEA/D and NSGA-II. IEEE Transactions on Evolutionary Computation **13**(2), 284–302 (2009)

245. Li, Y., Bhanu, B.: Utility-based Camera Assignment in a Video Network: A Game Theoretic Framework. Sensors Journal **11**(3), 676–687 (2011)

246. Liang, C.J.M., Liu, J., Luo, L., Terzis, A., Zhao, F.: RACNet: A High-Fidelity Data Center Sensing Network. Proceedings of the 7th ACM Conference on Embedded Networked Sensor Systems pp. 15–28 (2009)

247. Liu, J., Zhong, L., Wickramasuriya, J., Vasudevan, V.: uWave: Accelerometer-based personalized gesture recognition and its applications. Pervasive and Mobile Computing **5**(6), 657–675 (2009)

248. Liu, Y., Yao, X.: Ensemble learning via negative correlation. Neural Networks **12**(10), 1399–1404 (1999)

249. Lübbers, E., Platzner, M.: Cooperative multithreading in dynamically reconfigurable systems. In: Proceedings of the International Conference on Field Programmable Logic and Applications (FPL), pp. 1–4. IEEE (2009)

250. Lübbers, E., Platzner, M.: ReconOS: Multithreaded programming for reconfigurable computers. ACM Transactions on Embedded Computing Systems **9** (2009)

251. Lucas, B.D., Kanade, T.: An iterative image registration technique with an application to stereo vision. In: Proceedings of the International Joint Conference on Artificial Intelligence (IJCAI), pp. 674–679 (1981)

252. Maggio, M., Hoffmann, H., Santambrogio, M.D., Agarwal, A., Leva, A.: A comparison of autonomic decision making techniques. Tech. Rep. MIT-CSAIL-TR-2011-019, Computer Science and Artificial Intelligence Laboratory, Massachusetts Institute of Technology (2011)

253. Makris, D., Ellis, T., Black, J.: Bridging the Gaps between Cameras. In: Proceedings of Conference on Computer Vision and Pattern Recognition, vol. 2 (2004)

254. Marler, R.T., Arora, J.S.: Function-transformation methods for multi-objective optimization. Engineering Optimization **37**(6), 551–570 (2005)

255. Marrow, P.: Nature-inspired computing technology and applications. BT Technology Journal **18**(4), 13–23 (2000)

256. Marsaglia, G., Bray, T.A.: A convenient method for generating normal variables. SIAM Review **6**(3), 260–264 (1964)

257. Masahiro, N., Takaesu, H., Demachi, H., Oono, M., Saito, H.: Development of an automatic music selection system based on runner's step frequency. In: Proceedings of the 2008 International Conference on Music Information Retrieval, pp. 193–8 (2008)

258. Massie, M.L., Chun, B.N., Culler, D.E.: The Ganglia distributed monitoring system: design, implementation, and experience. Parallel Computing **30**, 817–840 (2004)

259. Mathar, R., Mattfeldt, J.: Pulse-coupled decentral synchronization. SIAM Journal on Applied Mathematics **56**(4), 1094–1106 (1996)

260. Max [computer software]. URL http://cycling74.com. (Accessed March 8, 2016)

261. McKay, M.D., Beckman, R.J., Conover, W.J.: A comparison of three methods for selecting values of input variables in the analysis of output from a computer code. Technometrics pp. 55–61 (2000)

262. Mehta, N.R., Medvidovic, N.: Composing architectural styles from architectural primitives. In: Proceedings of the European Software Engineering Conference and ACM SIGSOFT Symposium on the Foundations of Software Engineering, pp. 347–350 (2003). URL http://dblp.uni-trier.de/db/conf/sigsoft/fse2003.html#MehtaM03

263. Menasce, D.A., Sousa, J.a.P., Malek, S., Gomaa, H.: QoS Architectural Patterns for Self-architecting Software Systems. In: Proceedings of the 7th International Conference on Autonomic Computing (ICAC), pp. 195–204. ACM (2010). DOI 10.1145/1809049.1809084

264. Metcalfe, J., Shimamura, A.P. (eds.): Metacognition: Knowing about knowing. MIT Press, Cambridge, MA, USA (1994)

265. Michalski, R.S.: A Theory and Methodology of Inductive Learning. In: Machine Learning, Symbolic Computation, pp. 83–134. Springer Berlin Heidelberg (1983)

266. Miettinen, K., Mäkelä, M.M.: Interactive bundle-based method for nondifferentiable multiobjective optimization: nimbus. Optimization Journal **34**(3), 231–246 (1995)

267. Minku, L.L.: Online ensemble learning in the presence of concept drift. Ph.D. thesis, School of Computer Science, University of Birmingham, Birmingham, UK (2010)

268. Minku, L.L., Yao, X.: DDD: A new ensemble approach for dealing with concept drift. IEEE Transactions on Knowledge and Data Engineering **24**(4), 619–633 (2012)

269. Minku, L.L., Yao, X.: Software Effort Estimation as a Multi-objective Learning Problem. ACM Transactions on Software Engineering and Methodology **22**(4), 35:1–32 (2013)

270. Miranda, E.R., Wanderley, M.: New Digital Musical Instruments: Control and Interaction Beyond the Keyboard. A-R Editions, Inc., Middleton, WI (2006)

271. Mirollo, R.E., Strogatz, S.H.: Synchronization of pulse-coupled biological oscillators. SIAM Journal on Applied Mathematics **50**(6), 1645–1662 (1990)

272. Mitchell, M.: Self-awareness and control in decentralized systems. In: Proceedings of the AAAI Spring Symposium on Metacognition in Computation (2005). Available at http://www.cs.pdx.edu/ mm/self-awareness.pdf

273. Modler, P.: Neural networks for mapping hand gestures to sound synthesis parameters, vol. 18, p. 14. IRCAM — Centre Pompidou (2000)

274. Moens, B., van Noorden, L., Leman, M.: D-Jogger: Syncing music with walking. In: Proceedings of the Sound and Music Computing Conference, pp. 451–456. Barcelona, Spain (2010)

275. Morin, A.: Levels of consciousness and self-awareness: A comparison and integration of various neurocognitive views. Consciousness and Cognition **15**(2), 358–71 (2006)

276. Morin, A., Everett, J.: Conscience de soi et langage interieur: Quelques speculations. [Self-awareness and inner speech: Some speculations]. Philosophiques **XVII**(2), 169–188 (1990)

277. Müller-Schloer, C., Schmeck, H., Ungerer, T.: Organic computing: a paradigm shift for complex systems. Springer (2011)

278. Nakashima, H., Aghajan, H., Augusto, J.C.: Handbook of ambient intelligence and smart environments. Springer (2009)

279. Narukawa, K., Tanigaki, Y., Ishibuchi, H.: Evolutionary many-objective optimization using preference on hyperplane. In: Proceedings of the 2014 Conference on Genetic and Evolutionary Computation Companion, pp. 91–92. ACM (2014)

280. Natarajan, P., Atrey, P.K., Kankanhalli, M.: Multi-camera coordination and control in surveillance systems: A survey. ACM Transactions on Multimedia Computing, Communications and Applications **11**(4), 57:1–57:30 (2015). DOI 10.1145/2710128

281. Nebehay, G., Chibamu, W., Lewis, P.R., Chandra, A., Pflugfelder, R., Yao, X.: Can diversity amongst learners improve online object tracking? In: Z.H. Zhou, F. Roli, J. Kittler (eds.) Multiple Classifier Systems, *Lecture Notes in Computer Science*, vol. 7872, pp. 212–223. Springer (2013). DOI 10.1007/978-3-642-38067-9_19

282. Nebehay, G., Pflugfelder, R.: Consensus-based matching and tracking of keypoints for object tracking. In: Proceedings of the Winter Conference on Applications of Computer Vision (WACV). IEEE (2014)

283. Nebro, A.J., Luna, F., Alba, E., Beham, A., Dorronsoro, B.: AbYSS: adapting scatter search for multiobjective optimization. Tech. Rep. ITI-2006-2, Departamento de Lenguajes y Ciencias de la Computación, University of Málaga, Malaga (2006)

284. Neisser, U.: The Roots of Self-Knowledge: Perceiving Self, It, and Thou. Annals of the NY AoS. **818**, 19–33 (1997)
285. netem. URL http://www.linuxfoundation.org/collaborate/workgroups/networking/netem. (Accessed March 8, 2016)
286. Nguyen, A., Satish, N., Chhugani, J., Kim, C., Dubey, P.: 3.5-D blocking optimization for stencil computations on modern CPUs and GPUs. In: Proceedings of the ACM/IEEE International Conference for High Performance Computing, Networking, Storage and Analysis, pp. 1–13 (2010)
287. Niezen, G., Hancke, G.P.: Evaluating and optimising accelerometer-based gesture recognition techniques for mobile devices. In: Proceedings of AFRICON, pp. 1–6. IEEE (2009)
288. Nishida, K.: Learning and detecting concept drift. Ph.D. thesis, Hokkaido University (2008). URL http://lis2.huie.hokudai.ac.jp/ knishida/paper/nishida2008-dissertation.pdf
289. Nishida, K., Yamauchi, K.: Detecting concept drift using statistical testing. In: Proceedings of the Tenth International Conference on Discovery Science (DS) - Lecture Notes in Artificial Intelligence, vol. 3316, pp. 264–269. Sendai, Japan (2007)
290. Niu, X., Chau, T.C.P., Jin, Q., Luk, W., Liu, Q.: Automating elimination of idle functions by run-time reconfiguration. In: Proceedings of the 21st IEEE Annual International Symposium on Field-Programmable Custom Computing Machines (FCCM), pp. 97–104 (2013)
291. Niu, X., Coutinho, J.G.F., Luk, W.: A scalable design approach for stencil computation on reconfigurable clusters. In: Proceedings of the 23rd International Conference on Field programmable Logic and Applications (FPL), pp. 1–4 (2013)
292. Niu, X., Jin, Q., Luk, W., Liu, Q., Pell, O.: Exploiting run-time reconfiguration in stencil computation. In: Proceedings of the 22nd International Conference on Field programmable Logic and Applications (FPL), pp. 173–180 (2012)
293. Niu, X., Tsoi, K.H., Luk, W.: Reconfiguring distributed applications in FPGA accelerated cluster with wireless networking. In: Proceedings of the 21st International Conference on Field Programmable Logic and Applications (FPL), pp. 545–550 (2011)
294. NVIDIA: Cuda zone. http://www.nvidia.com/object/cuda_home_new.html. (Accessed March 8, 2016)
295. Nymoen, K., Chandra, A., Glette, K., Tørresen, J.: Decentralized harmonic synchronization in mobile music systems. In: Proceedings of the International Conference on Awareness Science & Technology (iCAST), pp. 1–6 (2014)
296. Nymoen, K., Chandra, A., Glette, K., Tørresen, J., Voldsund, A., Jensenius, A.R.: PheroMusic: Navigating a Musical Space for Active Music Experiences. In: Proceedings of the International Computer Music Conference (ICMC) joint with the Sound and Music Computing Conference, pp. 1715–1718 (2014)
297. Nymoen, K., Song, S., Hafting, Y., Tørresen, J.: Funky Sole Music: Gait recognition and adaptive mapping. In: Proceedings of the International Conference on New Interfaces for Musical Expression (NIME), pp. 299–302 (2014)
298. Okuma, K., Taleghani, A., de Freitas, N., Little, J., Lowe, D.: A Boosted Particle Filter: Multitarget Detection and Tracking. In: Proceedings of 8th European Conference on Computer Vision, vol. 3021, pp. 28–39 (2004)
299. Olfati-Saber, R.: Distributed Kalman filtering for sensor networks. In: Proceedings of the Conference on Decision and Control, pp. 5492–5498 (2007). DOI 10.1109/CDC.2007.4434303
300. Olsson, R.A., Keen, A.W.: Remote procedure call. The JR Programming Language: Concurrent Programming in an Extended Java pp. 91–105 (2004)
301. Ong, Y.S., Nair, P.B., Keane, A.J.: Evolutionary optimization of computationally expensive problems via surrogate modeling. AIAA Journal **41**(4), 689–696 (2003)
302. Ontañón, S., Plaza, E.: Multiagent Inductive Learning: An Argumentation-based Approach. In: J. Fürnkranz, T. Joachims (eds.) Proceedings of the 27th International Conference on Machine Learning (ICML), pp. 839–846. Omnipress, Haifa, Israel (2010)
303. Oxford: Oxford dictionaries: Adapt. http://www.oxforddictionaries.com/definition/english/adapt. (Accessed March 8, 2016)

304. Oza, N.C.: Online bagging and boosting. In: Proceedings of the IEEE International Conference on Systems, Man and Cybernetics, pp. 2340–2345 (2005)

305. Oza, N.C., Russell, S.: Experimental comparisons of online and batch versions of bagging and boosting. In: Proceedings of the seventh ACM SIGKDD International Conference on Knowledge Discovery and Data Mining, pp. 359–364 (2001)

306. Özuysal, M., Calonder, M., Lepetit, V., Fua, P.: Fast Keypoint Recognition Using Random Ferns. IEEE Transactions on Pattern Analysis and Machine Intelligence **32**(3), 448–461 (2010). DOI 10.1109/tpami.2009.23

307. Page, I., Luk, W.: Compiling occam into Field-Programmable Gate Arrays. In: Proceedings of the International Conference on Field programmable Logic and Applications (FPL) (1991)

308. Papakonstantinou, A., Liang, Y., Stratton, J.A., Gururaj, K., Chen, D., Hwu, W.W., Cong, J.: Multilevel Granularity Parallelism Synthesis on FPGAs. In: Proceedings of the 19th Annual International Symposium on Field-Programmable Custom Computing Machines (FCCM), pp. 178–185. IEEE (2011)

309. Parashar, M., Hariri, S.: Autonomic computing: an overview. In: Proceedings of the International Conference on Unconventional Programming Paradigms, pp. 257–269. Springer-Verlag, Berlin (2005)

310. Parsons, S., Bahsoon, R., Lewis, P.R., Yao, X.: Towards a better understanding of self-awareness and self-expression within software systems. Tech. Rep. CSR-11-03, University of Birmingham, School of Computer Science, UK (2011)

311. Paul, C., Bass, L., Kazman, R.: Software Architecture in Practice. MA: Addison-Wesley (1998)

312. Paul, C., Kazman, R., Klein, M.: Evaluating Software Architectures: Methods and Case Studies. Addison-Wesley (2002)

313. Paulson, L.: DARPA creating self-aware computing. IEEE Computer **36**(3), 24 (2003). DOI 10.1109/MC.2003.1185213

314. Peleg, A., Wilkie, S., Weiser, U.C.: Intel MMX for Multimedia PCs. Communications of the ACM **40**(1), 24–38 (1997)

315. Perkowitz, M., Philipose, M., Fishkin, K., Patterson, D.J.: Mining models of human activities from the Web. In: Proceedings of the 13th International Conference on World Wide Web, pp. 573–582 (2004)

316. Perrone, M., Liu, L.K., Lu, L., Magerlein, K., Kim, C., Fedulova, I., Semenikhin, A.: Reducing Data Movement Costs: Scalable Seismic Imaging on Blue Gene. In: Proceedings of the 26th International Parallel & Distributed Processing Symposium (IPDPS), pp. 320–329 (2012)

317. Perrone, M.P., Cooper, L.N.: When networks disagree: Ensemble methods for hybrid neural networks. Neural Networks for Speech and Image Processing, Chapman-Hall, New York pp. 126–142 (1993)

318. Peskin, C.S.: Mathematical aspects of heart physiology. Courant Institute of Mathematical Sciences, New York University New York (1975)

319. Pflugfelder, R., Bischof, H.: People Tracking across Two Distant Self-calibrated Cameras. In: Proceedings of International Conference on Advanced Video and Signal-based Surveillance. IEEE Computer Society Press (2006)

320. Pflugfelder, R., Bischof, H.: Tracking across Non-overlapping Views Via Geometry. In: Proceedings of the International Conference on Pattern Recognition (2008)

321. Phelps, S., McBurney, P., Parsons, S.: Evolutionary mechanism design: A review. Autonomous Agents and Multi-Agent Systems **21**(2), 237–264 (2010)

322. Piciarelli, C., Esterle, L., Khan, A., Rinner, B., Foresti, G.: Dynamic Reconfiguration in Camera Networks: a short survey. IEEE Transactions on Circuits and Systems for Video Technology **PP**(99), 1–13 (2015). DOI 10.1109/TCSVT.2015.2426575. (early access)

323. Pilato, C., Loiacono, D., Tumeo, A., Ferrandi, F., Lanzi, P.L., Sciuto, D.: Speeding-up expensive evaluations in highlevel synthesis using solution modeling and fitness inheritance. In: Y. Tenne, C.K. Goh (eds.) Computational Intelligence in Expensive Optimization Problems, vol. 2, pp. 701–723. Springer (2010)

324. Polikar, R.: Ensemble based systems in decision making. IEEE Circuits and Systems Magazine **6**(3), 21–45 (2006)
325. Polikar, R., Udpa, L., Udpa, S., Honavar, V.: Learn++: An incremental learning algorithm for supervised neural networks. IEEE Transactions on Systems, Man, and Cybernetics, Part C: Applications and Reviews **31**(4), 497–508 (2001)
326. Puckette, M.: Pure Data (PD) (software). URL http://puredata.info. (Accessed March 8, 2016)
327. Pylvänäinen, T.: Accelerometer based gesture recognition using continuous HMMs. In: Pattern Recognition and Image Analysis, pp. 639–646. Springer (2005)
328. Quaritsch, M., Kreuzthaler, M., Rinner, B., Bischof, H., Strobl, B.: Autonomous Multicamera Tracking on Embedded Smart Cameras. EURASIP Journal on Embedded Systems **2007**(1), 35–45 (2007)
329. Rajko, S., Qian, G., Ingalls, T., James, J.: Real-time gesture recognition with minimal training requirements and on-line learning. In: Proceedings of the IEEE Conference on Computer Vision and Pattern Recognition (CVPR), pp. 1–8. IEEE (2007)
330. Ramamurthy, S., Bhatnagar, R.: Tracking recurrent concept drift in streaming data using ensemble classifiers. In: Proceedings of the Sixth International Conference on Machine Learning and Applications (ICMLA), pp. 404–409. Cincinnati, Ohio (2007)
331. Rammer, I., Szpuszta, M.: Advanced .NET Remoting. Springer (2005)
332. Ramos, C., Augusto, J.C., Shapiro, D.: Ambient intelligence - the next step for artificial intelligence. IEEE Intelligent Systems **23**(2), 15–18 (2008). DOI 10.1109/MIS.2008.19
333. Rasmussen, C., Williams, C.: Gaussian Processes for Machine Learning. MIT Press (2006)
334. Reason [computer software]. URL https://www.propellerheads.se. (Accessed March 8, 2016)
335. ReconOS: A programming model and OS for reconfigurable hardware (2013). URL http://www.reconos.de/. (Accessed March 8, 2016)
336. Reisslein, M., Rinner, B., Roy-Chowdhury, A.: Smart camera networks. IEEE Computer **47**(5), 26–28 (2014)
337. Reyes, R., de Sande, F.: Automatic code generation for gpus in llc. The Journal of Supercomputing **58**(3), 349–356 (2011)
338. Richter, U., Mnif, M., Branke, J., Müller-Schloer, C., Schmeck, H.: Towards a generic observer/controller architecture for organic computing. In: C. Hochberger, R. Liskowsky (eds.) INFORMATIK 2006 – Informatik für Menschen, *LNI*, vol. P-93, pp. 112–119. Bonner Köllen Verlag (2006)
339. Rietmann, M., Messmer, P., Nissen-Meyer, T., Peter, D., Basini, P., Komatitsch, D., Schenk, O., Tromp, J., Boschi, L., Giardini, D.: Forward and adjoint simulations of seismic wave propagation on emerging large-scale GPU architectures. In: Proceedings of the Conference on High Performance Computing Networking, Storage and Analysis (SC) (2012)
340. Rinner, B., Esterle, L., Simonjan, J., Nebehay, G., Pflugfelder, R., Fernandez, G., Lewis, P.R.: Self-Aware and Self-Expressive Camera Networks. IEEE Computer **48**(7), 33–40 (2015)
341. Rinner, B., Winkler, T., Schriebl, W., Quaritsch, M., Wolf, W.: The evolution from single to pervasive smart cameras. In: Proceedings of the Second ACM/IEEE International Conference on Distributed Smart Cameras (ICDSC), pp. 1–10 (2008). DOI 10.1109/ICDSC.2008.4635674
342. Rinner, B., Wolf, W.: Introduction to Distributed Smart Cameras. Proceedings of the IEEE **96**(10), 1565–1575 (2008). DOI 10.1109/JPROC.2008.928742
343. RNA: Recursive Network Architecture. URL http://www.isi.edu/rna. (Accessed March 8, 2015)
344. Rochat, P.: Five levels of self-awareness as they unfold in early life. Consciousness and Cognition **12**, 717–731 (2003)
345. Russell, S.J., Norvig, P.: Artificial Intelligence - A Modern Approach, 3 edn. Pearson Education (2010)
346. Saaty, T.L.: The Analytical Hierarchical Process. McGraw-Hill (1980)
347. Sakellari, G.: The cognitive packet network: A survey. The Computer Journal **53** (2010)
348. SanMiguel, J.C., Shoop, K., Cavallaro, A., Micheloni, C., Foresti, G.L.: Self-Reconfigurable Smart Camera Networks. IEEE Computer **47**(5), 67–73 (2014)

349. Santambrogio, M., Hoffmann, H., Eastep, J., Agarwal, A.: Enabling technologies for self-aware adaptive systems. In: 2010 NASA/ESA Conference on Adaptive Hardware and Systems (AHS), pp. 149–156. IEEE (2010)

350. Santner, J., Leistner, C., Saffari, A., Pock, T., Bischof, H.: PROST: Parallel robust online simple tracking. In: Proceedings of the IEEE Conference on Computer Vision and Pattern Recognition (CVPR), pp. 723–730 (2010)

351. Schaumeier, J., Jeremy Pitt, J., Cabri, G.: A tripartite analytic framework for characterising awareness and self-awareness in autonomic systems research. In: Proceedings of the Sixth IEEE Conference on Self-Adaptive and Self-Organizing Systems Workshops (SASOW), pp. 157–162 (2012)

352. Schlömer, T., Poppinga, B., Henze, N., Boll, S.: Gesture recognition with a Wii controller. In: Proceedings of the 2nd International Conference on Tangible and Embedded Interaction, pp. 11–14. ACM (2008)

353. Schmeck, H.: Organic computing - a new vision for distributed embedded systems. In: Proceedings of the Eighth IEEE International Symposium on Object-Oriented Real-Time Distributed Computing (ISORC), pp. 201–203 (2005)

354. Schmickl, T., Thenius, R., Moslinger, C., Timmis, J., Tyrrell, A., Read, M., Hilder, J., Halloy, J., Campo, A., Stefanini, C., Manfredi, L., Orofino, S.: CoCoRo–The Self-Aware Underwater Swarm. In: Proceedings of the International Conference on Self-Adaptive and Self-Organizing Systems Workshops (SASOW), pp. 120–126. IEEE Computer Society, Ann Arbor, MI, USA (2011)

355. Schnier, T., Yao, X.: Using negative correlation to evolve fault-tolerant circuits. In: Proceedings of the 5th International Conference on Evolvable Systems: From Biology to Hardware (ICES'2003) – Lecture Notes in Computer Science, vol. 2606, pp. 35–46. Springer-Verlag (2003)

356. Scholz, M., Klinkenberg, R.: Boosting classifiers for drifting concepts. Intelligent Data Analysis 11(1), 3–28 (2007)

357. Sharan, K.: Java remote method invocation. In: Beginning Java 8 APIs, Extensions and Libraries, chap. 7, pp. 525–548. Springer (2014)

358. Shaw, M.J., Sikora, R.: A distributed problem-solving approach to inductive learning. Tech. Rep. CMU-RI-TR-90-262, School of Computer Science, Carnegie Mellon University (1990)

359. Shipp, C.A., Kuncheva, L.I.: Relationships between combination methods and measures of diversity in combining classifiers. Information Fusion 3(2), 135–148 (2002)

360. Showerman, M., Enos, J., Pant, A., Kindratenko, V., Steffen, C., Pennington, R., mei Hwu, W.: QP: A heterogeneous multi-accelerator cluster. In: Proceedings of the International Conference on High-Performance Clustered Computing (2009)

361. Shukla, S.K., Yang, Y., Bhuyan, L.N., Brisk, P.: Shared memory heterogeneous computation on PCIe-supported platforms. In: Proceedings of the 23rd International Conference on Field programmable Logic and Applications (FPL), pp. 1–4 (2013)

362. Simonjan, J., Esterle, L., Rinner, B., Nebehay, G., Dominguez, G.F.: Demonstrating autonomous handover in heterogeneous multi-camera systems. In: Proceedings of the International Conference on Distributed Smart Cameras, pp. 43:1–43:3 (2014). DOI 10.1145/2659021.2669474

363. Sironi, F., Bartolini, D.B., Campanoni, S., Cancare, F., Hoffmann, H., Sciuto, D., Santambrogio, M.D.: Metronome: Operating System Level Performance Management via Self-adaptive Computing. In: Proceedings of the Design Automation Conference (DAC). ACM (2012)

364. Sironi, F., Cuoccio, A., Hoffmann, H., Maggio, M., Santambrogio, M.: Evolvable Systems on Reconfigurable Architecture via Self-aware Adaptive Applications. In: Proceedings of the NASA/ESA Conference on Adaptive Hardware and Systems (AHS) (2011). DOI 10.1109/AHS.2011.5963933

365. Sironi, F., Triverio, M., Hoffmann, H., Maggio, M., Santambrogio, M.: Self-aware Adaptation in FPGA-based Systems. In: Proceedings of the International Conference on Field Programmable Logic and Applications. IEEE (2010)

366. Smallwood, J., McSpadden, M., Schooler, J.: The lights are on but no one's home: meta-awareness and the decoupling of attention when the mind wanders. Psychonomic Bulletin and Review **14**(3), 527–533 (2007)

367. Song, S., Chandra, A., Tørresen, J.: An ant learning algorithm for gesture recognition with one-instance training. In: Proceedings of the International Congress on Evolutionary Computation (CEC), pp. 2956–2963. IEEE (2013)

368. SRC Computers, L.: SRC-7 MAPstation. Tech. rep., SRC Computers (2009)

369. Srinivas, N., Deb, K.: Multiobjective optimization using nondominated sorting in genetic algorithms. Evolutionary Computation **2**(3), 221–248 (1994)

370. Stanley, K.O.: Learning concept drift with a committee of decision trees. Tech. Rep. UT-AI-TR-03-302, Department of Computer Sciences, University of Texas at Austin (2003)

371. Sterritt, R., Parashar, M., Tianfield, H., Unland, R.: A concise introduction to autonomic computing. Advanced Engineering Informatics **19**(3), 181–187 (2005)

372. Steuer, R.E., Choo, E.U.: An interactive weighted Tchebycheff procedure for multiple objective programming. Mathematical Programming **26**(3), 326–344 (1983)

373. Stone, P.: Layered Learning in Multiagent Systems: A Winning Approach to Robotic Soccer. MIT Press (2000)

374. Strassen, V.: Gaussian elimination is not optimal. Numerische Mathematik pp. 13:354–356 (1969)

375. Street, W., Kim, Y.: A streaming ensemble algorithm (SEA) for large-scale classification. In: Proceedings of the Seventh ACM International Conference on Knowledge Discovery and Data Mining (KDD), pp. 377–382. New York (2001)

376. Strenski, D.: The Cray XD1 computer and its reconfigurable architecture. Tech. rep., Cray Inc. (2005)

377. Strey, A., Bange, M.: Performance Analysis of Intel's MMX and SSE: A Case Study. In: Proceedings of 7th International Euro-Par Conference on Parallel Processing (Euro-Par), pp. 142–147. Manchester, UK (2001)

378. Susanto, K.W., Todman, T., Coutinho, J.G.F., Luk, W.: Design Validation by Symbolic Simulation and Equivalence Checking: A Case Study in Memory Optimization for Image Manipulation, *LNCS*, vol. 5404, pp. 509–520. Springer (2009)

379. Sutter, H.: The free lunch is over: A fundamental turn toward concurrency in software. Dr. Dobb's Journal (2005)

380. Sutton, R.S., Barto, A.G.: Reinforcement Learning: An Introduction. MIT Press (1998)

381. Taj, M., Cavallaro, A.: Distributed and decentralized multi-camera tracking. IEEE Signal Processing Magazine **28**(3), 46–58 (2011)

382. Tawney, G.A.: Feeling and self-awareness. Psyc. Rev. **9**(6), 570 – 596 (1902)

383. Tesauro, G.: Reinforcement learning in autonomic computing: A manifesto and case studies. IEEE Internet Computing **11**(1), 22–30 (2007)

384. Thomas, D., Luk, W.: Non-uniform random number generation through piecewise linear approximations. In: Proceedings of the International Conference on Field Programmable Logic and Applications (FPL), pp. 1–6 (2006)

385. Thomas, D.B., Luk, W.: Credit Risk Modelling using Hardware Accelerated Monte-Carlo Simulation. In: Proceedings of the 16th IEEE International Symposium on Field-Programmable Custom Computing Machines (FCCM), pp. 229–238 (2008)

386. Todman, T., Boehm, P., Luk, W.: Verification of streaming hardware and software codesigns. In: Proceedings of the International Conference on Field Programmable Technology (ICFPT), pp. 147–150. IEEE (2012)

387. Todman, T., Stilkerich, S.C., Luk, W.: Using Statistical Assertions to Guide Self-Adaptive Systems. International Journal of Reconfigurable Computing **2014**, 1–8 (2014). DOI 10.1155/2014/724585

388. Tong, X., Ngai, E.: A ubiquitous publish/subscribe platform for wireless sensor networks with mobile mules. In: Proceedings of the IEEE Eighth International Conference on Distributed Computing in Sensor Systems (DCOSS), pp. 99–108 (2012)

389. Tørresen, J., Hafting, Y., Nymoen, K.: A new Wi-Fi based platform for wireless sensor data collection. In: Proceedings of the International Conference on New Interfaces for Musical Expression, pp. 337–340 (2013)
390. Tørresen, J., Plessl, C., Yao, X.: Special Issue on "Self-Aware and Self-Expressive Systems". IEEE Computer **48**(7), 45–51 (2015)
391. Touch, J., Pingali, V.: The RNA Metaprotocol. In: Proceedings of the International Conference on Computer Communications and Networks, pp. 1–6 (2008). DOI 10.1109/ICCCN.2008.ECP.46
392. Trucco, E., Plakas, K.: Video Tracking: A Concise Survey. Journal of Oceanic Engineering **31**(2), 520–529 (2006)
393. Tse, A.H.T., Chow, G.C.T., Jin, Q., Thomas, D.B., Luk, W.: Optimising performance of quadrature methods with reduced precision. In: Proceedings of the International Conference on Reconfigurable Computing: Architectures, Tools and Applications (ARC), *Lecture Notes in Computer Science*, vol. 7199, pp. 251–263. Springer (2012). DOI 10.1007/978-3-642-28365-9_21
394. Tse, A.H.T., Thomas, D.B., Tsoi, K.H., Luk, W.: Dynamic scheduling Monte-Carlo framework for multi-accelerator heterogeneous clusters. In: Proceedings of the International Conference on Field-Programmable Technology (FTP), pp. 233–240 (2010)
395. Tsoi, K.H., Luk, W.: Axel: A Heterogeneous Cluster with FPGAs and GPUs. In: Proceedings of the 18th annual ACM/SIGDA International Symposium on Field Programmable Gate Arrays, pp. 115–124 (2010)
396. Tsymbal, A., Pechenizkiy, M., Cunningham, P., Puuronen, S.: Dynamic integration of classifiers for handling concept drift. Information Fusion **9**(1), 56–68 (2008)
397. Vassev, E., Hinchey, M.: Knowledge representation and awareness in autonomic service-component ensembles – state of the art. In: 14th IEEE International Symposium on Object/Component/Service-oriented Real-time Distributed Computing, pp. 110–119 (2011)
398. Vasudevan, S.: What is assertion-based verification? SIGDA E-News **42**(12) (2012)
399. Vermorel, J., Mohri, M.: Multi-Armed Bandit Algorithms and Empirical Evaluation. In: Proceedings of the European Conference on Machine Learning, pp. 437–448. Springer (2005)
400. Vickrey, W.: Counterspeculation, auctions, and competitive sealed tenders. The Journal of Finance **16**(1), 8–37 (1961)
401. Vinoski, S.: CORBA: Integrating diverse applications within distributed heterogeneous environments. IEEE Communications Magazine **35**(2), 46–55 (1997)
402. Volker, L., Martin, D., El Khayaut, I., Werle, C., Zitterbart, M.: A Node Architecture for 1000 Future Networks. In: Proceedings of the IEEE International Conference on Communications (ICC), pp. 1–5 (2009). DOI 10.1109/ICCW.2009.5207996
403. Volker, L., Martin, D., Werle, C., Zitterbart, M., El-Khayat, I.: Selecting Concurrent Network Architectures at Runtime. In: Proceedings of the IEEE International Conference on Communications (ICC), pp. 1–5 (2009). DOI 10.1109/ICC.2009.5199445
404. Wang, J., Brady, D., Baclawski, K., Kokar, M., Lechowicz, L.: The use of ontologies for the self-awareness of the communication nodes. In: Proceedings of the Software Defined Radio Technical Conference (SDR), vol. 3 (2003)
405. Wang, S., Minku, L.L., Yao, X.: A learning framework for online class imbalance learning. In: Proceedings of the IEEE Symposium on Computational Intelligence and Ensemble Learning (CIEL), pp. 36–45 (2013)
406. Wang, S., Minku, L.L., Yao, X.: Online class imbalance learning and its applications in fault detection. International Journal of Computational Intelligence and Applications **12**(1340001), (1–19) (2013)
407. Wang, S., Minku, L.L., Yao, X.: A multi-objective ensemble method for online class imbalance learning. In: Proceedings of the International Joint Conference on Neural Networks (IJCNN), pp. 3311–3318. IEEE (2014). DOI 10.1109/IJCNN.2014.6889545
408. Wang, S., Minku, L.L., Yao, X.: Resampling-based ensemble methods for online class imbalance learning. In: IEEE Transactions on Knowledge and Data Engineering, vol. 27, pp. 1356–1368. IEEE (2015). DOI 10.1109/TKDE.2014.2345380

409. Wang, Z., Tang, K., Yao, X.: A memetic algorithm for multi-level redundancy allocation. IEEE Transactions on Reliability **59**(4), 754–765 (2010)
410. Watson, R.: The Delta-t Transport Protocol: Features and Experience. In: Proceedings 14th Conference on Local Computer Networks, pp. 399–407 (1989). DOI 10.1109/LCN.1989.65288
411. Werner-Allen, G., Tewari, G., Patel, A., Welsh, M., Nagpal, R.: Firefly-inspired sensor network synchronicity with realistic radio effects. In: Proceedings of the 3rd International Conference on Embedded Networked Sensor Systems, pp. 142–153 (2005)
412. Weyns, D., Schmerl, B., Grassi, V., Malek, S., Mirandola, R., Prehofer, C., Wuttke, J., Andersson, J., Giese, H., Gäschka, K.M.: On patterns for decentralized control in self-adaptive systems. In: R. Lemos, H. Giese, H. Müller, M. Shaw (eds.) Software Engineering for Self-Adaptive Systems II, *Lecture Notes in Computer Science*, vol. 7475, pp. 76–107. Springer Berlin Heidelberg (2013)
413. Wikipedia: Adaptation (computer science). http://en.wikipedia.org/wiki/Adaptation. (Accessed March 8, 2016)
414. Winfield, A.: Robots with internal models: a route to self-aware and hence safer robots. In: J. Pitt (ed.) The Computer After Me. Imperial College Press / World Scientific Book (2014)
415. Wolf, W., Ozer, B., Lv, T.: Smart Cameras as Embedded Systems. IEEE Computer **35**(9), 48–53 (2002)
416. Wright, M.: Open Sound Control: an enabling technology for musical networking. Organised Sound **10**(3), 193–200 (2005)
417. Xiao, L., Zhu, Y., Ni, L., Xu, Z.: GridIS: An Incentive-Based Grid Scheduling. In: Proceedings of the 19th IEEE International Parallel and Distributed Processing Symposium, p. 65b (2005). DOI 10.1109/IPDPS.2005.237
418. Xilinx: SDAccel Development Environment. http://www.xilinx.com/products/design-tools/sdx/sdaccel.html. (Accessed March 8, 2016)
419. Ye, J., Dobson, S., McKeever, S.: Situation identification techniques in pervasive computing: A review. Pervasive and Mobile Computing **8**(1) (2012)
420. Yiannacouras, P., Steffan, J.G., Rose, J.: VESPA: portable, scalable, and flexible FPGA-based vector processors. In: Proceedings of the International Conference on Compilers, Architecture, and Synthesis for Embedded Systems, pp. 61–70 (2008)
421. Yilmaz, A., Javed, O., Shah, M.: Object Tracking: A Survey. ACM Computing Surveys **38**(4), 1–45 (2006)
422. Yin, F., D., M., Velastin, S.: Performance evaluation of object tracking algorithms. In: Proceedings of the International Workshop on Performance Evaluation of Tracking and Surveillance (2007)
423. Yin, L., Dong, M., Duan, Y., Deng, W., Zhao, K., Guo, J.: A high-performance training-free approach for hand gesture recognition with accelerometer. Multimedia Tools and Applications pp. 1–22 (2013)
424. Yu, X., Tang, K., Chen, T., Yao, X.: Empirical analysis of evolutionary algorithms with immigrants schemes for dynamic optimization. Memetic Computing **1**(1), 3–24 (2009)
425. Zadeh, L.: Optimality and non-scalar-valued performance criteria. IEEE Transactions on Automatic Control **8**(1), 59–60 (1963)
426. Zagal, J.C., Lipson, H.: Towards self-reflecting machines: Two-minds in one robot. In: Advances in Artificial Life. Darwin Meets von Neumann, *Lecture Notes in Computer Science*, vol. 5777, pp. 156–164. Springer (2011)
427. Zambonelli, F., Bicocchi, N., Cabri, G., Leonardi, L., Puviani, M.: On self-adaptation, self-expression, and self-awareness in autonomic service component ensembles. In: Proceedings of the Fifth IEEE Conference on Self-Adaptive and Self-Organizing Systems Workshops (SASOW), pp. 108–113 (2011)
428. Zarezadeh, A.A., Bobda, C.: Hardware Middleware for Person Tracking on Embedded Distributed Smart Cameras. Hindawi International Journal of Reconfigurable Computing (2012)
429. Zeppenfeld, J., Bouajila, A., Stechele, W., Bernauer, A., Bringmann, O., Rosenstiel, W., Herkersdorf, A.: Applying ASoC to Multi-core Applications for Workload Management. In:

C. Müller-Schloer, H. Schmeck, T. Ungerer (eds.) Organic Computing – A Paradigm Shift for Complex Systems, *Autonomic Systems*, vol. 1, pp. 461–472. Springer Basel (2011)

430. Zhou, A., Qu, B.Y., Li, H., Zhao, S.Z., Suganthan, P.N., Zhang, Q.: Multiobjective evolutionary algorithms: A survey of the state of the art. Swarm and Evolutionary Computation **1**(1), 32–49 (2011)

431. Ziliani, F., Velastin, S., Porikli, F., Marcenaro, L., Kelliher, T., Cavallaro, A., Bruneaut, P.: Performance evaluation of event detection solutions: the CREDS experience. In: Proceedings of the International Conference on Advanced Video and Signal Based Surveillance, pp. 201–206 (2005)

432. Zitzler, E., Deb, K., Thiele, L.: Comparison of Multiobjective Evolutionary Algorithm: Empirical Results. Evolutionary Computation **8**(2), 173–195 (2000)

433. Zitzler, E., Künzli, S.: Indicator-Based Selection in Multiobjective Search. In: Proceedings of the International Conference on Parallel Problem Solving from Nature (PPSN), vol. 3242, pp. 832–842 (2004)

434. Zitzler, E., Laumanns, M., Thiele, L.: SPEA2: Improving the Strength Pareto Evolutionary Algorithm. Tech. Rep. 103, Computer Engineering and Networks Laboratory (TIK), Swiss Federal Institute of Technology (ETH), Zurich (2001)

435. Zitzler, E., Thiele, L.: Multiobjective evolutionary algorithms: a comparative case study and the strength Pareto approach. IEEE Transactions on Evolutionary Computation **3**(4), 257–271 (1999)

436. Zitzler, E., Thiele, L., Laumanns, M., Fonseca, C.M., da Fonseca, V.G.: Performance assessment of multiobjective optimizers: An analysis and review. IEEE Transactions on Evolutionary Computation **7**(2), 117–132 (2003)

Index

Printed in the United States
By Bookmasters